bnas

Field Responsive Polymers

ACS SYMPOSIUM SERIES **726**

Field Responsive Polymers

Electroresponsive, Photoresponsive, and Responsive Polymers in Chemistry and Biology

Ishrat M. Khan, EDITOR
Clark Atlanta University

Joycelyn S. Harrison, EDITOR
National Aeronautics and Space Administration

American Chemical Society, Washington, DC

Library of Congress Cataloging-in-Publication Data

Field responsive polymers: electroresponsive, photoresponsive, and responsive polymers in chemistry and biology/Ishrat M. Khan, editor, Joycelyn S. Harrison, editor.

    p.   cm.—(ACS symposium series ; 726)

Includes bibliographical references and indexes.

ISBN 0-8412-3598-8

1. Smart materials. 2. Polymers—Optical properties. 3. Conducting polymers. 4. Photorefractive materials. 5. Photobiochemistry.

I. Khan, Ishrat, 1956– . II. Harrison, Joycelyn S. III. Series.

TA418.9.S62F54   1999
620.1'92—dc21

99–19603
CIP

The paper used in this publication meets the minimum requirements of American National Standard for Information Sciences—Permanence of Paper for Printed Library Materials, ANSI Z39.48–1984.

Foreword

THE ACS SYMPOSIUM SERIES was first published in 1974 to provide a mechanism for publishing symposia quickly in book form. The purpose of the series is to publish timely, comprehensive books developed from ACS sponsored symposia based on current scientific research. Occasionally, books are developed from symposia sponsored by other organizations when the topic is of keen interest to the chemistry audience.

Before agreeing to publish a book, the proposed table of contents is reviewed for appropriate and comprehensive coverage and for interest to the audience. Some papers may be excluded in order to better focus the book; others may be added to provide comprehensiveness. When appropriate, overview or introductory chapters are added. Drafts of chapters are peer-reviewed prior to final acceptance or rejection, and manuscripts are prepared in camera-ready format.

As a rule, only original research papers and original review papers are included in the volumes. Verbatim reproductions of previously published papers are not accepted.

ACS BOOKS DEPARTMENT

Contents

vii

PHOTORESPONSIVE POLYMERS: NONLINEAR OPTICAL AND PHOTOREFRACTIVE

RESPONSIVE POLYMERS IN CHEMISTRY AND BIOLOGY

INDEXES

Preface

In the next millennium, the class of polymers labeled as smart or field responsive polymers will play a key role in the emergence of a wide array of new devices, such as those manufactured by aerospace, communications, and the expanding biomedical industry. A responsive or smart polymer is one that responds to an external stimulus in a controlled, reproducible, and reversible manner. This external stimulus may be optical, electrical, mechanical, or environmental, such as a change in temperature or pH. The promise of smart polymeric materials results from their versatility and the relative ease by which structural changes can be synthesized into a polymer to create a desired functionality. Moreover polymers possess the processability that facilitates their incorporation into a variety of device configurations. In the past decade there has been a dramatic increase in the amount of research dedicated to the synthesis of novel responsive polymers. There are several markets just waiting to take advantage of these new materials. For example, in the telecommunications industry, low-loss, inexpensive optical wave guides and very-high-bandwidth, integrated modulators are two areas where improved NLO polymeric materials will provide breakthrough capabilities. Other technological concepts such as optical computing, microwave shielding, and transmitting for stealth systems, synthetic enzymes, and targeted drug delivery have been theorized, but the materials for making them have not yet been optimized. In the paragraphs below, we have highlighted the enormous and exciting potential of field responsive polymers in the development of emerging technologies. We have illustrated by examples the breadth of applicability for responsive polymeric systems.

A wealth of recent literature indicates that photorefractive polymeric materials are key to the future technology of optical data storage and image processing. It is envisioned that photorefractive polymer technology may one day enable the storage of the entire *Encyclopedia Britannica* on a disk the size of a dime [see Dagani, R. *C&E News*, 28, February, **1995**]. Such storage capability has the potential to revolutionize the manner and effectiveness of remote sensing and earth observation science missions by providing the capability to store large quantities of high-resolution digital images at high speed in a very compact device. The photorefractive effect is the spatial modulation of the index of refraction due to charge redistribution in an optically nonlinear material. The effect arises when charge carriers, generated by modulated light, separate via drift and diffusion mechanisms and become trapped to produce a nonuniform space–charge distribution. The resulting internal space–charge electric field modulates the refractive index to create a phase grating that can diffract light. This three-dimensional diffraction grating is a hologram that can be read with a reference laser beam and that will remain in the material when the light source is turned off. Several holograms can be stored in the same volume by tilting the sample or the laser source for each new hologram. The hologram can be readily erased by exposing the material to uniform light to evenly redistribute the charges. The hologram can also be inadvertently erased or altered if there is migration of trapped charges with time or variations in temperature.

The exciting possibilities of responsive polymers have been demonstrated by many groups, and among the more fascinating examples were shown by Hoffman and co-workers [see Slayton, P. S.; Shimboji, T.; Long, C.; Chilkoti, A.; Chen, G.; Harris, J. M.; Hoffman, A. S. *Nature*, **1995**, *378*, 472]. The work involved is controlling the biotin–ligand recognition process by conjugation of a tempera-ture-responsive polymer in the vicinity of the protein-binding pocket. The recogni-tion process may be turned off and on by changing the temperature of the system; a change in the temperature causes the polymer to collapse and block the access of the biotin to the pocket. Such polymers have also been utilized in the important area of organic catalysis where Bergbreiter and co-workers have cleverly manipu-lated temperature-responsive polymers. Bergbreiter successfully prepared smart catalysts whose reactivity may be regulated by a temperature-sensitive polymer support [see chapter 20]. By increasing the temperature, the polymer collapses, thereby decreasing the accessibility of reactants to the active catalysts bound to the polymer, and hence the overall activity is decreased. Ostensibly the product may be easily recovered and the catalyst activity regenerated by decreasing the temperature.

It is in the electronics industry where the research in the area of responsive polymers is currently most significant and visible. A number of recent studies have focused on the use of polymeric materials in smart microwave shielding or reflecting devices, which are important in stealth technology. These studies indi-cate that conducting polymers are capable of variable microwave transmission [see Rupich, M. W.; Liu, Y. P.; Kon, A. B. *Mat. Res. Soc. Symp. Proc.* **1993**, *293*, 163]. The setup used to demonstrate this consists of a multilayer structure where two conductive films are separated by a solid polymer electrolyte. The active conductive polymer is highly microwave transmitting when reduced and attenuat-ing when oxidized. The passive conductive polymer is transmitting both in the reduced and the oxidized states. The microwave transmittance can be modulated by applying a voltage across the two conductive polymer layers. Responsive polymers whose microwave properties may be tuned upon the application of small bias voltages raises the possibility of developing newer types of responsive materials for use in the development of "smart skins", which are of interest to the avionics/defense industry [see Khan, S. M.; Negi, S.; Khan, I. M. *Polym. News*, **1997**, *22*, 414]. The aim of "smart skins" is to integrate antennas, sensors, transmit/receive (T/R) modules, preprocessors, and signal processors into the platform of the skin during the structural design of an aircraft to yield a highly integrated, multifunctional, tunable structure [see Lockyer, A. J.; Kudva, J. N.; Kane, D. M.; Hill, B. P.; Martin, C. A.; Goetz, A. C.; Tuss, J. *SPIE Proc.*, **1994,** *2189*, 172].

Multilayer polymeric structures have also been used to fabricate synthetic muscles or actuators [see Pei, Q.; Inganas, O. *Synth. Met.* **1993,** *55–57,* 3718]. A synthetic muscle or actuator functions by converting chemical energy into mechanical energy. For example, a multilayer structure capable of carrying out this function consists of a polyethylene layer, a thin gold layer, and a polypyrrole layer immersed in an electrolyte solution. Oxidation of the polypyrrole results in diffusion of the perchlorate ions from the electrolyte into the polymer and a net increase in the volume of the polymer, which causes deflection of the biopolymer strip. Undoping causes the perchlorate ions to be expelled from the polymer, and the multilayer structure returns to the original position. Thus, this electrochemical process has resulted in mimicking muscles of living organisms. Obviously, in this setup the liquid electrolyte is a drawback, and ideally an all-solid-state device is desirable. Wallace and co-workers have reported such an all-polymer solid-state electrochemical actuator [see Lewis, T. W.; Spinks, G. M.; Wallace, G. G.; De Rossi, C. E.; Pachetti, M. *Polym. Prepr.* (*Am. Chem. Soc. Div. Polym. Chem.*) **1997,** *38(2),* 520]. Large degrees of bending were obtained with this actuator composed of two polypyrrole layers separated by a solid polymer electrolyte. In order for the full potential of these actuators to be realized, some of the drawbacks, such as slow response times and stability to recycling, must be solved. One method to decrease the response times of all polymer solid-state actuators may be to bring the solid polymer electrolyte and the electronic conductive polymer in intimate contact or within a few hundred angstroms of each other. This will significantly reduce the time required for the diffusion-controlled doping–undoping process because the distance of ion transport will be decreased from the micrometer level to a few hundred angstom level. The microphase-separated mixed (ionic and electronic) conducting polymer system reported recently, in theory, will permit the fabrication of multilayer actuating structures with much faster response times [see Khan, I. M.; Li, J.; Arnold, S.; Pratt, L. in *Electrical and Optical Polymer Systems*; Wise, D. L. et al. Eds.; Marcell Dekker, New York, **1998;** p 331].

The future is indeed exciting and is dependent on the development of appropriate polymer systems to overcome drawbacks associated with some of the reported smart devices or structures. In addition to developing appropriate polymer systems with the required properties, the processing properties of the polymers must be addressed to allow fabrication of complementary polymers into appropriate smart device configurations.

Polymers that respond to an external field in a controlled, reproducible, and reversible manner have tremendous application potential. As we have highlighted in the preceding examples, the diverse utility of such polymers spans the gamut from biological to stealth to electro-optic systems. This highly interdisciplinary area includes research in chemical synthesis, property characterization, device development, and system integration. Chapters featured in this book divide the broader classification of responsive polymers into three more specialized categories: electroresponsive, photoresponsive, and responsive polymers in chemistry and biology. Within each of these categories some of the chapters offer an overview of specific types of responsive polymeric materials such as photorefractive and amorphous piezoelectric polymers. Other chapters present leading-edge research that encompasses a variety of materials, experimental techniques, theories, processes, and applications. The breadth of the information presented in this book should make it useful for materials scientists, polymer chemists, physicists, and engineers in a broad spectrum of industries.

Acknowledgments

Funding to support the symposium upon which this book is based was provided by the Polymer Chemistry Division of the American Chemical Society and PRF Grant # 32468-SE. We also greatly appreciate the help of Anne Wilson, Kelly Dennis, and Tracie Barnes of the ACS Books Department who provided the guidance and support to bring this book to a fruitful completion. Finally, we thank the contributors for their timely efforts and for sharing their valuable research results in this forum.

ISHRAT M. KHAN
Department of Chemistry
Clark Atlanta University
Atlanta, GA 30314

JOYCELYN S. HARRISON
Langley Research Center
National Aeronautics and Space Administration
Hampton, VA 23681–0001

ELECTRORESPONSIVE POLYMERS

Chapter 1

Microwave and Optical Properties of Conducting Polymers: From Basic Research to Applications

P. Hourquebie, P. Buvat, and D. Marsacq

CEA, Le Ripault, BP 16-37260 Monts, France

Microwave absorbing materials can advantageously be obtained using conducting polymers. Among those polymers, soluble ones are of interest as they can be processed and mixed with insulating polymers using classical techniques such as spin-coating, molding or extrusion. Nevertheless, final electrical properties and stability over time will depend on structural characteristics of the polymer (monomer, comonomer, defects content, chain length, crystallinity) and also on the details of processing techniques. Effect of structural properties of polymers on electrical or optical properties have been studied in detail in order to go further in applications. Main results of these studies are presented in this paper.

The unusual dielectric properties of conducting polymers make them a unique new class of microwave absorbing materials (1). We have previously studied the links between structural (crystal structure, counter-anion size, molar mass, length of the alkyl chain on the substituted monomer) and electrical properties (σ_{dc} and ε^* values) (2). The key parameters affecting conduction properties are clearly the interchain distance and the delocalization length as it was proposed by Wang et. al. a few years ago (3). These parameters are closely related to the method by which the polymer is synthesized, and doped.

As an extension of our earlier work we have studied soluble conductive polymers (sulfonic acids doped polyaniline (4), poly(3-alkyl thiophenes) (5)) for the design of a new generation of microwave absorbing materials (6). These studies have shown that microwave properties were very dependent on minor changes in experimental procedures. Therefore it was necessary to determine how synthetic conditions effects influence structural parameters of polymers, and how these structural parameters influence electrical, optical properties and stability over time. In the case of poly(alkylthiophene)s, solubility of polymers permits a complete structural characterization and then a better understanding of links between structural and electrical properties. In a recent publication (7) we have shown that quasi-metallic properties already obtained with polyaniline (8) or polypyrrole (9) are now obtained with poly(alkylthiophene)s especially in the field of optical properties in the infrared domain.

In this paper we will point out the main results of these studies and we will show how the control of structural parameters of polymers leads to materials with improved electrical properties suitable for applications. Four points will be discussed, the modification of structural parameters at molecular level, the modification of properties at supramolecular level (chain conformation, crystallinity, homogeneity of blends), the effects of structural parameters on electrical properties and aging phenomena. In the last part we will draw some conclusions and open perspectives for future studies.

Modification of structural properties at molecular level

In order to modify structural properties at molecular scale, synthesis parameters were varied and the effects on polymers structure determined.

Synthesis and characterization of polyaniline. Polymerization of aniline was carried out according to a general method (Figure 1) described by Y. Cao et. al (10) but with some changes of experimental parameters.

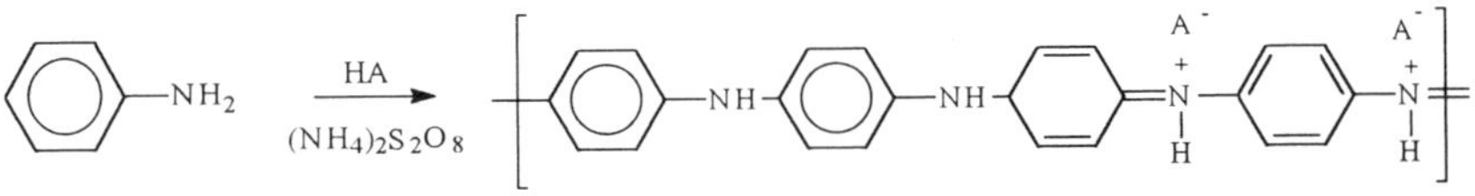

Figure 1:   Synthesis of polyaniline, oxidation in acidic medium (HA).

Parameters such as, synthesis medium (HCl, camphor sulfonic acid (CSA) or dodecylbenzene sulfonic acid (DBSA)), molar ratio aniline/oxidant, method for oxidant addition, temperature, polymerization duration, dedoping conditions will have effects on polymer properties in terms of yield, chain length, effective conjugation, defects rate, and electrical properties. Temperature was kept to 0°C, -30°C or -40°C and reactions were stopped after various durations in order to control the molecular weight of samples. Polymerization durations were varied from 1 hour to 5 days and depended on the reaction temperature. The inherent viscosities of the polyanilines (Pani) were determined at 25 °C in 0.1%w solutions in concentrated sulfuric acid (95 %), using an Ubbelohde viscometer. For instance, high viscosity samples (1.4 dl/g) were obtained after 3 days at -40 °C while low viscosity samples (0.6 dl/g) were obtained after 1 hour at 0 °C.

A relationship was found between polymerization duration and inherent viscosity for polymers synthesized at low temperature (- 40 °C). Inherent viscosity increases from 0.6 to 1.4 when duration of polymerization increases from one to five days. In the case of synthesis at 0°C no correlation was obtained between duration of polymerization and inherent viscosity. A careful control of other parameters (synthesis medium, molar ratio aniline/oxidant, method for oxidant addition) have permitted to get samples with inherent viscosities ranging from 0.55 to 2.1 dl/g in a reproducible way.

Synthesis and characterization of poly(3-alkylthiophene)s. The basic reactions used for the synthesis of poly(3-alkylthiophene)s are shown in Figure 2.

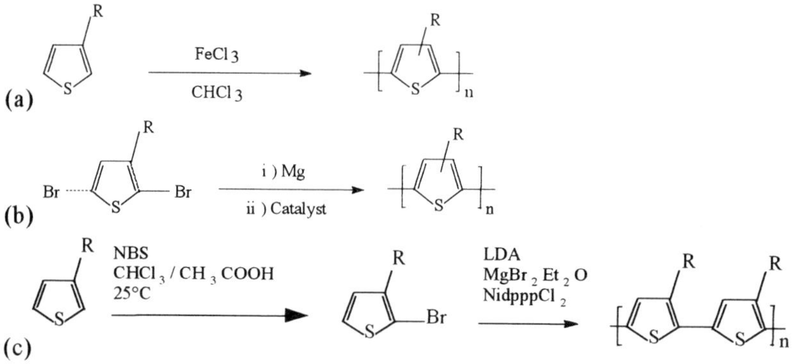

Figure 2 :   Synthesis routes for poly(alkylthiophene)s
a) oxidation of 3-n-alkylthiophenes
b) polycondensation of the dibrominated monomer
c) regioregular poly(3-alkylthiophenes)
according to Mc Cullough et al. (5)

The first step is the synthesis of the monomer. All 3-n-alkylthiophenes were prepared according to Kumada (11). The second step is the synthesis of the polymer. A simple and inexpensive method to produce P3AT is by direct oxidation of 3-n-alkylthiophenes using $FeCl_3$ as the oxidant/catalyst according to Sugimoto et al (12). The problem with this synthesis is the remaining Fe which is not easy to remove and leads to soon aging of the polymer and then to poor processability. A second route is the polycondensation of the dibrominated monomer (13). In this case no Fe is present in the materials and the stability of properties is increased. This synthesis was found to be very convenient in order to modify the molecular weight of polymers. In fact, by introducing a few percents of the monobrominated monomer we can vary the molecular weight from 3500 to 5500 (Figure 3). If more than five percents is added polymerization doesn't occur.
The third route leads to regioregular poly(3-alkylthiophenes). They were prepared according to Mc Cullough et al (14). Only the bromination of monomer step was changed. Regioselective bromination of all 3-n-alkylthiophenes was performed using N-bromosuccinimide in $CHCL_3/AcOH$ 1/1. This procedure gives almost no 5-bromo-3-n-alkylthiophene and therefore purification is greatly simplified. All operations were carried out under inert atmosphere conditions using pure nitrogen and dried and distilled solvents. This point makes the synthesis more complicated but permits to control the defects content in the polymer in terms of head to head couplings.
With these three kinds of syntheses we have been able to get a series of polymers with various alkylchain length on the monomer, various molecular weight and various defect rates. The effect of defects rate on conjugation length can be seen for example with the evolution of the maximum of absorption on UV-Vis spectra. For the first synthesis route the maximum absorption in chloroform solution was at 430 nm for poly(3-butylthiophene), it was at 435 nm for the second route and 452 nm for the third one. More generally a correlation was found between defects content

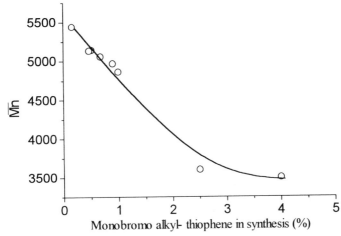

Figure 3 :   Effect of monobromo butyl-3 thiophene concentration on molecular
weight (Mn) of polymer obtained through polycondensation of the
dibrominated monomer. Molecular weight is obtained by Gel
Permeation Chromatography calibrated with polystyrene standards.

(% of head to head couplings) and the location of the maximum absorption on UV-
vis spectrum (Figure 4). This correlation was in agreement with results already
published on poly(3-hexylthiophene) (5,15-18) This correlation was found to be
largely independent of the alkylchain length on the monomer and of molecular
weight of the polymers.

Modification of structural properties at supramolecular level

When control of structural parameters of the polymer is achieved, properties of the
material will also depend on processing routes. Processing routes will have an
impact on polymer chain conformation and then on localization of charge carriers.
Polymers are mainly processed from solutions or from their molten state. We will
show here the extreme importance of the conformation of polymers in solution
when one is concerned with final electrical properties. The conformation of polymer
in solution can be tailored by a careful choice of the solvent type, i.e. dielectric
constant and solubility parameters or some added salts in the solution. Even in the
case of extrusion, when the polymer is processed in the molten state, the prior
conformation in solution plays an important role on the final properties of
conductive polymer based blends.
    As an example, solutions of emeraldine base and CSA in m-cresol, or DBSA in
xylene (4) can be cast or spin-cast onto various substrates. The properties of the
material will then depend on the structural parameter of the polymer but also on the
experimental details.
    UV-vis spectra of DBSA doped Polyaniline solutions in m-xylene were found
to be dependent on heat treatment in the case of high viscosity samples (Figure 5).
During heat treatment the band at 750 nm attributed to localised carriers is gradually
shifted toward low energies and its intensity is decreased while an absorption
increases at low energy leading to a more "metallic" (19) spectrum.

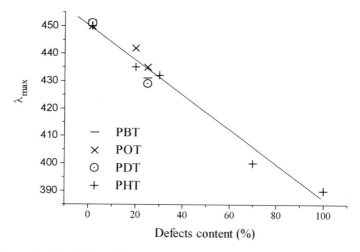

Figure 4 :   Evolution of λ_{max}(nm) with defects content (% of head to head couplings measured by ^{1}H NMR) for poly(3-butylthiophène) (PBT), poly(3-octylthiophène) (POT), poly(3-dodecyllthiophène) (PDDT) synthesized in this work and for poly(3-hexylthiophène) (PHT) of references 5, 15-18.

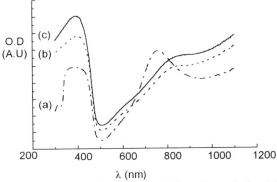

Figure 5 : Evolution of UV-visible spectrum of the solution before(a), after 1 hour
(b) and after 3 hours (c) of heat treatment (Synthesis -40 °C, 3 days).

On the contrary, spectra of low viscosity samples show weak evolution with heat treatment of the solution.
Moreover the UV-Vis spectra were found to be dependent on the isomer of xylene used and on duration of doping process (Figure 6-7). It is surprising to see that for the same polymer, depending on the xylene isomer used, opposite evolution of properties is obtained. In o-xylene (Figure 6) the maximum absorption is gradually shifted toward low energy when in m-xylene (Figure 7) the shift is toward high energy. In the first case the polymer is more delocalised whereas in the second case localisation increases.

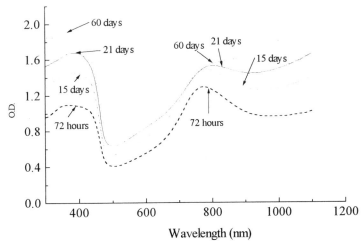

Figure 6 : Evolution of UV-Vis spectra with time for PANI/DBSA solutions in
o-xylene.

8

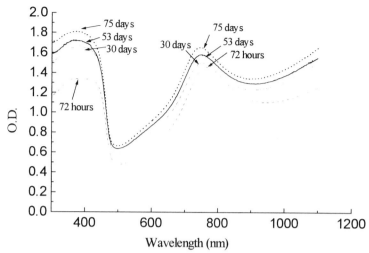

Figure 7 :   Evolution of UV-Vis spectra with time for Pani/DBSA solutions in m-xylene.

The difference in dielectric constant of solvent between o-xylene (ε_r = 2.57) and m-xylene (ε_r = 2.37) could be an explanation for more metallic like spectra in the case of o-xylene. This effect of dielectric constant seems to be quite general as we have been able to extend it to poly(3-alkylthiophene) solutions. The absorption spectra of head to tail polymerized poly(3-dodecylthiophene) (HTpDDT) doped in solution in different solvents have been obtained for a polymer concentration of 0.4 mg/ml and a doping level of 25%. All solutions present the same general features: two absorption bands in the NIR, characteristic of the bipolaron bands. In Figure 8 we have plotted the wavelength of the first and the second bipolaronic transition as a function of dielectric constant of solvent. The effect on electronic spectra of changing the solvent is dramatic. The spectrum in a polar solvent, nitrobenzene (ε_r=35.7) is significantly shifted toward low energy compared with spectra measured in relative non polar solvent, such as dichloromethane (ε_r=9.1), chloroform (ε_r=4.8) and toluene (ε_r=2.4).

In the case of toluene, it is postulated that there is little dissociation of ions (positive charges on the polymer and negative $FeCl_4^-$ counter-anions). The positive and negative ions associated with the chain tend to form ion pairs inhibiting electrostatic repulsion between the positive charges on the chain. This effect tends to favor a compact coil conformation of the polymer. As the dielectric constant of the solvent increases, dissociation of negative (counter-anion) and positive (polymer) ions increases, resulting in coulombic repulsion of positive charges on the polymer chain and in an expansion of the initial compact coil conformation of the poly(alkylthiophene) chain, and thus a greater conjugation length.

This effect is similar to results already published in the case of polyaniline (8). In this case, using m-cresol instead of chloroform, changes the molecular conformation of the polyaniline from compact coil to expanded coil, resulting in an enhancement of the free carrier feature and a decrease of the localized polarons band. We show here a similar phenomenon for poly(3-alkylthiophene)/FeCl₃ system.

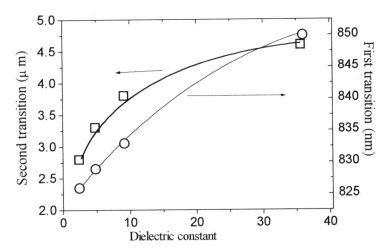

Figure 8:   Wavelength of the first and second bipolaronic transition as a function of dielectric constant of the solvent.

Polyaniline based materials can be obtained by dissolving the doped powder in a solvent containing an insulating polymer (PMMA, PE, EVA) or by extrusion of the insulating polymer and doped polyaniline powder. In the case of extrusion of DBSA doped polyaniline with low density polyethylene we found out that homogeneity of blends and final properties are strongly dependent on the polymer structural properties and the conformation in solution. Best results were obtained with low viscosity polymers and polyaniline powder recovered from metallic type solutions. Powders recovered from solution exhibiting localised absorption usually lead to inhomogenious blends with some aggregates of polyaniline. Aggregates may be due to incomplete doping of polyaniline aggregates leading to poor compatibility with polyethylene. Moreover after the extrusion process the conformation of the chain surprisingly remains the same as it can be seen on UV-vis spectra of thin films of diluted Pani/DBSA-PE films.

Effect of structural parameters on electrical properties

Electrical and optical characterization. A four probe apparatus (JANDEL JA 010) was used for σ_{dc} measurements. The electric current was provided by a KEITHLEY 224 current source and a KEITHLEY 617 electrometer was used to measure the voltage. σ_{dc} measurements were carried out on free standing films or deposits. Film thicknesses were determined using a Dektak ST surface profiler. In order to get reliable values of conductivity surface resistance R of deposits on glass slides were measured as a function of the thickness of the film (Figure 9). Conductivity was then given by the slope of the curve of the inverse of R as a function of thickness. Correlation coefficients better than 0.99 insure reliable values of conductivity.

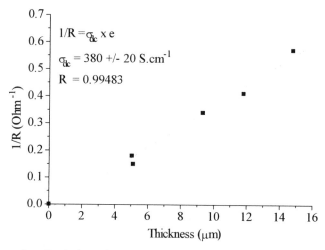

Figure 9 : Evolution of the inverse of surface resistance with thickness of polyaniline deposits.

ε^* measurements were carried out using a network analyzer (HP 8720) working between 130 MHz and 20 GHz with a 50 Ω line. Powder samples are pressed or molded in order to give die castings compatible with APC7 standard. S parameters were then measured and values of ε^* were calculated. Precision of the measurements was about 5 % for ε^* values.

The transmitted, reflected and absorbed part of the incident electromagnetic power were measured on deposits or thin films using a network analyzer (HP 8720) with a rectangular wave guide.

Specular reflectance measurements ($q = 7°$) were carried out between 6000 and 400 cm^{-1} using a Bruker IFS88 FTIR (resolution 8 cm^{-1}). A gold mirror was used as reference.

Effect of molecular structural parameters on electrical and optical properties. In the case of low temperature synthesis polyaniline samples, conductivity was found to be related to inherent viscosity (Figure 10). Such a correlation is not found for samples synthesized at 0°C. This may be related to the fact that low temperature synthesis leads to defect free materials where conjugation length increases with molecular weight of the polymer. In the case of 0 °C samples, defects content is high and limits the conjugation length whatever the molecular weight is.

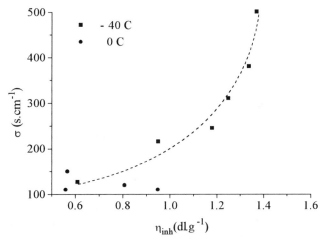

Figure 10 : Evolution of conductivity with inherent viscosity for polyaniline samples synthesized at high (0 C) or low (- 40 C) temperature.

The decrease of defect rates (low temperature synthesis) and the increase of molecular weight permits to get high values of reflectivity in infrared domain (about 80 %). Similar results have been obtained with poly(alkylthiophene)s going from randomly (RpDDT) to head to tail (HTpDDT) polymerized poly(3-dodecylthiophene) (Figure 11).

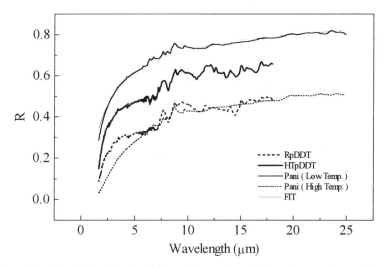

Figure 11 : Evolution of reflectivity with wavelength for various polymers (see text for details) and fit to Hagen Rubens equation in the case of high reflectivity polyaniline sample.

In the case of polyaniline reflectivity is high enough to be described by the Hagen Rubens equation. The reflection coefficient R at a wavelength λ for a given resistivity ρ is given by the equation ($1 - R = 4\left(\dfrac{\pi \varepsilon_0 c \rho}{\lambda}\right)^{0.5}$). An estimation of microscopic conductivity can then be obtained. This conductivity, 1950 S.cm^{-1} is about four times higher than the static conductivity measured. In the case of the other materials, the levels are to low and application of this equation doesn't fit experimental results showing that we are too far from metallic behavior.

In the case of polyaniline samples the increase in reflection coefficient is correlated to the increase in static conductivity. In the case of poly(alkylthiophene)s, we found out that control of molecular structure permits to increase conductivity at molecular level but is not sufficient to insure high level of static conductivity of the material. Moreover no correlation was found between static conductivity and reflection coefficient in the infrared domain. In fact static conductivity is not only influenced by molecular parameters but also by structural parameters at supramolecular scale. The effect of structural parameters on microwave properties of low conductivity samples (10^{-4}-5 S.cm^{-1}) has already been reported (2). By increasing the conductivity we have shown that microwave properties for high conductivity samples will only depend on the static conductivity and thickness of samples in term of absorbed, reflected and transmitted power for a given frequency. Different materials such as polypyrrole, polyaniline or indium tin oxide deposits were found to give same results when the surface resistance of the deposits was kept the same.

In summary it is shown that control of molecular properties leads to better conductivity at microscopic level. Depending on the material, this increase at molecular level is more or less discernible at macroscopic level (static conductivity). For microwave properties, if static conductivity is increased, levels of ε^* are also increased. Those levels are no more dependent on molecular structure if static conductivity is sufficiently high (more than 5 S.cm^{-1}).

Effects of structural parameters at supramolecular scale on electrical properties. The UV-Vis spectra evolution of DBSA doped Pani in m-xylene is correlated with modification of the conductivity of the two polymers before and after heat treatment of the solution. In the case of low viscosity sample the initial conductivity is 24 S.cm^{-1} and remains about the same value after heat treatment of the solution in agreement with no change in UV-Vis spectrum. On the contrary the initial conductivity of high viscosity sample is about 90 S.cm^{-1} and is increased up to 150 S.cm^{-1} after heat treatment of the solution leading to a more metallic like UV-Vis spectrum.

This illustrates the fact that electric properties strongly depend on processing routes and that the effects can be more or less important depending on the structural characteristics of the polymer.

Also in the case of poly(3-alkyl thiophene) samples we found out that a more expanded coil like conformation would lead to higher reflectivity in the infrared domain (7) provided that the defects rate is sufficiently low.

Interesting ε^* levels for microwave absorption can be reached with low conducting polymer contents (1-5%) (6). However, those levels, for a given conducting polymer are highly dependent on the synthesis parameters used to get the polymer (Temperature, duration, ...), on the way the doping process was carried out (solution concentration, evaporation temperature, kind of solvent,...) and on the way the final material was obtained (evaporation of the two polymers in solution, extrusion of the two polymers). This point is illustrated by the evolution of dielectric properties before and after the extrusion process. In this case we note an increase of dielectric properties which means that the quality of the blend is increased and that no aging phenomena occur during the extrusion process (Figure 12). Moreover, in this case the low viscosity sample was found to give better results after extrusion because of a better compatibility with polyethylene and then better homogeneity of the blend.

It was found that materials obtained through solution evaporation lead to a better reproducibility and high values of ε^* while extruded materials could present wide dispersion of results. This point is essential for applications and shows that if conducting polymer based materials, with low content of conducting polymers (1-5 %$_w$) permit to get suitable values for microwave applications, works remain to be done when ε^* values have to be severely controlled such as in the case of stealth applications (1).

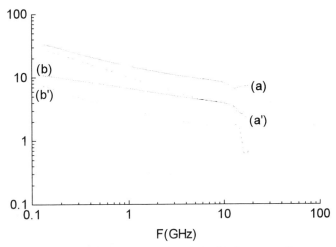

Figure 12 : Evolution of ε'(a,b) and ε''(a',b') with frequency for polyaniline(5%w)/Polyethylene blend samples, before (b,b') and after extrusion (a,a') process.

Stability of properties

The stability of properties also depends on the polymer structural properties. In the case of low viscosity polyaniline samples, heating at 120°C in air, for 24 hours, leads to a decrease of conductivity by a factor of 100 while only a factor 10 is found for the high viscosity sample. IR and UV spectra were recorded during aging and only slight variations where obtained on spectra. In fact a very low defect content due to aging has a strong influence on dc conductivity even if it is not detected by IR or UV-Vis spectroscopy. On the contrary loss in infrared reflectivity was low when compared to the decrease of conductivity. For example reflectivity in the infrared is decreased by 5 % when conductivity is divided by a factor 3. This may be explained by the fact that reflectivity is due to the concentration of metallic domains and is less influenced by connectivity between metallic islands. This results from a preferential degradation of the amorphous phase between metallic domains where diffusion of oxygen is easier.

It was also observed that after aging the polymer has lost its solubility and then that cross linking reactions may occur in the material.

Aging experiments on thin deposits (3-5 μm) of Pani on glass substrates where performed at various temperatures for the two polyaniline samples doped with DBSA and for the high viscosity polyaniline sample doped with CSA. Conductivity level was 500 S.cm^{-1} for high viscosity CSA doped sample, and 150 and 25 S.cm^{-1} respectively for high and low viscosity samples (6). Decrease in conductivity as a function of time was well described by first order laws. It was possible than to determine the activation energy of the phenomenon and to deduce, at room temperature the time needed for the conductivity to be divided by a factor two. The results are that the time needed is about 10 years for CSA-doped samples, three years for the high viscosity DBSA doped sample and only three months for the low viscosity DBSA doped sample. We can see here that the higher the conductivity, the higher the stability of the material is. In fact a higher conductivity is due to a material with higher crystallinity and organization leading to a better stability. CSA doped samples are interesting for their stability properties but are not liable to be extruded with polyethylene. So work remains to be done in order to get extruded polyaniline with expected values of radioelectric parameters and enough stability.

Conducting polymers for radar absorbing materials.

We have recently reviewed the use of conducting polymers in radar absorbing materials (1). The aim of this application is to generate materials able to decrease the reflected microwave energy of a metallic target illuminated by a radar. We have shown that such an application requires a careful control not only of dc conductivity but also of radioelectric parameters that are influenced by details of conduction mechanism at microscopic level (2). The great advantage carried on by conducting polymers is the wide diversity of synthesis routes and the possibility to tune the radioelectric parameters through doping or control of structural parameters. Different examples of such materials are given in reference (1) and will not be developed here. The challenge is now to generate new absorbers with conducting polymers having better controlled structural parameters leading to a wide choice of processing routes and better stability.

Conclusion

Control of synthesis and process parameters permit the tailoring of electrical properties. Concerning the static conductivity, we have shown that the decrease of defects and the increase of molecular weight permit to obtain high levels of conductivity and better stability. Concerning microwave properties, achievement of homogeneous blends permits to obtain materials with interesting properties at very low concentrations of conductive polymer. For IR properties the decrease of defects rate and the control of chain conformation permit to obtain high levels of reflectivity in the infrared range.

Soluble conducting polymers are very attractive materials in order to realize microwave absorbing materials. However, structural parameters of the polymer and details of preparation techniques have to be severely controlled in order to get reproducible and expected values. More specifically, DBSA doped polyaniline based materials are promising materials if two major problems are fixed : homogeneity of properties after extrusion and stability over time. Work is in progress now in order to increase properties stability of the polymer.

References

1. Olmedo, L.; Hourquebie, P.; Jousse, F. « *Microwave properties of conducting polymers* » in « *Handbook of Organic and Conductive Molecules and Polymers* », H. S Nalwa Ed., John Wiley, 1997, vol. 3, pp 367-428.
2. Hourquebie, P.; Olmedo, L. *Synth. Met.* **1994**, *65*, pp 19-26.
3. Wang, Z.H.; Ray, A.; Mc Diarmid, A.G.; Epstein, A.J. *Phys. Rev. B.* **1991**, *43*, 5, pp 4373-4384.
4. Cao, Y.; Smith, P.; Heeger, A.J. *Synth. Met.* **1992**, *48*, pp 91-97.
5. Mc. Cullough, R.D.; Lowe, R.D. *J. Chem. Soc., Chem. Commun.* **1992**, pp 70-72.
6. Hourquebie, P.; Blondel, B. *Synth.Met.* **1997**, *85*, pp 1437-1438.
7. Buvat P.; Hourquebie P. *Macromolecules* **1997**, *30*, pp 2685-2692.
8. Mc Diarmid, A.G.; Epstein, A.J. *Synth. Met.* **1994**, *65*, pp 103-116.
9. Kohlman, R.S.; Joo, J.; Wang, Y.Z. *Physical Review Letters* **1995**, *74*, 5, pp 773-776.
10. Cao, Y.; Andreatta, A.; Heeger, A.J. *Polymer* **1989**, *30*, pp 2305-2311.
11. Kumada, M.; Nakajima, I.; Kodama, S. *Tetrahedron* **1982**, *38*, 22, pp 3347-3354.
12. Yoshino, K.; Nakajima, S.; Onoda, M. *Synth. Met.* **1989**, *28*, pp C349-C357.
13. Österholm, J.E.; Laakso, J. Patent WO 89/01015, 09.02.89).
14. Mc Cullough, R.D.; Lowe, R.D.; Anderson, D.L. *J. Org. Chem.* **1993**, *58*, pp 904-912.
15. Souto Maior, R. M.; Hinkelmann, K; Eckert, H.; Wudl, F. *Macromolecules* **1990**, *23*, pp 1268-1279.
16. Yamamoto, T.; Morita, A.; Miyazaki, Y. *Macromolecules* **1992**, *25*, pp 1214-1223.
17. Leclerc, M.; Martinez, F.; Wegner, G. *Makromol. Chem.* **1989**, *190*, pp 3105-3116.
18. Inganäs, O.; Sulaneck, W.R. *Synth. Met.* **1988**, *22*, pp 395-406.
19. Xia, Y.; Joo, J.; Mc Diarmid, A.G.; Epstein, A.J. *Chem. Mater.* **1995**, *7*, pp 443-445.

Chapter 2

Separation and Concentration of Anionic Organic Electrolytes by Electrotransport through Polyethylene Films Grafted with Cationic Polymers

Kazunori Yamada, Koki Sasaki, and Mitsuo Hirata

Department of Industrial Chemistry, College of Industrial Technology, Nihon University, Narashino, Chiba 275–8575, Japan

Electrotransport of anionic and neutral phenyl compounds such as benzoic acid (BA), benzenesulfonic acid (BSA), and phenyl-1,2-ethanediol (PhED) was studied using polyethylene films photografted with 2-(dimethylamino)ethyl methacrylate (DMAEMA).   The BA and BSA permeabilities of DMAEMA-grafted PE (PE-g-PDMAEMA) films were considerably increased by the application of the direct current at pH 6.   On the other hand, the permeability of phenyl-1,2-ethanediol (PhED) increased slightly.   BA and BSA could be selectively permeated from the respective binary BA/PhED and BSA/PhED mixtures by making use of the difference in the permeabilities between the anionic and neutral phenyl compounds.   In addition, anionic phenyl compounds could be concentrated by 1.8 to 1.9 times the initial concentrations by the continuous application of the direct current in the aqueous BA and BSA solution systems.

It is possible to add new properties to polymer substrates by grafting various monomers onto the surfaces or throughout the bulk of the polymeric substrates such as polyethylene (PE) (1-5), polypropylene (PP) (6-9), or poly(tetrafluoroethylene) (PTFE) (10-13) by using UV radiation (1-5), Co γ rays (7,10), plasma (6), or electron beam (8,9) as an energy source or by the combined use of the plasma treatment and photografting (11-13).   We reported in previous papers (2,3) that the wettabilities and adhesivities of polyethylene (PE) plates were enhanced without affecting any bulk properties by the photograftings of hydrophilic monomers.   Furthermore, it was found that the grafted layers formed on the PE substrates could absorb a significant amount of water.   When the photograftings of hydrophilic monomers were carried out throughout the bulk of PE films used in place of PE plates, the grafted PE films possessed reasonable mechanical properties in the swollen state (4,14).   They are suitable for use in hydrogel systems or functional membranes (5).   Among these grafted PE films, 2-(dimethylamino)ethyl methacrylate (DMAEMA) grafted PE (PE-g-PDMAEMA) films showed good electrical conductivities and water-absorptivities even at lower grafted amounts (4,14).   Generally, polyelectrolytes undergo conformational changes with a change in the environmental conditions such as pH, ionic strength, and electric field of the medium (15,16).   Therefore, if such polyelectrolytes are covalently bonded to polymeric films as grafted polymer chains, their permeabilities to neutral and ionic small molecules can be controlled in response to one or more of the above mentioned environmental changes. Our work has been geared toward studying the application of grafted PE films as functional membranes for the separation, concentration, and recovery of anionic and cationic compounds. In fact, many kinds of ionic membranes have been designed for the concentration and selective separation of anionic and cationic compounds by several researchers (17-19). We reported in previous papers that ionic phenyl compounds such as benzoic acid (BA), benzenesulfonic acid (BSA), and p-aminobenzoic acid could be uphill transported against the concentration gradient by the use of the pH difference across the PE-g-PDMAEMA films. The concentration of each ionic phenyl compound in the acidic side increased by about 1.8 times the initial concentration (20,21). However, it took much longer time to concentrated the concentration of the ionic phenyl compounds against the concentration gradient. Thus, we tried to examine electrotransport properties of PE-g-PDMAEMA films to some anionic phenyl compounds.

In this study, the control of permeation of anionic phenyl compounds and their selective separation and concentration by electrotransport were examined using PE-g-PDMAEMA films with the aim of applying them to various types of functional membranes.

Experimental Section

Photografting.   A film of PE (thickness = 30 μm, density = 0.924 g/cm^3) supplied from Tamapoly Co., Ltd., in Japan was used as a polymer substrate. DMAEMA was purified by distillation under reduced pressure.   Other chemicals were used as supplied.   The PE films, cut strips of 6.0 cm length and 3.0 cm width, were dipped for 1 min in a 50 ml of an acetone solution containing 0.25 g benzophenone (BP) as a sensitizer to coat the PE surfaces with BP (1-3).   Next, the pH value of an aqueous DMAEMA monomer solution at 1.0 mol/dm^3 was adjusted to 8.0 with concentrated HCl to prepare PE-g-PDMAEMA films.   Photografting of DMAEMA onto the BP-coated PE films was carried out by applying UV rays emitted from a 400 W high pressure mercury lamp to the aqueous DMAEMA monomer solutions adjusted to pH 8 in the Pyrex glass tubes at 60 ℃ using a Riko RH400-10W rotary photochemical reactor (4).   The grafted amount was calculated from the weight increase of the samples in mmol/g.

Permeation Control in Response to the Direct Current.   Two platinum mesh electrodes and PE-g-PDMAEMA films swollen in NH4Cl/HCl or NH4Cl/NaOH buffer solutions of pH 4 to 10 housed in the permeation cells were arranged in order of one platinum electrode, the PE-g-PDMAEMA film, and another platinum electrode (22,23).   A 100 ml of buffer solution containing BA, BSA, or PhED (concentration = 2.0 mmol/dm^3) was put in one side of the cell and a 100 cm^3 of the buffer solution in the other side of the cell, and then the direct current was turned on and off in a stepwise manner with stirring.   During applying the direct current, 0.5 ~ 2 mol/dm^3 HCl or NaOH was added to the feed and permeate solutions to keep their initial pH values.   The amounts of permeated BA, BSA, and PhED were determined by measuring the absorbance of the aliquots (λ = 219 nm for BA and BSA and λ = 205 nm for PhED), and then each aliquot was immediately returned to the permeate solution.

Selective Permeation by Electrotransport.   The binary mixtures of BA/PhED and BSA/PhED in a NH4Cl/HCl buffer solution of pH 6 (100 cm^3) were put in the feed side, and the direct current of 10 mA was applied with stirring.   The amounts of two phenyl compounds, designating one phenyl compound as A and another as B,

Table I. Determination of the amount of permeated phenyl compounds for the binary BA/PhED and BSA/PhED mixture systems.

Binary component	λ_1 (nm)	$\log \varepsilon$ (dm³/mol·cm)	λ_2 (nm)	$\log \varepsilon$ (dm³/mol·cm)
BA/PhED $<$ BA	212	3.76	219	3.87
PhED		3.76		3.10
BSA/PhED $<$ BSA	209	3.86	220	3.67
PhED		3.86		2.85

permeated from the binary mixtures through PE-g-PDMAEMA films were determined by measuring the absorbance of the aliquots taken from the permeate solutions at two different wavelengths, λ_1 and λ_2, as shown in Table I. The absorbances were determined both at one wavelength λ_1 at which the molar absorption coefficient of A was equal to that of B ($\varepsilon_{A1} = \varepsilon_{B1}$) and at another wavelength λ_2 at which the molar absorption coefficients of two phenyl compounds were considerably different ($\varepsilon_{A1} \neq \varepsilon_{B1}$). The amounts of permeated phenyl compounds, A and B, were calculated using equations 1 and 2:

$$Abs_2 = C_A \cdot \varepsilon_{A2} \cdot L + (C_{A+B} - C_A) \cdot \varepsilon_{B2} \cdot L \qquad (1)$$

$$C_{A+B} = C_A + C_B \qquad (2)$$

In addition, the permselectivities were estimated from the separation factors calculated using equation 3:

$$\text{Separation factor} = \frac{Y_{BA\ or\ BSA} / Y_{PhED}}{X_{BA\ or\ BSA} / X_{PhED}} \qquad (3)$$

where X and Y denote the molar fractions of phenyl compounds in the feed and in the permeate solutions, respectively, and the subscripts are the abbreviations of the phenyl compounds used here.

Concentration by Electrotransport. An aqueous BA or BSA solution at 2 mmol/dm³ was put in both anode and cathode sides of the cell, and then the direct current of 10 mA was continuously applied. The concentrations of BA or BSA in the feed and permeate sides were spectroscopically measured. The transport fraction was calculated from the maximum concentration in the anode side, C_{max}, and the initial concentration, C_{int}, using equation 4:

$$\text{Transport fraction} = \frac{C_{max} - C_{int}}{C_{int}} \cdot 100 \qquad (4)$$

Results and Discussion

The dependence of the BA and BSA permeabilities on the amount of grafted PDMAEMA and the pH value of the buffer solution during the application of the direct current was followed up using PE-g-PDMAEMA films. First, the BA and BSA permeabilities were examined during applying the direct current of 5 mA at pH 6 using PE-g-PDMAEMA films with different grafted amounts. Figure 1 shows the variations in the BA and BSA permeabilities of PE-g-PDMAEMA films with the grafted amount. The BA and BSA permeabilities showed the maximum values about 7 ~ 8 mmol/g. No ungrafted layers in any PE-g-PDMAEMA films could be observed more than 5 mmol/g (4). Since the grafted amount leads to the increase in the amount of dimethylamino groups as a positively charged functional group, the BA and BSA permeabilities increase with an increase in the grafted amount. However, a further increase in the grafted amount leads to the increase in the total thickness of the PE-g-PDMAEMA film, so the BA and BSA permeabilities decrease. Secondly, the BA and BSA permeabilities were examined in the buffer solutions of different pH values using the PE-g-PDMAEMA film with 7.7 mmol/g which showed the maximum permeabilities in Figure 1. Figure 2 shows the variation in the BA and BSA permeabilities with the pH value. The BA and BSA permeabilities passed through the maximum value at pH 6. This result can be explained in terms of the viscometric behavior of aqueous PDMAEMA solutions (ionic strength = 0.01 mol/dm^3, NaCl) (4) and the protonation behavior of dimethylamino groups affixed to PDMAEMA as a function of pH value (21). The maximum value of the reduced viscosity of PDMAEMA in the aqueous solutions was also observed at the same pH value. Therefore, it can be suggested that the maximum permeabilities of PE-g-PDMAEMA film to BA and BSA at pH 6 is due to the expansion of grafted PDMAEMA chains caused by the protonation of dimethylamino groups.

On the basis of the above results, electrotransports of BA and BSA were followed up at pH 6 using the PE-g-PDMAEMA film with 7.7 mmol/g. Figure 3 shows the BA and BSA permeabilities of PE-g-PDMAEMA film with 7.7 mmol/g on repeatedly turning on and off the direct current between 0 and 5 mA at pH 6. When the platinum mesh electrode in the permeate side was used as the anode, the BA and BSA permeabilities considerably increased by the application of the direct current. When the platinum mesh electrode in the permeate side was used as the cathode, the BA and BSA permeabilities were depressed. These results indicate that increased permeabilities of BA and BSA were caused not only by the concentration gradient but also by the migration of benzoate and benzenesulfonate anions toward the anode. In addition, electrotransport of BA was examined using a methacrylamide (MAAm, neutral) grafted PE film with 31.0 mmol/g and a methacrylic acid (MAA, anionic) grafted PE film with 13.8 mmol/g under the same conditions as shown in Figure 3(a). Both grafted PE films had no grafted layers. Figure 4 shows the BA permeabilities of the PE-g-PMAAm and PE-g-PMAA films on repeatedly turning on and off the direct current between 0 and 5 mA at pH 6. It is apparent that the BA permeabilities slightly increased for the PE-g-PMAAm film and little for the PE-g-PMAA film on applying the direct current. These results indicate that grafted polymer chains with positively charged functional groups are required to obtain a high permeability during the application of the direct current. The on-off regulation of permeation observed by turning on and off the direct current could be repeated in shorter intervals than that repeated by the temperature and pH changes (5).

In addition, the permeabilities of PhED as a neutral phenyl compound were investigated under the same electrical field as above described. Figure 5 shows the PhED permeabilities of the PE-g-PDMAEMA film at pH 6 by repeatedly turning on and off the direct current of 10 mA. The PhED permeabilities during turning on the direct current only slightly increased compared to those during turning off the direct current, even when the platinum mesh electrode in the permeate side was used as the anode. Therefore, it seems from this result that grafted PDMAEMA chains with positively charged dimethylamino groups orient toward the cathode in the feed solution. This behavior also exerts a favorable influence on the increase in the permeation of BA and BSA molecules on applying the direct current in addition to the above-mentioned two factors.

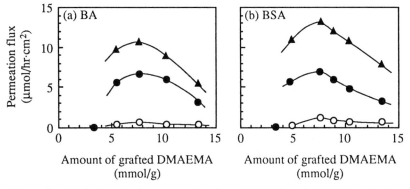

Figure 1. Variations in the permeation flux of (a) BA and (b) BSA with the amount of grafted DMAEMA in a buffer solution of pH 6. Direct current (mA): O: 0, ●: 5, ▲:10.

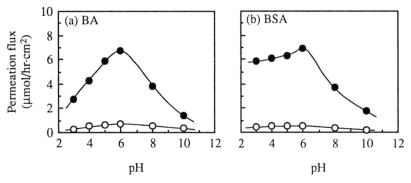

Figure 2. Variations in the permeation flux of (a) BA and (b) BSA with the pH value of the buffer solution for PE-g-PDMAEMA film with 7.7 mmol/g. Direct current (mA): O : 0, ● : 5.

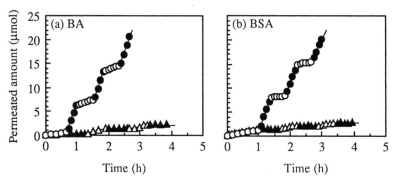

Figure 3. Reversible permeation of (a) BA and (b) BSA through PE-g-PDMAEMA film with 7.7 mmol/g at pH 6 by repeatedly turning on and off the direct current between 0 (O , △) and 5 mA (● , ▲). Electrode in the permeate side : O , ● : anode, △ , ▲ : cathode.

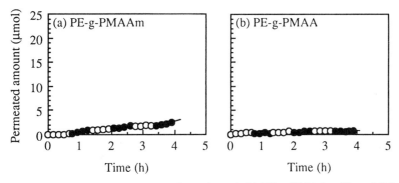

Figure 4. Permeation behavior of BA through (a) PE-g-PMAAm film with 3 1.0mmol/g and (b) PE-g-PMAA with 13.8 mmol/g at pH 6 by repeatedly turning on and off the direct current 0 (O ,△) and 5 mA (● ,▲). Electrode in the permeate side : O , ● : anode, △ , ▲ : cathode.

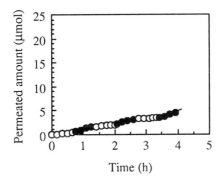

Figure 5. Reversible permeation of PhED through PE-g-PDMAEMA film with 7.7 mmol/g at pH 6 be repeatedly turning on and off the direct current between 0 (O) and 10 mA (●). Electrode in the permeate side was used as the anode.

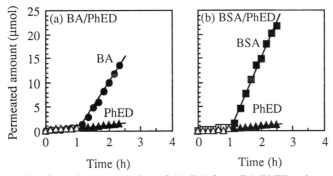

Figure 6. Selective permeation of (a) BA from BA/PhED mixture and (b) BSA from BSA/PhED mixture (C = 2.0 mmol/dm^3) through PE-g-PDMAEMA film with 7.7 mmol/g at pH 6 on applying the direct current of 10 mA. Electrode in the permeate side : O,●: cathode, △, ▲: anode. Molar fraction of anionic phenyl compounds (BA and BSA) was 0.50.

The permselectivities of PE-g-PDMAEMA film by electrotransport were followed up from in the BA/PhED and BSA/PhED systems at pH 6.   Figure 6 shows the permselectivities of the PE-g-PDMAEMA film with 7.7 mmol/g from the binary BA/PhED and BSA/PhED mixtures of pH 6 (mole % = 50/50) on applying the direct current.   While applying the direct current, benzoate and benzenesulfonate anions were selectively transported to the anode side through the PE-g-PDMAEMA film. The separation factors for the binary BA/PhED and BSA/PhED mixtures were calculated to be 17.8 and 23.8, respectively, from equation 3.   It is clear from Figure 5 that PE-g-PDMAEMA films show high permselectivities to anionic phenyl compounds.   In addition, selective permeation experiments were carried out using the BA/PhED or BSA/PhED mixtures with different molar fractions of BA or BSA at a constant total concentration of 2.0 mmol/dm^3.   Figure 7 shows the effect of feed composition of BA and BSA on the separation of BA/PhED and BSA/PhED mixtures and separation factor.   Separation factors increased with a decrease in the molar fraction of BA or BSA.   It is apparent from Figure 6 (b) that the selective separation of anionic phenyl compounds such as BA and BSA was attained through the PE-g-PDMAEMA films and the permeate solutions enriched with BA and BSA could be obtained even from a low BA and BSA feed solution.   Such a high permselectivity to BA or BSA was considered to be attributed to the migration of benzoate or benzenesulfonate anions toward the anode and electrostatic attraction between benzoate and benzenesulfonate anions and grafted PDMAEMA chains with positively charged dimethylamino groups.   It can be considered that the BSA fraction is higher than BA fraction in the permeate in spite of the molar fractions of anionic phenyl compounds in the feed, since most of BSA molecules dissociate into anions.

The concentration of anionic phenyl compounds through the PE-g-PDMAEMA films by electrotransport was studied.   An aqueous BA or BSA solution at 2 mmol/dm^3 was put in both sides of the cell, and then the direct current of 10 mA was continuously applied.   Figure 8 shows the changes in the concentrations of BA and BSA and the pH values in the feed and permeate solutions with the time.   The concentrations of BA and BSA in the anode side increased with the passage of the time and went up to over 3.8 mmol/dm^3.   The electrotransport of anionic phenyl anions to the anode side progressed concurrently with a sharp increase in the pH value in the cathode side.   It becomes apparent from these results that benzoate and benzenesulfonate anions are uphill transported to the anode side against their concentration gradient with simultaneous transfer of hydrogen ions generated by electrolysis of water molecules.   Bubbles produced in the vicinity of the electrodes during the application of the direct current were promptly removed because the presence of the bubbles resulted in a decrease in the migration of phenyl anions due to a significant increase in the electrical resistance.   The transport fraction of BA and BSA calculated from equation 4 amounted to 87 % and 95 %, respectively. Although the concentration of BA more rapidly increased than that of BSA, the maximum concentration of BA was lower than that of BSA.   It is understandable from the dissociation behavior of BA molecules calculated using the dissociation constant of pKa = 4.20 and the protonation behavior of dimethylamino groups affixed to DMAEMA chains determined from the colloid titration with potassium poly(vinyl alcohol) sulfate (KPVS) [17,21] that the degree of dissociation of BA molecules is low and most of dimethylamino groups are protonated in the pH range below pH 4. Since the pH value in the anode side is lower than that in the cathode side, benzoate anions incorporated into the PE-g-PDMAEMA film by electrostatic attraction with positively charged dimethylamino groups are readily released to the anode medium. For the aqueous BSA solution, however, the pH range below pH 4 was maintained for 3 hrs so that the higher maximum concentration was obtained.   The concentration of BA and BSA by electrotransport could be attained in more shorter

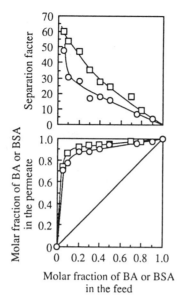

Figure 7. Effect of the feed composition on the separation factor and selective permeation of BA from BA/PhED mixture (○) and BSA from BSA/PhED mixture (□) at pH 6 on applying the direct current of 10 mA for PE-g-PDMAEMA film of 7.7 mmol/g. Electrode in the permeate side was used as the cathode.

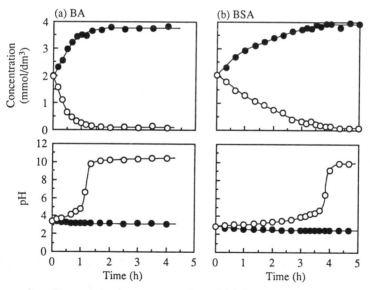

Figure 8. Changes in the concentration of (a) BA and (b) BSA and pH values in the anode (○) and cathode (●) sides on contituously applying the direct current of 10 mA using a PE-g-PDMAEMA film of 7.7 mmol/g.

24

time than that by the uphill transport using the pH difference across the PE-g-PDMAEMA films as a driving force and the maximum concentrations obtained here was a little higher than those by the uphill transport. From the above results, the PE-g-PDMAEMA films are expected to be applied to electrodialysis membranes for anionic organic compounds.

Literature Cited

1. Ogiwara, Y.; Takumi, M.; Kubota, H. *J. Appl. Polym. Sci.* **1982**, *27*, 3743.
2. Yamada, K.; Kimura, T.; Tsutaya, H.; Hirata, M. *J. Appl. Polym. Sci.* **1992**, *44*, 993.
3. Yamada, K.; Tsutaya, H.; Tatekawa, S.; Hirata, M. *J. Appl. Polym. Sci.* **1992**, *46*, 1065.
4. Yamada, K.; Tatekawa, S.; Hirata, M. *J. Colloid Interface Sci.* **1994**, *162*, 144.
5. Yamada, K.; Sato, T.; Tatekawa, S.; Hirata, M. *Polym. Gels Networks* **1994**, *2*, 323.
6. Gupta, B.D.; Tyagi, P.K.; Ray, A. R.; Singh, H. *J. Macromol. Sci. Chem.* **1990**, *A27*, 831.
7. Dogué, I. L. J.; Mermilliod, N.; Gandini, A. *J. Appl. Polym. Sci.* **1995**, *56*, 33.
8. Epailland, F. P.; Chevet, B.; Brosse, J. C. *J. Appl. Polym. Sci.* **1994**, *53*, 1291.
9. Yun, L.; Zhu, Z.; Hanmin, Z. *J. Appl. Polym. Sci.* **1994**, *53*, 405.
10. Hegazy, E. A.; El-Assy, N. B.; Taher, N. H.; Desouki, A. M. *Radiat. Phys. Chem.* **1989**, *33*, 539.
11. Yamada, K.; Hayashi, K.; Sakasegawa, K.; Onodera, H.; Hirata, M. *Nippon Kagaku Kaishi* [*J. Chem. Soc. Jpn., Chem. Ind. Chem.*] **1994**, 424.
12. Yamada, K.; Ebihara, T.; Gondo, T.; Sakasegawa, K.; Hirata, M. *J. Appl. Polym. Sci.* **1996**, *61*, 1899.
13. Matsuda, K.; Shibata, N.; Yamada, K.; Hirata, M. *Nippon Kagaku Kaishi* [*J. Chem. Soc. Jpn., Chem. Ind. Chem.*] **1997**, 575.
14. Yamada, K.; Hirata, M. *Interfacial Aspects of Multicomponent Polymer Materials*; Sperling, L. H., Ed.; Plenum: Orlando, Florida, in press.
15. Gregor, H. P.; Gold, D. H.; Frederick, M. J. Polym. Sci. **1957**, *23*, 467.
16. Hirata, M.; Yamada, K.; Matsuda, K.; Kokufuta, E. *ACS Symp. Ser.* **1994**, *548*, 493.
17. Uragami, T.; Yoshida, F.; Sugihara, M. *Separation Sci. Technol.* **1988**, *23*, 1067.
18. Yoshikawa, M.; Suzuki, M.; Sanui, K.; Ogata, N.; *J. Membrane Sci.* **1987**, *32*, 235.
19. Nonaka, T.; Takeda, T.; Egawa, H. *J. Membrane Sci.* **1993**, *76*, 193.
20. Yamada, K.; Sato, K.; Hirata, K. *Polyelectrolytes Potsdam '95 First International Symposium on Polyelectrolytes and International Bunsen-Discission-Meeting, Polyelectrolytes in Solution and at interfaces*, **1995**, 166, No. P5.26.
21. Yamada, K.; Sato, K.; Hirata, M. *J. Mater. Sci.* submitted for publication.
22. Sasaki, K.; Yamada, K.; Hirata, M. *Polym. Prepr. Jpn.* **1996**, *45(3)*, 401, No. II Pc 023.
23. Yamada, K.; Hirata, M. *Int. Symp. High-Tech Polymers and Polymeric Complexes Prepr.* **1996**, 79.

Chapter 3

Ionic Polymer–Metal Composites as Biomimetic Sensors and Actuators–Artificial Muscles

M. Shahinpoor[1], Y. Bar-Cohen[2], T. Xue[2], Joycelyn S. Harrison[3], and J. Smith[3]

[1]Artificial Muscles Research Institute, University of New Mexico, Albuquerque, NM 87131
[2]NASA Jet Propulsion Laboratory (JPL), California Institute of Technology, Pasadena, CA, 91109–8099
[3]Composites and Polymers Branch, NASA Langley Research Center, Hampton, VA 23681–0001

ABSTRACT

This chapter presents an introduction to ionic-polymer-metal composites and some mathematical modeling pertaining to them. It further discusses a number of recent findings in connection with ion-exchange polymer metal composites (IPMC) as biomimetic sensors and actuators. Strips of these composites can undergo large bending and flapping displacement if an electric field is imposed across their thickness. Thus, in this sense they are large motion actuators. Conversely by bending the composite strip, either quasi-statically or dynamically, a voltage is produced across the thickness of the strip. Thus, they are also large motion sensors. The output voltage can be calibrated for a standard size sensor and correlated to the applied loads or stresses. They can be manufactured and cut in any size and shape. In this paper first the sensing capability of these materials is reported. The preliminary results show the existence of a linear relationship between the output voltage and the imposed displacement for almost all cases. Furthermore, the ability of these IPMC's as large motion actuators and robotic manipulators is presented. Several muscle configurations are constructed to demonstrate the capabilities of these IPMC actuators. This paper further identifies key parameters involving the vibrational and resonance characteristics of sensors and actuators made with IPMC's. When the applied signal frequency is varied, so does the displacement up to a point where large deformations are observed at a critical frequency called resonant frequency where maximum deformation is observed. Beyond which the actuator response is diminished. A data acquisition system was used to measure the parameters involved and record the results in real time basis. Also the load characterization of the IPMC's were measured and showed that these actuators exhibit good force to weight characteristics in the presence of low applied voltages. Finally, reported are the cryogenic properties of these muscles for potential utilization in an outer space environment of few Torrs and temperatures of the order of -140 degrees Celsius. These muscles are shown to work quite well in such harsh cryogenics environments and thus present a great potential as sensors and actuators that can operate at cryogenic temperatures.

Keywords: Ionic Polymer-Metal Composite Sensor, Soft Actuator, Artificial Muscles, Biomimetic Sensor, Vibrations, Resonance.

1. INTRODUCTION

Ion-exchange polymer-metal composites (IPMC) are active actuators that show large deformation in the presence of low applied voltage and exhibit low impedance. They operate best in a humid environment and can be made as a self-contained encapsulated actuators to operate in dry environments as well. They have been modeled as both capacitive and resistive element actuators that behave like biological muscles and provide an attractive means of actuation as artificial muscles for biomechanics and biomimetics applications. Grodzinsky[1], Grodzinsky and Melcher[2,3] and Yannas, Grodzinsky and Melcher[4] were the first to present a plausible continuum model for electrochemistry of deformation of charged polyelectrolyte membranes such as collagen or fibrous protein and were among the first to perform the same type of experiments on animal collagen fibers essentially made of charged natural ionic polymers and were able to describe the results through electro-osmosis phenomenon. Kuhn[5] and Katchalsky[6], Kuhn, Kunzle, and Katchalsky[7], Kuhn, Hargitay, and Katchalsky[8], Kuhn, and Hargitay[9], however, should be credited as the first investigators to report the ionic chemomechanical deformation of polyelectrolytes such as polyacrylic acid (PAA), polyvinyl chloride (PVA) systems. Kent, Hamlen and Shafer[10] were also the first to report the electrochemical transduction of PVA-PAA polyelectrolyte system. Recently revived interest in this area concentrates on artificial muscles which can be traced to Shahinpoor and co-workers and other researchers [11-14, 22-53], Osada[15], Oguro, Asaka and Takenaka[16], Asaka, Oguro, Nishimura, Mizuhata and Takenaka[17], Guo, Fukuda, Kosuge, Arai, Oguro and Negoro[18], De Rossi, Parrini, Chiarelli and Buzzigoli[19] and De Rossi, Domenici and Chairelli[20]. More recently De Rossi, Chiarelli, Osada, Hasebe, Oguro, Asaka, Tanaka, Brock, Shahinpoor, Mojarrad[11-69] have been experimenting with various chemically active as well as electrically active ionic polymers and their metal composites as artificial muscle actuators.

Essentially polyelectrolytes possess ionizable groups on their molecular backbone. These ionizable groups have the property of dissociating and attaining a net charge in a variety of solvent medium. According to Alexanderowicz and Katchalsky[17] these net charge groups which are attached to networks of macromolecules are called polyions and give rise to intense electric fields of the order of 10^{10} V/m. Thus, the essence of electromechanical deformation of such polyelectrolyte systems is their susceptibility to interactions with externally applied fields as well as their own internal field structure. In particular if the interstitial space of a polyelectrolyte network is filled with liquid containing ions, then the electrophoretic migration of such ions inside the structure due to an imposed electric field can also cause the macromolecular network to deform accordingly. Shahinpoor[18,22,25,26,28,29,31-,36] and Shahinpoor and co-workers[21,23,24,27,30] have recently presented a number of plausible models for micro-electro-mechanics of ionic polymeric gels as electrically controllable artificial muscles in different dynamic environments. The reader is referred to these papers for the theoretical and experimental results on dynamics of ion-exchange membranes - platinum composite artificial muscles.

The IPMC muscle used in our investigation is composed of a perfluorinated ion exchange membrane (IEM), which is chemically composited with a noble metal such as gold or platinum. A typical chemical structure of one of the ionic polymers used in our research is

$$[-(CF_2-CF_2)_n-(CF-CF_2)_m-]$$
$$|$$
$$O-CF-CF_2-O-CF_2-SO_3^-.....M^+$$
$$|$$
$$CF_3$$

where n is such that 5<n<11 and m ~ 1, and M^+ is the counter ion (H^+, Li^+ or Na^+). One of the interesting properties of this material is its ability to absorb large amounts of polar solvents, i.e. water. Platinum, Pt, metal ions, which are dispersed through out the hydrophilic regions of the polymer, are subsequently reduced to the corresponding metal atoms. This results-in the formation of a dendritic type electrodes.

Metallization of Ion-Exchange Membranes

In Metalizing this material there is a first stage of in-depth molecular metallization and a second stage of surface plating and electroding. Thus, the important stage of compositing is the first stage which can be postulated to take place according to the following chemical reacctions :

$$[Pt(NH_3)_4]^{2+} + 2e^- \Rightarrow Pt^o + 4NH_3 \tag{1.1}$$

$$LiBH_4 + 8OH^- \Rightarrow BO_2^- + Li^+ + 6H_2O + 8e^- \tag{1.2}$$

From equations (1.1) and (2.1), it is possible to draw the following:

$$LiBH_4 + 4[Pt(NH_3)_4]^{2+} + 8OH^- + 8e^- \Rightarrow 4Pt^o + 16NH_3 + BO_2^- + Li^+ + 6H_2O + 8e^-$$
$$= LiBH_4 + 4[Pt(NH_3)_4]^{2+} + 8OH^- \Rightarrow 4Pt^o + 16NH_3 + BO_2^- + Li^+ + 6H_2O \tag{1.3}$$

Also, the solid form of $LiBO_2$ occasionally precipitates. Therefore, the overall reaction may be,

$$LiBH_4 + 4[Pt(NH_3)_4]^{2+} + 8OH^- \Rightarrow 4Pt^o(s) + 16NH_3(g) + LiBO_2(s?) + 6H_2O(l) \tag{1.4}$$

Now, the biggest question is the source of hydroxyl ions. Apparentely, the following reaction may be possible.

$$LiBH_4 + 4H_2O \Rightarrow 4H_2 + LiOH(Li^+ + OH^-) + B(OH)_3(s?) \tag{1.5}$$

This indicates 9 moles of $LiBH_4$ are required for reducting 4 moles of $Pt(NH_3)_4^{2+}$.

2-THEORETICAL CONSIDERATION

A simple one-dimensional model of electrically-induced deformation of ionic polymeric gels is such that :

$$\sigma = (1/3)E(C_0, C_i)(\lambda - \lambda^{-2}), \tag{2.1}$$

$$\sigma = \kappa(C_0, C_i)E^{*2} \tag{2.2}$$

where σ is the stress, λ is the stretch, $E(C_0, C_i)$ is the corresponding Young's modulus of hyper-elasticity, C_0 is the polymer solid concentration, C_i, $(i=1,2,....,N)$'s are the molal concentration of various ionic species in the aqueous medium, κ (C_0, C_i) is an electromechanical coefficient and E^* is the local electric field. Thus bending can occur due to differential contraction and expansion of outer most remote regions of a strip if an electric field is imposed across its thickness as shown below in Figure 1. Since ionic polyelectrolytes are for the most part three dimensional network of macromolecules cross-linked nonuniformly, the concentration of ionic charge groups are also nonuniform within the polymer matrix. Therefore the mechanism of bending is partially related to the redistribution of fixed ions and migration of mobile ions within the network due to the imposition of an electric field. However, recent modeling effort on the sensing and actuation have revealed that this effect may play an insignificant role on the actuation which may be dominated by surface charge interactions. This subject is currently under investigation.

A simple one-dimensional model of electrically-induced dynamic deformation or vibration of a cantilever beam made with such IPMC artificial muscle strips is given by the following equations :

$$\rho \frac{\partial^2 y}{\partial t^2} = \frac{\partial \sigma}{\partial x} + F(x, t), \tag{2.3}$$

$$\varepsilon = \varepsilon_c + \kappa_E \eta, \quad -C < \eta < C, \tag{2.4}$$

$$\lambda = 1 + \varepsilon, \tag{2.5}$$

$$\lambda_+ - \lambda_- = 2\kappa_E C, \tag{2.6}$$

where F is the body force per unit volume of the muscle, ρ is the density, ε is the strain, subscript c indicates values at the neutral axis of the cross-section of the strip, C is the distance of the outer-most remote fibers, κ_E is the local curvature due to an imposed electric field, η is a cross-sectional parameter, E^* is the local electric field, x and t are axial location and time variables and subscripts + and -, respectively indicate the values of variable at the outermost remote fibers.

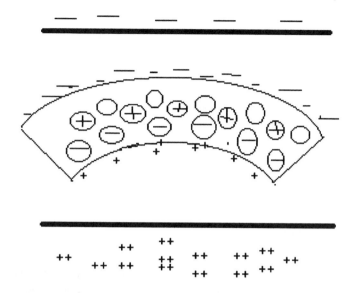

Figure 1. General redistribution of charges in an ionic polymer due to an imposed electric field.

Thus bending can occur due to differential contraction and expansion of outer most remote fibers of a strip if an electric field is imposed across its thickness as shown below in Figures 1 and 2. Numerical solutions to the above set of dynamic equations are presently underway and will be reported later. However, it must be mentioned that the governing equations (1)-(6) display a set of highly non-linear dynamic equations of motion for the IPMC artificial muscles.

Presently attempts are under way to establish existence and uniqueness of dynamic solutions to the above equations mathematically. However, experimental observations in our laboratory clearly indicate the non-linear motion characteristics of such muscles as well as unique vibrational response and resonance characteristics.

For detailed dynamics description and analysis of the continuum theory of ionic polymeric gel the reader is referred to Segalman, Witkowski, Adolf and Shahinpoor[25]. Since polyelectrolytes are for the most part three dimensional network of macromolecules cross-linked nonuniformly, the concentration of ionic charge groups are also nonuniform within the polymer matrix. Therefore the mechanism of swelling and contraction are intimately related to osmotic diffusion of solvent, ions and counterions into and out of the gel. One possible way to describe this mechanism is to model the system by the governing continuum mechanics equations and Neo-Hookean deformation theory. In the next section an analytical relation is presented as described by Segalman, Witkowski, Adolf and Shahinpoor[25].

3-ION TRANSPORT MECHANISMS

Let $c(X,t)$ be the solvent concentration, $H(X,t)$ be the ionic concentration, $x(X,t)$ be the position vector of a typical gel element, X be the reference material coordinate, and t be the time such that the governing continuum mechanics equation takes the following forms:

$$\frac{\partial c}{\partial t} = \nabla \cdot \left[D_{1,1}(c,H)\nabla c + D_{1,2}(c,H)\nabla H \right] - \nabla \cdot (c\dot{x}) \tag{3.1}$$

$$\frac{\partial H}{\partial t} = \nabla \cdot \left[D_{2,1}(c,H)\nabla c + D_{2,2}(c,H)\nabla H \right] - \nabla \cdot (H\dot{x}) + \dot{H}_s \tag{3.2}$$

$$\frac{\partial \rho_g}{\partial t} + \nabla(\rho \dot{x}) = 0 \tag{3.3}$$

$$\rho_g \ddot{x} = \nabla : S + \rho_g f_b \tag{3.4}$$

$$\frac{\partial \varepsilon}{\partial t} = S : \nabla \dot{x} + \nabla \cdot q + \nabla \cdot q_c + J \cdot E + \rho_g h \tag{3.5}$$

where x is the displacement, a superposed dot stands for a differentiation with respect to time, $D_{i,j}$ is diffusion coefficient, $\dot{H}_s$ is the source term for the production of ions in the gel, ρ_g is the gel density, S is the stress tensor, f_b is the body force vector which includes electromagnetic and gravitational terms, ε is the specific internal energy of the ionic polymeric gel, q is the heat flux vector, q_c is the chemical energy flux vector, J is the electric current flux vector, E is the electric field vector, and h is the specific source of energy production in the gel. The stress tensor S, is related to deformation gradient field by means of Neo-Hookean type constitutive equation which may be represented by the following equation:

$$S = G(c)\left[F^{-1}(F^{-1})^T - I \right] + pI \tag{3.6}$$

where $F = (\partial x / \partial X)$, I is the identity matrix, superposed T stands for transpose, G(c) is the Young's modulus, p is an unknown Lagrangian multiplier to be found by solving system of equations 1-12. The solution to this model will enable one to electrically control the polymeric muscle bending and therefore the motion of the swimming robotic structure. For additional references on modeling of IPMC artificial muscles the reader is referred to references [11]-[14] and [22]-53]

4-BIOMIMETIC SENSING CAPABILITY OF IPMC

Investigations of the use of ion-exchange-membrane materials as sensors can be traced to Sadeghipour, Salomon, and Neogi[58] where they used such membranes as a pressure sensor/damper in a small chamber which constituted a prototype accelerometer. However, it was Shahinpoor [39] who first discussed the phenomenon of flexogelectric effect in connection with dynamic sensing of ionic polymeric gels. In this paper the focus is on the application of the IPMC sensor on quasi-static or dynamic displacement sensing where the response of the sensor against large imposed displacements was investigated. To get a better understanding of the mechanism of sensing, more explanation must be given about the general nature of the ionic polymers.

As shown in Figures 1 and 2, IPMC strips generally bend towards the anode and if the voltage signal is reversed they also reverse their direction of bending. Conversely by bending the material, shifting of mobile charges become possible due to imposed stresses. Consider Figure 2 where a rectangular strip of the composite sensor is placed between two electrodes. When the composite is bent a stress gradient is built on the outer fibers relative to the neutral axis (NA). The mobile ions therefore will shift toward the favored region where opposite charges are available. The deficit in one charge and excess in the other can be translated into a voltage gradient which is easily sensed by a low power amplifier.

4.1-Quasi- Static Sensing

The experimental results showed that a linear relationship exists between the voltage output and imposed quasi-static displacement of the tip of the IPMC sensor as shown in Figure 3. The experimental set up was such that the tip of the cantilevered IPMC strip as shown in Figure 2 was mechanically moved and the corresponding output voltage recorded. The results are shown in Figure 3.

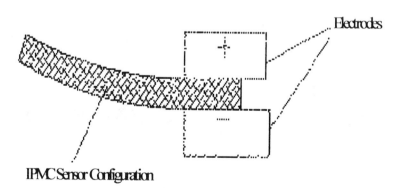

Electrodes

IPMC Sensor Configuration

Figure 2. Simple IPMC sensor placed between two electrodes.

32

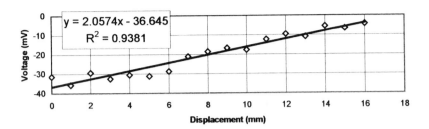

Sensor Response (+ Displacement)
Membrane Face Down

$y = 2.0574x - 36.645$
$R^2 = 0.9381$

Figure 3. Inverted IPMC film sensor response for positive displacement input.

4.2-Dynamic Sensing

When strips of IPMC are dynamically disturbed by means of a dynamic impact or shock loading, a damped electrical response is observed as shown in Figure 4. The dynamic response was observed to be highly repeatable with a fairly high band width to 100's of Hz.. This particular property of IPMC's may find a large number of applications in large motion sensing devices for a variety of industrial applications. Since these muscles can also be cut as small as one desires, they present a tremendous potential to micro-electro-mechanical systems (MEMS) sensing and actuation applications.

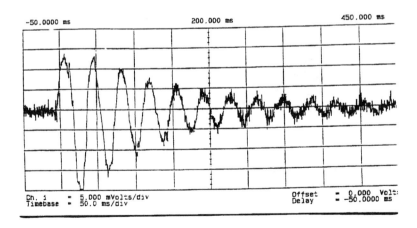

Figure 4- Dynamic sensing response in the form of output voltage of strips (40mmx5mmx0.2mm) of IPMC subject to a dynamic impact loading as a cantilever.

5-BIOMIMETIC ACTUATION PROPERTIES OF IPMC's

5.1- General Considerations

As mentioned before, IPMCs are large motion actuators that operate under a low voltage compared to other actuators such as peizocerams or shape memory alloys. Table 1 shows a comparison between the capability of IPMC materials and both electroceramics and shape memory alloys. As shown in Table 1, IPMC materials are lighter and their potential striction capability can be as high as two orders of magnitude more than EAC materials. Further, their response time is significantly higher than Shape Memory Alloys (SMA). They can be designed to emulate the operation of biological muscles and have unique characteristics of low density as well as high toughness, large actuation strain and inherent vibration damping.

TABLE 1: Comparison of the properties of IPMC, SMA and EAC

Property	Ionic polymer-Metal Composites (IPMC)	Shape Memory Alloys (SMA)	Electroactive Ceramics (EAC)
Actuation displacement	>10%	<8% short fatigue life	0.1 - 0.3 %
Force (MPa)	10 - 30	about 700	30-40
Reaction speed	μsec to sec	sec to min	μsec to sec
Density	1- 2.5 g/cc	5 - 6 g/cc	6-8 g/cc
Drive voltage	4 - 7 V	NA	50 - 800 V
Power consumption	watts	watts	watts
Fracture toughness	resilient, elastic	elastic	fragile

These muscles are manufactured by a unique chemical process in which a noble metal (Pt) is deposited within the molecular network of the base ionic polymer. Equations (1.1) through (1.5) depict the essence of such chemical compositing which is followed by a surface plating and electroding process. One of the interesting properties of IPMC artificial muscles is its ability to absorb large amounts of polar solvents, i.e. water. Platinum salt ions, which are dispersed through out the hydrophilic regions of the polymer, are subsequently chemically reduced to the corresponding metal atoms. This results-in the formation of dendritic type electrodes. In Figure 5, scanning electron micrographs are shown in two magnifications, with an order of magnitude difference. On the left, a view is given of the edge of an electroded muscle. The Pt metal covers each surface of the film with some of the metal penetrating the subsurface regions of the material. A closer view with x10 magnification is shown in Figure 5 on the right.

34

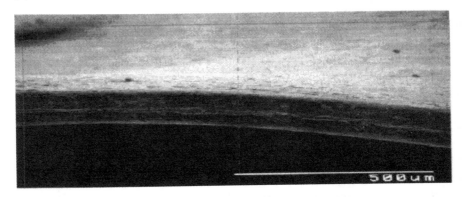

(a)

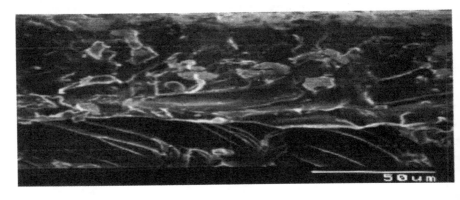

(b)

Figure 5: Scanning Electron Micrographs of the Structure of IPMC, (a) displays the the thickness edge of the muscle while (b) depicts the metal particle deposition on the network inside the muscle

When an external direct voltage of 2 volts or higher is applied on a IPMC film, it bends towards the anode. An increase in the voltage level (up to 6 or 7 volts) causes a larger bending displacement. When an alternating voltage is applied, the film undergoes swinging movement and the displacement level depends not only on the voltage magnitude but also on the frequency. Lower frequencies (down to 0.1 or 0.01 Hz) lead to higher displacement (approaching 25mm) for a 0.5cmx2cmx0.2mm thick strip. Thus, the movement of the muscle is fully controllable by the applied electrical source. The muscle performance is also strongly dependent on the water content which serves as an ion transport medium and the dehydration rate gradient across the film leads to a pressure difference. The frequency dependence of the ionomer deflection as a function of the applied voltage is shown in Figure 6. A single film was used to emulate a miniature bending arm that lifted a mass weighing a fraction of a gram. A film-pair weighing 0.2-g was configured as a linear actuator and using 5V and 20 mW successfully induced more than 11% contraction displacement. Also, the film-pair displayed a significant expansion capability, where a stack of two film-pairs 0.2 cm thick expanded to about 2.5 cm wide (see Figure 7).

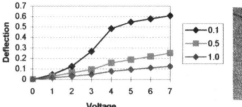 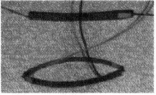

Figure 6: The deflection of a IPMC strip as a function of the frequency (0.1, 0.5 and 1 Hz) and the applied voltage.

Figure 7: IPMC film-pair in expanded mode. A reference pair (top) and an activated pair (bottom).

5.2- Muscle actuators for soft robotic applications

IPMC films have shown remarkable displacement under relatively low voltage, using very low power. Since the IPMC films are made of a relatively strong material with a large displacement capability, we investigated their application to emulate fingers. In Figure 8, a gripper is shown that uses IPMC fingers in the form of an end-effector of a miniature low-mass robotic arm.

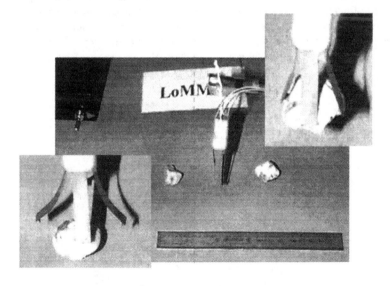

Figure 8: An end-effector gripper lifting 10.3-g rock under 5-V, 25-mW activation using four 0.1-g fingers made of IPMC's.

The fingers are shown as vertical gray bars and the electrical wiring, where the films are connected back-to-back, can be seen in the middle portion of Figure 8. Upon electrical activation, this wiring configuration allows the fingers to bend either inward or outward similar to the operation of a hand and thus close or open the gripper fingers as desired. The hooks at the end of the fingers represent the concept of nails and secure the gripped object that is encircled by the fingers.

To date, multi-finger grippers that consist of 2- and 4-fingers were produced, where the 4-finger gripper shown in Figure 8 was able to lift 10.3-g. This gripper prototype was mounted on a 5-mm diameter graphite/epoxy composite rod to emulate a light weight robotic arm. This gripper was driven by a 5 volts square wave signal at a frequency of 0.1 Hz to allow sufficient time to perform a desirable demonstration of the capability of the Gripper -- opening the gripper fingers, bringing the gripper near the collected object, closing the fingers and lifting an object with the arm. The demonstration of this gripper capability to lift a rock was intended to pave the way for a future potential application of the gripper to planetary sample collection tasks (such as Mars Exploration) using ultra-dexterous and versatile end-effector.

5.3- Linear and Platform type actuators

For detailed dynamics description and analysis of the dynamic theory of ionic polymeric gels the reader is referred to Shahinpoor and co-workers [11-14,22-70]. Since polyelectrolytes are for the most part three dimensional network of macromolecules cross-linked nonuniformly, the concentration of ionic charge groups are also nonuniform within the polymer matrix. Therefore the mechanism of bending is partially related to migration of mobile ions within the network due to imposition of an electric field as shown in Figure 1. However, recent investigation by the author and his co-workers point to a stronger effect due to surface charge interactions which will be reported later. Figure 9 depicts the bending deformation of a typical strip with varying electric field, while Figure 10 displays the variation of deformation with varying frequency of alternating electric field.

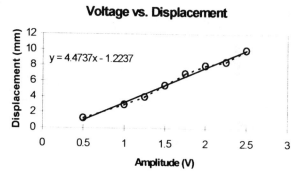

Voltage vs. Displacement

$y = 4.4737x - 1.2237$

Figure 9-Bending Displacement versus Voltage for a typical IPMC strip of 5mmx0.20mmx20mm under a frequency of 0.5Hz.

Based on such dynamic deformation characteristics, linear and platform type actuators can be designed and made dynamically operational. These types of actuators are shown in Figure 11.

6-LARGE AMPLITUDE VIBRATIONAL RESPONSE OF IPMC's

6.1-General Considerations

Strips of IPMC were used to study their large amplitude vibration characteristics. The IPMC strips were chemically composited with Platinum. A small function generator circuit was designed and built to produce approximately ±4.0V amplitude alternating wave at varying frequency. In order to study the feasibility of using IPMC artificial muscles as vibration damper, a series of muscles made from IPMC's were cut into strips and attached either end-to-end or to one fixed platform and another movable platform in a cantilever configuration. By applying a low voltage the movement of the free end of the beam could be calibrated and its response measured, accordingly. Typical data for the frequency-dependence of amplitude of lateral oscillations of the muscle strips subjected to alternating voltages of various forms such as sinusoidal, rectangular, saw-tooth or pulsed were investigated. Furthermore, the static deformation of the strip with voltage as well as the frequency dependence of deflection-voltage curves were evaluated.

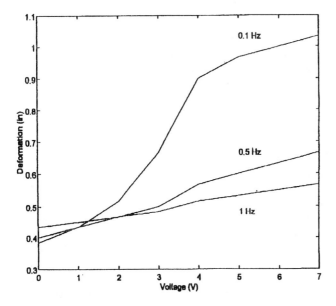

Figure 10-Frequency dependence of bending deformation of IPMC composite muscles

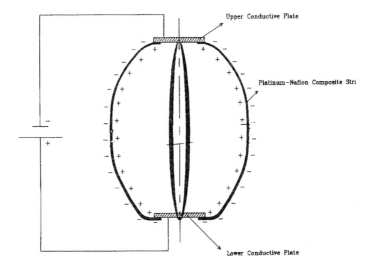

Figure 11- A typical linear-type robotic actuators made with IPMC legs

6.2--EXPERIMENTAL OBSERVATIONS

IPMC artificial muscle strips of about 2-4cmx4-6mm were cut and completely swollen in a suitable solution such as water to swell. The IPMC muscle strip typically weighed 0.1-0.4 grams and its thickness measured about 0.2mm. The strip was then held by a clamping setup between two platinum plate terminals which were wired to a signal amplifier and generator apparatus driven by Labview software through an IBM compatible PC containing an analog output data acquisition board. The amplifier (Crown model D-150A) was used to amplify the signal output of a National Instrument data acquisition card (AT-AO-10). Software was written to produce various waveforms such as sinusoid, square, triangular and saw tooth signals at desired frequencies up to 100 Hz and amplitudes up to 10 volts. When a low direct voltage was applied, the membrane composite bent toward the anode side each time. So by applying an alternating signal we were able to observe alternating bending of the actuator that followed the input signal very closely up to 35 Hz. At voltages higher than 2.0 volts, degradation of displacement output of the actuator was observed which may be due to dehydration. Water acts as the single most important element for the composite bending by sequentially moving within the composite depending on the polarity of the electrodes. The side facing the anode dehydrated faster than the side facing the cathode leading to a differential stresses which ultimately leads to bending of the composite. So, prior to each experiment, the composite was completely swollen in water. The displacement of the free end of a typical 2cmx4mm composite membrane was then measured for the frequency range of 0.1-35 Hz for sinusoid input voltage at 2.0 volts amplitude (Figure 12).

39

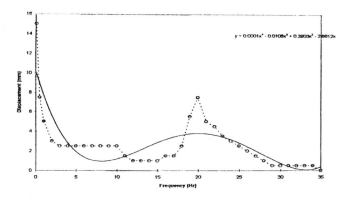

Figure 12- Amplitude of displacement versus the imposed frequency for a voltage of 2 volts for a 2cmx4mmx0.2mm sample.

Resonance was observed at about 20 Hz where the associated displacement was observed to be 7.5mm. It should be noted that as the actuator dehydrated the resonance frequency and maximum displacement varied accordingly. By encapsulating the strips in a plastic membrane such as Saran[R], the deterioration in the amplitude of oscillation decreased with time. However, the initial amplitude of oscillation for the same level of voltage was smaller than the unwrapped case due to increased rigidity of the strip. For our sample actuator the resonance occurred in the frequency range of 12 to 28 Hz.

Based on such dynamic deformation characteristics, noiseless swimming robotic structures as shown in Figure 13 and cilia assembly-type robotic worlds, similar to coral reefs, as shown in Figure 14, were constructed and tested for collective vibrational dynamics. Furthermore, wing flapping flying machines, schematically shown in Figure 15, can be equipped with these muscles.

Figure 13. Robotic swimmer with muscle undulation frequency of 3 Hz (frame time interval, 1/3 second)

Figure 14-Cilia-Type assembly of IPMC-Pt Muscles Simulating Collective Dynamic Vibrational Response Similar to Coral Reefs and could create anti-biofouling surfaces

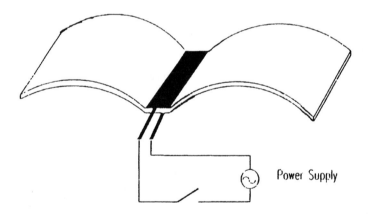

Power Supply

Figure 15-Wing-flapping flying machines design depicted schematically.

7-LOAD AND FORCE CHARACTERIZATION OF IPMC's

7.1-General Considerations

In order to measure the force generated by strips of these muscles in a cantilever form an experimental set up was designed using a load cell. A load cell (Transducer Techniques, model GS-30, 30 grams capacity) and corresponding signal conditioning module (Transducer Techniques, model TMO-1) together with a power supply was setup and connected to a PC-platform data acquisition and signal generation system composed of a 12-bit analog output board (National Instrument AT-AO-10) and a 16-bit multi-input-output board (National Instrument AT-MIO-16XE-50). A Nicolet scope was used to monitor the input and output waveform. Labview™ software was used to write a program to generate various waveform such as sinusoid, square, saw tooth, and triangular signals at desired frequencies and amplitudes. The effective length of the membrane was 10mm. . This made the effective weight of the muscle producing a force to be about 20 milligrams. The resulting graphs were then adjusted for initial noise and pre-load and plotted over 5 second period (2.5 cycles). The force capability of these muscles , on average was measured to be about 400 N/Kgm indicating that these muscles can lift almost 40 times their own weight. Figures 16 depict such general trends.

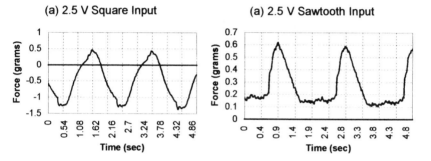

Figure 16. IPMC actuator response for square and saw tooth wave input at 2.5 Volts rms and a current of about 20 milliamps

8-CRYOGENIC PROPERTIES OF IPMC ARTIFICIAL MUSCLES

In this section are reported a number of recent experimental results pertaining to the behavior of ionic polymer metal composites (IPMC) under low pressure (few Torrs) and low temperature (-140 degrees Celsius). These experimental results have been obtained in a cryogenic chamber at NASA/JPL as well as a cryogenic chamber at the Artificial Muscles Research Institute at UNM. The interest at NASA/JPL was to study the actuation properties of these muscles in a harsh space environment such as one Torr of pressure and -140 degrees Celsius temperature. While at UNM the electrical properties , sensing capabilities as well as actuation properties of these muscles were tested in an atmospheric pressure chamber with a low temperasture of -80 degrees Celsius.

In general the results show that these materials are still capable of sensing and actuation in such harsh conditions as the following Figures 17 through 24 display. Furthermore, these IPMC artificial muscles become less conductive, i.e., their electrical resistance increases with decreasing temperature. This result appears to defy the generally accepted fact that resistance of metallic conductors increases /decreases with increasing/decreasing temperature, respectively.

42

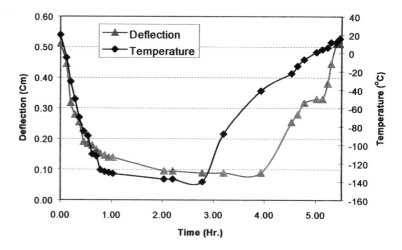

Figure 17- Deflection characteristics of IPMC as a function of time and temperature

Figure 21 (a) clearly shows a remarkable trend which is opposite to the normal trend of resistance-temperature variations in conductors. The graph is showing that as the temperature decreases in IPMC artificial muscles the resistance increases . For any given temperature, there is a range of linear response of V vs. I, which indicates a close to a pure resistor response. This rather remarkable effect is presently under study. However, one plausible explanation is that the colder the temperature the less active are the ionic species within the network of IPMC and thus the less ionic current activities. Since current is voltage over the resistance R, i.e., I=V/R, thus R has to increase to accommodate the decreasing ionic current due to decreasing temperature.

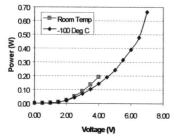

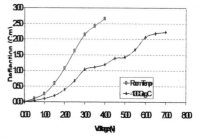

Figure 18: Power consumption of the IPMC strip bending actuator as a function of activation voltage.

Figure 19: Deflection of the bending IPMC strip as function of voltage

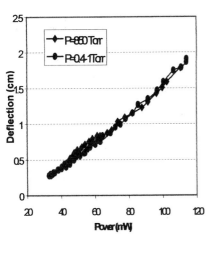
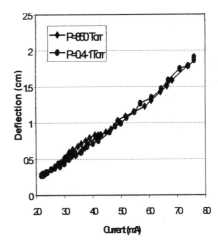

a. View of the deflection vs. power

b. View of the deflection vs. current

Figure 20- Deflection versus power and current under a constant voltage of 3 volts And a frequency of 0.1-Hz. For two different pressures

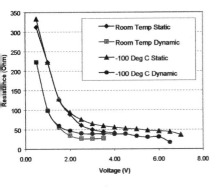
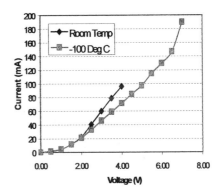

(a)-IPMC strip static (V/I) and dynamic (V/I) resistance at various temperature.

(b)-The relation between voltage and current for an IPMC strip that was exposed to RT and to -100°C.

Figure 21- Effect of temperature on the electrical resistance.

Figures 22 , 23 and 24 show the relationship between the temperature, voltage , current, power and displacement in a typical IPMC strips. Note that the behaviour of this material at low temperatures resembles more a semi-conductor type response to colder temperatures rather than a typical metalic conductor.

44

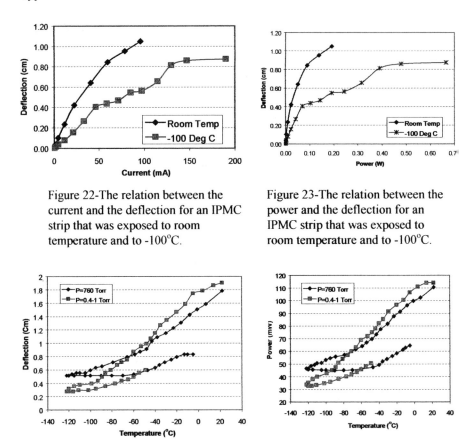

Figure 22-The relation between the current and the deflection for an IPMC strip that was exposed to room temperature and to -100°C.

Figure 23-The relation between the power and the deflection for an IPMC strip that was exposed to room temperature and to -100°C.

Figure 24-Deflection and power consumption of the IPMC muscle as a function of temperature with pressure as a parameter. Vpeak=3 V, Freq=0.1Hz.

9-SUMMARY

An introduction to ionic polymer metal composites as biomimetic sensors and actuators were presented. Some theoretical modeling on the mechanisms of sensing and actuation of such polymer composites were given. Highly Dynamic sensing characteristics of IPMC strips were remarkable in accuracy and repeatability and found to be superior to existing motion sensors and micro sensors. A new type of soft actuator and multi-fingered robotic hand were made from IPMC artificial muscles and found to be quite superior to conventional grippers and multi-fingered robotic hands. The feasibility of designing linear and platform type robotic actuators made with IPMC artificial muscle were presented. By applying a low voltage the movement of free end of the actuator could be calibrated and its response could be measured, accordingly. The feasibility of designing dynamic vibrational systems of artificial muscles made with IPMC artificial muscle were presented. Our experiments confirmed that these types of composite muscles show remarkable bending displacement that follow input signal very closely. When the applied signal frequency is varied, so did the displacement up to a point where large deformations were observed at a critical frequency called resonant frequency where maximum deformation was observed, beyond which the actuator response was diminished. A data acquisition system was used to measure the parameters involved and record the results in real time basis. The observed remarkable vibrational characteristics of IPMC composite artificial muscles clearly point to the potential of these muscles for biomimetics applications such as swimming robotic structures, wing-flapping flying machines, slithering snakes, heart and circulation assist devices, peristaltic pumps and dynamic robotic cilia-worlds. The cryogenic properties of these materials were quite unique. The fact that they still operated at very low temperatures such as -140 degrees Celsius shows their potential as cryogenic sensors and actuators. Their resistance increased with decreasing temperature, a property that is opposite to all metallic conductors.

9-ACKNOWLEDGMENT

The results reported in this manuscript were obtained partially under the NASA LoMMAs Task that is sponsored by NASA HQ, Code Q as part of the TeleRobotics program. Mr. David Lavery and Dr. Chuck Weisbin are the NASA HQ and JPL TeleRobotics program managers, respectively. This research was also partially supported by the Artificial Muscles Research Institute (AMRI) of UNM as well as Environmental Robots Incorporated through a grant from NRL. Thanks are also due to Dr. Kwang Kim for chemical reaction analysis.

10-REFERENCES

1. Grodzinsky, A.J., "Electromechanics of Deformable Polyelectrolyte Membranes", Sc.D. Dissertation, Dept. of Elec. Eng., MIT, Cambridge, June 1974.
2. Grodzinsky, A. J. and Melcher, J. R., "Electromechanics of Deformable, Charged Polyelectrolyte Membranes", Proc. 27th Annu. Conf. Engineering in Medicine and Biology, Vol. 16, 1974, paper 53.2.
3. Grodzinsky, A. J., Melcher, J. R., "Electromechanical Transduction with Charged Polyelectrolyte Membranes", IEEE Transactions on Biomedical Engineering, Vol. BME-23, No. 6, pp421-433, November 1976.
4. Yannas, I. V., Grodzinsky, A., J., "Electromechanical Energy Conversion with Collagen Fibers in an Aqueous Medium", Journal of Mechanochemical Cell Motility, vol. 2, pp113-125, 1973.
5. Kuhn,W.,"Reversible Dehnung und Kontraktion bei Anderung der Ionisation eines Netzwerks Polyvalenter Fadenmolekulionen", Experientia, Vol.V, pp318-319, 1949.
6. Katchalsky, A., "Rapid Swelling and Deswelling of Reversible Gels of Polymeric Acids by Ionization", Experientia, Vol. V, pp319-320, 1949.

7. Kuhn,W., Kunzle, O., Katchalsky, A., "Verhalten Polyvalenter Fadenmolekelionen in Losung", Halvetica Chemica Acta, vol. 31, pp1994-2037, 1948.

8. Kuhn, W., Hargitay, B., Katchalsky, A., Eisenberg, H., "Reversible Dilation and Contraction by Changing the State of Ionization of High-Polymer Acid Networks", Nature, vol. 165, pp514-516, 1950.

9. Kuhn, W., Hargitay, B., "Muskelahnliche Kontraktion und Dehnung von Netzwerken Polyvalenter Fadenmolekulionen", Experientia, vol. VII, pp1-11, 1951.

10. Hamlen, R. P., Kent, C. E., Shafer, S. N., "Electrolytically Activated Contractile Polymer", Nature, Vol. 206, pp1149-1150, 1965.

11. Shahinpoor, M.,, "Continuum Electromechanics of Ionic Polymeric Gels as Artificial Muscles for Robotic Applications", Smart Material and Structures Int. J., Vol. 3, pp. 367-372, 1994.

12. Shahinpoor, M.,, "Microelectro-Mechanics of Ionic Polymeric Gels as Artificial Muscles for Robotic Applications", Proceeding of the IEEE Robotics & Automation Conf., vol. , pp. , 1993.

13. Shahinpoor, M., Mojarrad, M., "Active Musculoskeletal Structures Equipped with a Circulatory System and a Network of Ionic Polymeric Gel Muscles", Proceedings of the 1994 International Conference on Intelligent Materials, pp. 1079-1085, 1994.

14. Shahinpoor, M., Wang, G., Mojarrad, M., "Elctro-Thermo-Mechanics of Spring-Loaded Contractile Fiber Bundles with Applications to Ionic Polymeric Gel and SMA Actuators", Proceedings of the International Conference on Intelligent Materials" ICIM'94, Williamsburg, VA., pp. 1105-1116, 1994.

15. Osada, Y., "Electro-Stimulated Chemomechanical System Using Polymer Gels (An Approach to Intelligent Artificial Muscle System)", Proceeding of the International Conference on Intelligent Materials, pp155-161, 1992.

16. Oguro, K., Asaka, K., Takenaka, H., "Polymer Film Actuator Driven by Low Voltage", Proceedings of 4th International Symposium on Micro Machine and Human Science at Nagoya, pp39-40, 1993.

17. Asaka, K., Oguro, K., Nishimura, Y., Mizuhata, M., Takenaka, H., "Bending of Polyelectrolyte Membrane-Platinum Composites by Electric Stimuli, I. Response Characteristics to Various Waveforms", Polymer Journal, Vol. 27, No. 4, pp436-440, 1995.

18. Guo, S., Fukuda, T., Kosuge, K., Arai, F., Oguro, K., Negoro, M., "Micro Catheter System with Active Guide Wire Structure, Experimental Results and Characteristic Evaluation of Active Guide wire Using ICPF Actuator", Osaka National Research Institute, Japan, pp191-197, 1994.

19. De Rossi, D., P. Parrini, P. Chiarelli and G. Buzzigoli, " Electrically-Induced Contractile Phenomena In Charged Polymer Networks : Preliminary Study on the Feasibility of Muscle-Like Structures,", Transaction of American Society of Artificial Internal Organs, vol. XXXI, pp. 60-65, (1985)

20. De Rossi, D., C. Domenici and P. Chiarelli, " Analog of Biological Tissues for Mechanoelectrical Transduction : Tactile Sensors and Muscle-Like Actuators,", NATO-ASI Series, Sensors and Sensory Systems for Advanced Robots, vol. F43,, pp. 201-218, (1988)

21. Alexanderowicz, A.,Katchalsky, A., "Colligative Properties of Polyelectrolyte Solutions in Excess of Salt", Journal of Polymer Science, Vol. 1A, pp3231-3260, 1963.

22. Shahinpoor, M.,, "Nonhomogeneous Large Deformation Theory of Ionic Polymeric Gels in Electric and pH Fields", Proceedings of the 1993 SPIE Conference on Smart Structures and Materials, Feb. 1-4, Albuquerque, Vol. 1916, pp. 40-50, 1993.

23. Shahinpoor, M.,, "Micro-Electro-Mechanics of Ionic Polymeric Gels as Electrically Controlled Artificial Muscles," Proc. 1994 Int. Conf on Intelligent Materials, ICIM'94, June 1994, Williamsburg, VA, pp. 1095-1104, 1994

24. Shahinpoor, M., "Conceptual Design, Kinematics and Dynamics of Swimming Robotic Structures Using Ionic Polymeric Gel Muscles", Smart Materials and Structures Int. J., Vol. 1, pp. 91-94, 1992.

25. Segalman, D., Witkowsky, W., Adolf, D., Shahinpoor, M., "Electrically Controlled Polymeric Muscles as Active Materials used in Adaptive Structures", Proceedings of ADPA/AIAA/ASME/SPIE Conference on Active Materials and Adaptive Structures, Alexandria, VA, November 1991.

26. Shahinpoor, M.,, "Micro-Electro-Mechanics of Ionic Polymeric Gels As Electrically-Controllable Artificial Muscles,", Int. J. Intelligent Material Systems, vol. 6, no. 3, pp. 307-314, 1995

27. Mojarrad, M., and Shahinpoor, M., "Noiseless Propulsion for Swimming Robotic Structures Using Polyelectrolyte Ion-Exchange Membranes,", Proc. SPIE 1996 North American Conference on Smart Structures and Materials, February 27-29, 1996, San Diego, California, vol. 2716, paper no. 27, 1996

28. Shahinpoor, M., and M. Mojarrad, "Ion-Exchange Membrane-Platinum Composites As electrically Controllable Artificial Muscles,", Proc. 1996 Third International Conference on Intelligent Materials, ICIM'96, and Third European Conference on Smart Structures and Materials, Lyon, France, SPIE Publication No. ICIM'96 , pp. 1171-1184, June 1996

29. Shahinpoor, M.,, "Electro-Mechanics of Bending of Ionic Polymeric Gels as Synthetic Muscles for Adaptive Structures,", ASME Publication AD-Vol. 35, Adaptive Structures and Material Systems, edited by G.P. Carman and E. Garcia, Vol. AD-35, pp.11-22, 1993

30. Shahinpoor, M.,"Electro-Mechanics of Resilient Contractile Fiber Bundles with Applications To Ionic Polymeric Gel and SMA Robotic Actuators" Proc. 1994 IEEE International Conference on Robotics & Automation , vol. 2, pp. 1502-1508, San Diego,   California, May 1994

31. Shahinpoor, M.,, "The Ionic Flexogelectric Effect" Proc. 1996 Third International Conference on Intelligent Materials, ICIM'96, and Third European Conference on Smart Structures and Materials, June 1996, Lyon, France

32. Shahinpoor, M.,, "Design and Development of Micro-Actuators Using Ionic Polymeric Micro-Muscles,", Proc. ASME Design Engn. Technical Conference, Boston, MA, September (1995)

33. Shahinpoor, M., and M.S. Thompson, "The Venus Flytrap As A Model For Biomimetic Material With Built-In Sensors and Actuators," J. Materials Science & Engineering, vol.C2, pp. 229-233, (1995)

34. Shahinpoor, M.,, "Design and Modeling of A Novel Spring-Loaded Ionic Polymeric Gel Actuator," Proc. SPIE 1994 North American Conference on Smart Structures and Materials ., February 94, Orlando, Florida, vol. 2189, paper no. 26, pp.255-264, (1994)

35. Shahinpoor, M.,, "Microelectro-Mechanics of Ionic Polymeric Gels As Synthetic Robotic Muscles," Proc. SPIE 1994 North American Conference on Smart Structures and Materials, February 94, Orlando, Florida, vol. 2189, paper no. 27, pp.265-274, (1994)

36. Shahinpoor, M.,, "Micro-Electro-Mechanics of Ionic Polymeric Gels as Electrically Controlled Synthetic Muscles,", Biomedical Engineering Recent Advances, Editor : J.Vossoughi, University of District of Columbia Press, Washington, D.C., April 1994, vol.1, pp.756-759, (1994)

37. Shahinpoor, M.,"Electro-Mechanics of Resilient Contractile Fiber Bundles with Applications To Ionic Polymeric Gel and SMA Robotic Actuators" Proc. 1994 IEEE International Conference on Robotics & Automation , vol. 2, pp. 1502-1508, San Diego, California, May (1994)

38. Shahinpoor, M.,, "Electro-Thermo-Mechanics of Spring-Loaded Contractile Fiber Bundles with Applications To Ionic Polymeric Gel and SMA Actuators," Proc. 1994 Int. Conf. on Intelligent Materials, ICIM'94, June 1994, Williamsburg, VA, pp. 1105-1116, (1994)

39. Shahinpoor, M., "A New Effect in Ionic Polymeric Gels : The Ionic "Flexogelectric Effect," Proc. SPIE 1995 North American Conference on Smart Structures and Materials, February 28-March 2, 1995, San Diego, California, vol. 2441, paper no. 05, (1995) .

40. Shahinpoor, M.,, "Active Polyelectrolyte Gels as Electrically-Controllable Artificial Muscles and Intelligent Network Structures,", Book Paper, in Active Structures, Devices and Systems, edited by H.S. Tzou, G.L. Anderson and M.C. Natori, World Science Publishing, Lexington, Ky., (1995)

41. Shahinpoor, M.,, " Ionic Polymeric Gels As Artificial Muscles For Robotic and Medical Applications, Int. Journal of Science & Technology vol. 20, no. 1, Transaction B, pp. 89-136, (1996)

42. Shahinpoor, M., and Y. Osada, "Heart tissue Replacement with Ionic Polymeric Gels" Proc. 1996 ASME Winter Annual Meeting, San Francisco, California, November 12-18, (1995)

43. Shahinpoor, M.,," Design, Modeling and Fabrication of Micro-Robotic Actuators with Ionic Polymeric Gel and SMA Micro-Muscles,", Proc. 1995 ASME Design Engineering Technical Conference, Boston, MA, September (1995)

44. Mojarrad, M., and Shahinpoor, M., "Noiseless Propulsion for Swimming Robotic Structures Using Polyelectrolyte Ion-Exchange Membranes,", Proc. SPIE 1996 North American Conference on Smart Structures and Materials, February 27-29, 1996, San Diego, California, vol. 2716, paper no. 27, (1996)

45. Salehpoor, K., Shahinpoor, M., and M. Mojarrad, "Electrically Controllable Ionic Polymeric Gels As Adaptive Lenses," , Proc. SPIE 1996 North American Conference on Smart Structures and Materials, February 27-29, 1996, San Diego, California, vol. 2716, paper no. 18, (1996)

46. Salehpoor, K.,Shahinpoor, M., and M. Mojarrad, "Electrically Controllable Artificial PAN Muscles,", Proc. SPIE 1996 North American Conference on Smart Structures and Materials, February 27-29, 1996, San Diego, California, vol. 2716, paper no. 07, (1996)

47. Shahinpoor, M., and M. Mojarrad, "Ion-Exchange Membrane-Platinum Composites As electrically Controllable Artificial Muscles,", Proc. 1996 Third International Conference on Intelligent Materials, ICIM'96, and Third European Conference on Smart Structures and Materials, pp. 1012-1017, June 1996, Lyon, France

48. Shahinpoor, M.,, "The Ionic Flexogelectric Effect" Proc. 1996 Third International Conference on Intelligent Materials, ICIM'96, and Third European Conference on Smart Structures and Materials, pp. 1006-1011, June 1996, Lyon, France

49. Shahinpoor, M., and M. Mojarrad, "Biomimetic Robotic Propulsion Using Ion-Exchange Membrane Metal Composite Artificial Muscles, ", Proceedings of 1997 IEEE Robotic and Automation Conference, Albuquerque, NM, April (1997)

50. Shahinpoor, M., , Salehpoor, K., and Mojarrad, M.," Some Experimental Results On The Dynamic Performance of PAN Muscles,", Smart Materials Technologies, SPIE Publication No. vol. 3040, pp. 169-173, (1997)

51. Shahinpoor, M.,, Salehpoor, K., and Mojarrad, M., "Linear and Platform Type Robotic Actuators Made From Ion-Exchange Membrane-Metal Composites,", Smart Materials Technologies, SPIE Publication No. vol. 3040, pp.192-198, (1997)

52. Shahinpoor, M. and Mojarrad, M.," Ion-Exchange-Metal Composite Sensor Films,", Proceedings of 1997 SPIE Smart Materials and Structures Conference, vol. 3042-10, San Diego, California, March (1997)

53. Shahinpoor, M. and Mojarrad, M.," Electrically-Induced Large Amplitude Vibration and Resonance Characteristics of Ionic Polymeric Membrane-Metal Composites,", Proceedings of 1997 SPIE Smart Materials and Structures Conference, vol. 3041-76, San Diego, California, March (1997)

54. Osada, Y., Hasebe, M., "Electrically Activated Mechanochemical Devices Using Polyelectrolyte Gels", Chemistry Letters, pp1285-1288, 1985.

55. Kishi, R., Hasebe, M., Hara, M., Osada, Y., "Mechanism and Process of Chemomechanical Contraction of polyelectrolyte Gels Under Electric Field", Polymers for Advanced Technologies, vol. 1, pp19-25, 1990.

56. Brock, D., Lee, W., Segalman, D., Witkowski, W., "A Dynamic Model of a Linear Actuator Based on Polymer Hydrogel", Proceedings of the International Conference on Intelligent Materials, pp210-222, 1994.

57. Mojarrad, M., Shahinpoor, M., "Ion-exchange-Metal Composite Artificial Muscle Load Characterization And Modeling", Smart Materials Technologies, SPIE Publication No. vol. 3040, pp. 294-301, (1997)

58. Sadeghipour, K., Salomon, R., Neogi, S., "Development of A Novel Electrochemically Active Membrane and 'Smart' Material Based Vibration Sensor/Damper", Smart Materials and Structures, Vol. 1, pp 172-179, 1992.

59. Tzou, H. S., Fukuda, T., "Precision Sensors, Actuators and Systems", Kluwer Academic Publishers 1992

60. Rieder, W. G., Busby, H. R., "Introductory Engineering Modeling Emphasizing Differential Models and Computer Simulations", Robert E. Krieger Publishing Company, Malabar, Florida, 1990.

61. Ugural, A. C., Fenster, S. K., "Advanced Strength and Applied Elasticity", Elsevier, New York, 1987.

62. Bar-Cohen, Y., T. Xue, B. Joffe, S.-S. Lih, Shahinpoor, M.,, J. Simpson, J. Smith, and P. Willis, " Electroactive polymers (IPMC) low mass muscle actuators, " Proceedings of 1997 SPIE Conference on Smart Materials and Structures, March-5, San Diego, California, (1997)

63. Shahinpoor, M., "Artificial Muscles," ERI Press, Albuquerque, New Mexico, Pending Publications, (1997)

64. Furukawa and J. X. Wen, "Electrostriction and Piezoelectricity in Ferroelectric Polymers," Japanese Journal of Applied Physics, Vol. 23, No. 9, pp. 677-679, 1984.

65. I. W. Hunter and S. Lafontaine, "A comparison of muscle with artificial actuators," IEEE Solid-State Sensor and Actuator Workshop, pp. 178-165, 1992.

66. Shahinpoor, M.,, "Continuum electromechanics of ionic polymeric gels as artificial muscles for robotic applications," Smart Materials and Structures, Vol. 3, pp. 367-372, 1994.

67. Kornbluh, K., R. Pelrine and J. Joseph, " Elastomeric dielectric artificial muscle actuators for small robots," Proceeding of the 3rd IASTED International Conference, June, 14-16, 1995.

68. Pelrine, R., R. Kornbluh, J. Joseph and S. Chiba, "Artificial muscle actuator," Proc. of the First International Micromachine Sym., Nov. 1-2, pp. 143-146, 1995.

69. Heitner-Wirguin, C., "Recent advances in perfluorinated ionomer membranes: Structure, properties and applications," Journal of Membrane Science, V 120, No. 1, pp. 1-33, 1996.

70. Bar-Cohen, Y., T. Xue, B. Joffe, S.-S. Lih, P. Willis, J. Simpson, J. Smith, M. Shahinpoor, and P. Willis, "Electroactive Polymers (EAP) Low Mass Muscle Actuators," Proceedings of SPIE, Vol. SPIE 3041, Smart Structures and Materials 1997 Symposium, Enabling Technologies: Smart Structures and Integrated Systems, Marc E. Regelbrugge (Ed.), ISBN 0-8194-2454-4, SPIE, Bellingham, WA (June 1997), pp. 697-701.

Chapter 4

Nonaqueous Polymer Electrolytes for Electrochromic Devices

J. R. Stevens, W. Wieczorek, D. Raducha, and K. R. Jeffrey

**Department of Physics, University of Guelph, Guelph,
Ontario N1G 2W1, Canada**

Nonaqueous alkali metal salt polyether electrolytes and nonaqueous H_3PO_4 proton conducting electrolyte gels have been synthesized for application in electrochromic devices. The ionic conductivity of these electrolytes has been optimized to be in the range 10^{-3} S/cm to 10^{-6} S/cm at room temperature. The most favourable, stable, polyether electrolytes are blends based upon EO/PO (= 50/50 or 60/40) copolymers with molecular weights in the range of 300-600 and $LiCF_3SO_3$ or $Li(CF_3SO_2)_2N$ concentrations of 25:1, PMMA concentrations in the range 8 to 40 vol% and PC < 15 vol%. The most favourable, stable nonaqueous gels were those with H_3PO_4 dissolved in PC (20 to 50 vol%) entrapped in a PMMA network using a crosslinking agent. These electrolytes are stable from -30° to 100°C.

According to the Report of the World Commission on Environment and Development (*1*) and the research and development being carried out under the International Energy Agency (*2*), interest in solar energy materials, like energy efficient window coatings, is growing. Materials with optical switching (dynamic) properties were reviewed by Lampert and Granqvist (*3*) and Granqvist (*4*) and consist of photochromic, thermochromic and electrochromic materials depending on whether their optical properties are changed due to irradiation, temperature, or electric potential difference.

Electrochromism is defined as a persistent but reversible optical change in absorption or reflection produced electrochemically in a medium by an applied electric field or current. An electrochromic device in which this process can be realized is schematically shown in Figure 1. Generally such a device contains transparent conductors, an ion storage layer, an ionic conductor and an optically active electrochromic layer. We will focus on properties of the ionic conductor.

Since the first reports that alkali metal salts could be dissolved in poly(ethylene oxide) (PEO) (*5-7*) much progress has been made in the synthesis of polymer electrolytes (*8*). The main role of an electrolyte in electrochromic devices is to allow

52

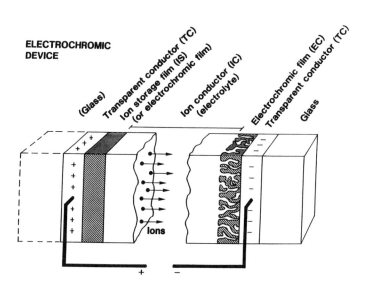

Figure 1.   Schematic sketch of an electrochromic device.

ions to be shuttled between an electrochromic film and an ion storage film also known as a "counter electrode". It is beneficial for polymer electrolytes working in electrochromic devices to have the following properties: wide operational temperature range, high exchange current density, ionic conductivity $> 10^{-5}$ S/cm, good to excellent adhesive properties, greater than 80% transparency in the visible region of the electromagnetic spectrum, stability against such environmental factors as moisture and temperature variation, and either be self-supporting films or have good, bubble free, coatability onto an electrochromic film substrate.

A wide range of solid and liquid electrolytes has been tested in various electrochromic devices and the results of these studies have been extensively discussed (4). The main electrolyte groups are: aqueous acidic electrolytes, nonaqueous lithium electrolytes, ceramic ionic conductors (β-alumina, Nasicon etc.), polyelectrolytes, polymer solid electrolytes (complexes of polyethers or polyimines with alkali metal salts and proton donors) and plasticized polymer ionic conductors.

Ambient temperature proton conducting systems studied so far are either not very stable (e.g. heteropolyacids) or their conductivity depends on the amount of water present (e.g. hydrated Nafion or Dow membranes). The presence of water is the main disadvantage of using aqueous or polyelectrolyte systems. It has been shown that most of the electrochromic oxides react with water molecules. This limits the lifetime of electrochromic devices containing electrolytes with even residual traces of water. The utilization of ceramic electrolytes leads to interfacial problems manifested by an increase in the overall resistance of the device which results in limiting current effects. Moreover most of the intercalated materials used as electrodes in electrochromic devices expand during the intercalation of ions; this often results in the cracking of the ceramic electrolyte.

Because of the above mentioned limitations flexible nonaqueous polymeric electrolytes have been used as electrolyte membranes in various electrochromic devices. Polymer electrolytes with ionic conductivities up to 10^{-3} S/cm at room temperature have been synthesized and tested in "smart window" and electrochromic display configurations. Generally these electrolytes are of two types; lithium salt doped polyether copolymers (9-17) and nonaqueous, proton conducting strong acid solutions (18-20); both types entrapped in a polymeric gel. Lithium bis(triflouromethanesulfonyl)imide (Li(CF$_3$SO$_2$)$_2$N) and lithium triflouromethanesulfonate (LiCF$_3$SO$_3$) are the favoured lithium salts and phosphoric, antimonic and sulphuric acids have been used as sources of protons. The gels are usually poly(methyl methacrylate) (PMMA) or a copolymer of PMMA and glycidal methacrylate (GMA), poly(vinyl acetate), poly(acrylamide) or poly(vinylidene fluoride).

The highest conductivities in lithium salt doped systems have been achieved in plasticized systems (13-17). These are characterized by relatively high ionic conductivities (up to 10^{-3} S/cm at room temperature) and transparency for up to 30 mass% of PMMA where PMMA has been used (13,14). A two step polymerization of glycidyl methacrylate (GMA) in the presence of propylene carbonate (PC) and a lithium salt (15), the use of small molecule cyclotriphosphazenes in poly[bis(methoxyethoxy) phosphazene] (MEEP)-LiCF$_3$SO$_3$ (16), and the use of fumed-silica particulates in low molecular weight poly(ethylene oxide) (17) with lithium salts (imide and triflate anions) are examples of placticized systems. In these systems anions are the dominant mobile species.

Scanning electron microscope studies indicate that polyether/PMMA/Li salt blends have a microphase separated morphology with the PMMA well dispersed in sizes of 2-20 μm (11). The effect of the molecular weight of the components on the "window of compatibility" is known from the literature for polymer blends such as poly(ethylene oxide) (PEO)/PMMA (9-12). The blending of PMMA with the polyether was found to stabilize the electrolyte against moisture uptake and provide adhesion to a glass substrate.

There has recently been widespread interest in the development of proton conducting polymeric electrolytes which can be used at ambient and moderate temperatures (18-28). The advantage of proton conducting systems in comparison with alkali metal electrolytes arises from their potentially higher conductivity. Thus we should expect a faster coloration-bleaching time in electrochromic devices compared with lithium salt electrolytes. Proton conductors are systems in which polar polymers with basic sites on a main polymer chain form compounds with strong acids such as H$_2$SO$_4$ or H$_3$PO$_4$. The polymeric films formed must be chemically and mechanically stable. Properties of these systems have been reviewed by Lassegues (21). Ambient temperature conductivities obtained for some of these proton polymeric electrolytes were higher than 10^{-3} S/cm (18-21,27,28). The C-O bond in ethers and alcohols is broken by strong acids; such degradation is accelerated by traces of water. Under laboratory conditions all necessary precautions have been made to keep these electrolytes "dry" (22). However it is not clear what the effect of residual traces of water on conductivity will be as well as the long time stability of these electrolytes in large area applications.

The main objectives of the work discussed here are to optimize the conductivity of nonaqueous electrolytes for electrochromic devices. The electrochemistry of nonaqueous alkali metal salt electrolytes is well developed because of the possible application of these systems in ambient temperature alkali metal batteries (29). However, little is known about the properties of nonaqueous concentrated solutions of strong acids which potentially have high protonic conductivity. Most of the studies of liquid proton conducting electrolytes described in the literature are devoted to aqueous electrolytes or to very dilute solutions ($<10^{-4}$ mol/dm^3) of moderately strong acids in nonaqueous solutions (29-30). (The comments in reference 29 regarding the effects of alkali salts in nonaqueous systems also apply to acids.) Nonaqueous proton conducting gels have attracted our attention due to the possibility of their application in various electrochemical devices (e.g. sensors or electrochromic windows) working at ambient temperatures.

Results and Discussion

The techniques and materials used are discussed in references 13, 14, 18 and 19.

PEG based electrolytes. Highly conductive solid polymer electrolytes based upon methyl capped poly(ethylene glycol) (PEG) of $M_w=350$ g/mol and random copolymers of poly(ethylene oxide-co-propylene oxide) (EO/PO) of $M_w=600$ g/mol (m.u. ratio = 1) blended with up to 50% by volume of PMMA and doped with $LiCF_3SO_3$ (O:Li = 25:1) have been synthesized (Table I (14)). Samples based on PEG and containing 13 vol% of PMMA or less (P1, P2) were found to be viscous liquids which crystallized at ~225 K while those containing more than 19 vol% of PMMA did not crystallize and could be prepared as thin films. For up to 38 vol% of PMMA all electrolytes were transparent and exhibited only one T_g in the region close to T_g for the PEG-$LiCF_3SO_3$ (O/M=25:1) system. For samples containing 38 vol% and 50 vol% of PMMA (P6, P7), phase separation was observed and a second glass transition appeared in the region appropriate for the T_g of PMMA. As calculated from the GPC spectra, molecular weights for PMMA polymerized in PEG-salt systems varied between 100000 and 130000 with a relatively narrow distribution $M_w/M_n=1.5$.

Ionic conductivity is a crucial parameter in the classification of polyether/PMMA/ $LiCF_3SO_3$ systems for use as polymer electrolytes in electrochromic "smart windows". For liquid samples (6 and 13vol% of PMMA) conductivities were even higher than for the pure PEG/$LiCF_3SO_3$ system (sample PO) and exceeded 10^{-4} S/cm at 25° C (Table I). Conductivity is observed to increase (up to 10^{-3} S/cm at around 100° C) with increase in temperature. In the temperature range 20° C to 100° C the temperature dependence of conductivity of the blend based electrolytes is VTF and follows

$$\sigma = AT^{-0.5} \exp[-E/k_B(T-T_o)]$$

where E is a pseudo-activation energy for ionic conductivity, A is a pre-exponential factor which is proportional to the number of charge carriers, and T_o (in Kelvins) is a quasi-equilibrium glass transition temperature 30-50 degrees below T_g. For temperatures below 20°C, conductivities of these systems are found to have either a VTF or an Arrhenius ($\sim\exp[-E/k_B T]$) temperature dependence (Figure 2).

The simplest method which suppresses the crystallization of PEG without diminishing chain flexibility is to prepare three component amorphous blends consisting of polymer, salt and a low molecular plasticizer such as propylene carbonate (PC). The temperature dependence of the conductivity for the PEG-PMMA-$LiCF_3SO_3$ system (sample P3) is compared in Figure 3 with similar amorphous electrolytes containing PC (samples T1 and T2). The conductivities of the T_1 and T_2 electrolytes are 3.6×10^{-4} S/cm and 1.6×10^{-4} S/cm at room temperature (T_R) and decrease to respectively 8.0×10^{-5} and 2.7×10^{-5} S/cm at 0° C but still exceed 5.0×10^{-6} S/cm at around -20° C. These three component electrolytes are stable up to 100° C. Values of T_g are slightly lower than those measured for similar electrolytes without PC.

Another method of increasing the polarity of electrolytes and hence their ionic conductivity is to introduce polar comonomers via copolymerization with

Table I. Sample[a] Characterization for Samples Based on PEG and EO/PO Copolymers

Sample code	polyether	vol % of PMMA	σ at 273 K [S/cm]	σ at 298 K [S/cm]
P0	PEG	0	3.1×10^{-5}	1.5×10^{-4}
P1	PEG	6	4.2×10^{-5}	2.2×10^{-4}
P2	PEG	13	3.6×10^{-5}	1.8×10^{-4}
P3	PEG	19	1.5×10^{-5}	1.1×10^{-4}
P4	PEG	26	1.2×10^{-5}	9.0×10^{-5}
P5	PEG	31	9.1×10^{-6}	6.5×10^{-5}
P6	PEG	38	6.6×10^{-6}	4.2×10^{-5}
P7	PEG	50	1.3×10^{-6}	9.5×10^{-6}
C1	EO:PO (85:15)	0	6.1×10^{-6}	1.1×10^{-4}
C2	EO:PO (60:40)	0	1.1×10^{-5}	1.4×10^{-4}
C3	EO:PO (50:50)	0	2.9×10^{-5}	1.2×10^{-4}
C4	EO:PO (95:5)	10	1.0×10^{-5}	8.9×10^{-5}
C5	EO:PO (85:15)	10	1.2×10^{-5}	8.0×10^{-5}
C6	EO:PO (60:40)	12	1.3×10^{-5}	8.5×10^{-5}
C7	EO:PO (50:50)	13	1.1×10^{-5}	7.0×10^{-5}
T1	PEG	17, (22, PC)	7.9×10^{-5}	3.6×10^{-4}
T2	PEG	27, (14, PC)	2.7×10^{-5}	1.6×10^{-4}
T3	PEG[c]	23, 10 EO-coAN	6.7×10^{-6}	5.4×10^{-5}

[a] All samples doped with $LiCF_3SO_3$, ether oxygen to metal ratio = 25:1.
[b] % Yield was calculated with respect to PMMA.
[c] EO-co-AN copolymer containing 10 molar % of AN monomeric units.
SOURCE: Adapted from ref. 13.

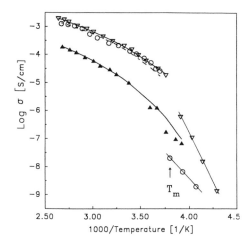

Figure 2.   A comparison of conductivities measured for PEG-PMMA-LiCF₃SO₃ electrolytes. Samples of various concentrations of PMMA denoted in vol %. Concentration of LiCF₃SO₃ is O:M = 25:1. (O) 0 vol % of PMMA; (▽) 13 vol % of PMMA; (▲) 50 vol % of PMMA. (Reproduced with permission from ref. 14. Copyright 1995 Elsevier.)

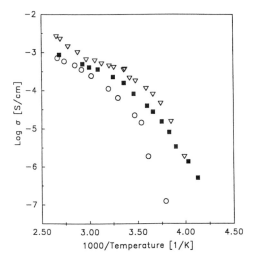

Figure 3.   A comparison of conductivities measured for PEG-PMMA-LiCF₃SO₃-PC electrolytes. Concentration of LiCF₃SO₃ is O:M = 25:1. (▽) 17 vol % of PMMA, 22 vol % of PC; (■) 27 vol % of PMMA, 14 vol % of PC; (O) 19 vol % of PMMA. (Reproduced with permission from ref. 14. Copyright 1995 Elsevier.)

methyl methacrylate. A random MMA-AN copolymer was prepared by copolymerization in the PEG-salt environment (Table I). Samples containing more than 20 vol% of AN were found to produce precipitates from the reaction mixture, thus resulting in opaque electrolytes. Similar conductivities were measured for transparent samples of lower AN concentrations and PEG-PMMA based blends with PMMA concentrations similar to those for AN (see Table I).

EO/PO based electrolytes. Noncrystallizable electrolytes of low PMMA concentration (10-13 vol%) can be obtained from random EO/PO copolymers containing 50 and 60 mol% of EO monomeric units (samples C2, C3, C6, C7) whereas blends of PMMA with EO/PO copolymers (85, 95 mol% of EO, samples C1, C4, C5) still exhibit a tendency to crystallize. For electrolytes based on EO/PO copolymers ambient temperature conductivities are slightly lower than the PEG-PMMA blend based described above (see Table I). This is a result of the weaker complexation abilities of the PO monomeric units. However at temperatures in the range of the melting point of the crystalline PEG phase, conductivities of amorphous EO/PO copolymer based systems remain in the range 5×10^{-7} to 5×10^{-6} S/cm whereas those of PEG-PMMA based blends drop to 10^{-9} to 10^{-8} S/cm (see Figure 2).

The salt concentration dependence of ionic conductivity was studied for copolymers of EO/PO (50:50) of molecular weights $M_w = 600$ g/mol and $M_w = 2600$ g/mol. Polymer electrolytes based on these polyethers were investigated with or without the presence of PMMA as shown in Table II in which ionic conductivities at 273 and 298K are listed (*13,14*). As evident from our studies, statistical copolymers with EO/PO molar ratio equal to (50:50) or (60:40) are the best for smart window application because of their amorphous character. The improvement in conductivity when going from molecular weight 2600 to 600 (see Table II (*14*)) is due to the lower viscosity and greater mobility of polyether chains which facilitates the movement of ions in the polyether matrix. Further it was possible to incorporate greater amounts of PMMA into these EO/PO based polymer electrolyte systems enhancing the mechanical properties while maintaining good conductivity. Electrolytes with up to 13 vol% PMMA are transparent and exhibit high ionic conductivity. The conductivity is further improved with the addition of a small amount (<15 vol%) of PC plasticizer.

Proton conducting gels. It is generally believed that the solvent should be the main conduction medium in a solvent/gel system; the conduction mechanism for gel electrolytes is similar to that for liquid systems. The polymer matrix simply acts as a "container" assuring good structural integrity. Also of importance is whether the solvent molecules are protonated by the protonic dopant in the system. The protophilic solvent itself can contribute to proton conduction leading to a Grotthus type of proton transport (*20*); the activation energy for proton conduction is lowered and an increase in conductivity is possible in the presence of a mixture of protonated and unprotonated solvents molecules. PC and dimethylformamide (DMF) seem to be the most appropriate solvents due to relatively high dielectric constants, low viscosities and a

Table II. Sample Characterization for EO/PO/PMMA/LiCF$_3$SO$_3$ Electrolytes

sample code	copolymer EO/PO M_w[g/mol]	LiCF$_3$SO$_3$ O/M ratio	PMMA vol%	T_g [K]	σ_{273} [S/cm]	σ_{298} [S/cm]
A1	600	50:1	0	197	1.1x10^{-6}	5.3x10^{-5}
A2	"	30:1	0	200	3.8x10^{-5}	9.7x10^{-5}
A3	"	25:1	0	204	2.4x10^{-5}	1.8x10^{-4}
A4	"	16:1	0	209	1.2x10^{-4}	1.5x10^{-4}
A5	"	25:1	10	204	3.5x10^{-5}	7.0x10^{-5}
A6	"	25:1	20	204	2.5x10^{-7}	3.0x10^{-5}
A7	"	25:1	30	205	2.4x10^{-7}	2.1x10^{-5}
B1	2600	50:1	0	212	1.6x10^{-7}	1.3x10^{-5}
B2	"	30:1	0	215	1.4x10^{-7}	1.1x10^{-5}
B3	"	25:1	0	219	1.2x10^{-7}	1.4x10^{-5}
B4	"	16:1	0	228	1.3x10^{-5}	1.0x10^{-5}
B5	"	8:1	0	252	2.0x10^{-9}	8.0x10^{-7}
B6	"	25:1	10	220	5.0x10^{-8}	5.7x10^{-6}

SOURCE: Reprinted with permission from ref. 14. Copyright 1995 Elsevier.

wide operational temperature range. We have measured the T_R conductivities of gels entrapping H_3PO_4 in other solvents and confirm our choice of PC and DMF as the most interesting solvents with gel/H_3PO_4 conductivities approaching 10^{-3} S/cm (*18,19*).

Various types of polymer networks were synthesized depending on the type of low molecular weight solvent used for the preparation of H_3PO_4 doped gels. Most of these networks contain monomeric units of methacrylic acid esters such as MMA and/or GMA. GMA is a difunctional monomer with a reactive unsaturated C=C bond and oxirane ring (*18*). Oxirane rings can react with the P-OH groups of H_3PO_4 and its mono or diesters leading to the formation of C-OH groups. This is followed by the addition reaction of oxirane rings with C-OH groups according to the "activated monomer mechanism" (*18*). The polymer networks for PC based electrolytes were obtained without GMA by using the free radical copolymerization of MMA with a crosslinking agent, triethylene glycol dimethacrylate, since precipitation products were observed when GMA was used. Homogeneous PC based gels can be obtained within broad solvent and H_3PO_4 concentration ranges.

DMF Based Electrolytes. Transparent gels incorporating DMF can be obtained in wide solvent and H_3PO_4 concentration ranges for polymer hosts based on GMA homopolymers as well as those based on copolymers with other acrylic monomers (e.g. methyl methacrylate, acrylonitrile, acrylamide). For DMF/GMA electrolytes the temperature dependence of the conductivity is Arrhenius for samples containing less than 25 mass % of H_3PO_4; two examples are shown in Figure 4. The activation energy for conduction calculated for these samples is in the range 0.1-0.25 eV which suggests the possibility of proton transport via a Grotthus type mechanism (*31*). Conductivities measured for these systems are higher than for the other DMF based gels over the entire temperature range studied. For DMF based electrolytes with higher H_3PO_4 concentrations the temperature behavior of the conductivity is similar to that discussed below for the PC based electrolytes (see Figure 5).

The intensity of vibrations characteristic of P-O(H) bonds in H_3PO_4 (~1000 cm^{-1}) increases with an increase in H_3PO_4 concentration for the DMF-H_3PO_4 based gel electrolytes. The position of the maximum of this band remains almost unchanged with an increase in the H_3PO_4 concentration. For the DMF based gel electrolyte containing 7.4 mass % of H_3PO_4 a shoulder is observed at ~971 cm^{-1} which we suggest is due to the formation of mono or diesters of H_3PO_4. According to previously published data (*32,33*) the formation of ester structures should be related to a down shift in the maximum of the P-(OH) vibration. However, for gel samples with higher H_3PO_4 concentration the presence of a peak due to the ester vibration is most probably masked by a stronger mode due to excess H_3PO_4.

Interaction between the DMF carbonyl groups and H_3PO_4 leading either to the protonation of the carbonyl group or to the formation of hydrogen bonds is confirmed by a shift of the maximum of the C=O IR band to lower frequencies (1674 to 1646 cm^{-1}) as the concentration of H_3PO_4 is increased from 0 to 45 mass%. The full width at half maximum (FWHM) of the C=O mode increases with an increase in H_3PO_4 concentration. The position of the band related to C-N vibrations is centered at 1382-

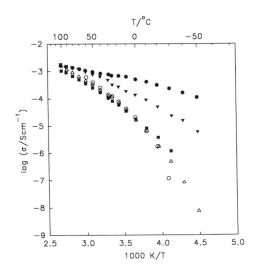

Figure 4.   The ionic conductivity of DMF-H$_3$PO$_4$ entrapped gel electrolytes as a function of inverse temperature. Samples containing 10.5 mass % of GMA based polymer matrix with 50 [○], 45 [Δ], 34 [■], 23 [▼] and 7.4 [●] mass % of H$_3$PO$_4$. (Reproduced with permission from ref. 18. Copyright ACS.)

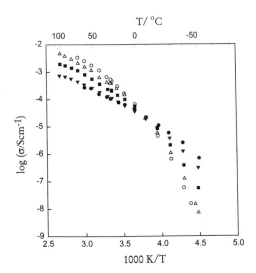

Figure 5.   The ionic conductivity of PC-H$_3$PO$_4$ gel electrolytes as a function of inverse temperature. Samples containing 10.5 mass % of MMA based polymer matrix with 50 [○], 45 [Δ], 34 [■], 23 [▾] and 17.5 [●] mass % of H$_3$PO$_4$. (Reproduced with permission from ref. 18. Copyright ACS.)

1388 cm^{-1} throughout the entire H_3PO_4 concentration range which indicates that interactions between H_3PO_4 and DMF occur mainly via the carbonyl groups of DMF. Electrolytes based on DMF exhibit higher ambient temperature conductivities (up to 8×10^{-4} S/cm) due to the possibility of solvent protonation and therefore the possibility of proton transport via a Grotthuss type mechanism. (See Figure 6.) FT-IR spectra confirm the possibility of the protonation of the carbonyl oxygens of DMF by H_3PO_4.

Figure 6 presents the conductivity isotherms obtained at 25°C as a function of H_3PO_4 concentration for gels containing DMF and PC. Opposite trends are observed for these systems in the H_3PO_4 concentration range 10-40 mass%. The explanation of this observation is, we believe, related to different conduction mechanisms and to the different nature of the solvents. At low H_3PO_4 concentrations in DMF there is a higher degree of dissociation according to the law of mass action as shown.

$$H-\overset{\overset{O}{\|}}{C}-N\overset{CH_3}{\underset{CH_3}{\diagdown}} \; + \; H_3PO_4 \; \rightleftharpoons \; H-\overset{\overset{+OH}{\|}}{C}-N\overset{CH_3}{\underset{CH_3}{\diagdown}} \; + \; H_2PO_4^-$$

At higher concentrations of H_3PO_4 the formation of ion pairs and possibly higher multiplets occurs in DMF which has a lower dielectric constant (37) than PC (65) (29,30). The formation of neutral associated species lowers the conductivity in DMF as the H_3PO_4 concentration increases.

PC Based Electrolytes. Figure 5 shows the temperature dependence of the ionic conductivity for gel electrolytes entrapping PC with various H_3PO_4 concentrations. Two temperature regions with opposite trends in conductivity versus H_3PO_4 concentration can be seen. Above ~ -10°C conductivity increases with an increase in the H_3PO_4 concentration. Below ~ -10°C conductivities decrease with an increase in H_3PO_4 concentration. For PC-H_3PO_4 gel electrolytes conductivities are lower than those measured for liquid PC-H_3PO_4 electrolytes due to the "dilution effect" of the polymer matrix phase (19).

The change in the temperature dependence of the conductivity observed for PC based gel electrolytes at ~ -10°C (see Figure 5) is most likely related to changes in the local viscosity of these electrolytes. An increase in T_g with an increase in the H_3PO_4 concentration was observed (19) and confirms the possibility of H_3PO_4 interactions with both the solvent and polymer matrix which stiffen the protonic gels. The T_g's measured for the gel electrolytes are higher than those measured for the liquid-H_3PO_4 electrolytes of similar concentrations of H_3PO_4. This difference increases with an increase in H_3PO_4 concentration.

The bands formed due to vibrations in the ions formed on dissociation of H_3PO_4 can be seen in the FT-IR and FT-Raman spectra. For example in the region of the 1790 cm^{-1} (C = 0) band in PC there appears three new bands due to the acid at ~1764, 1743 and 1698 cm^{-1}. The intensity of these "acid" lines relative to the intensity of the 1790

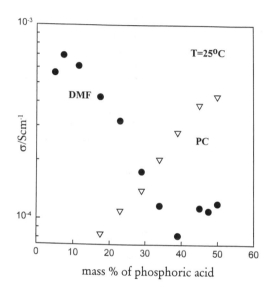

Figure 6. Isotherms of ionic conductivity of DMF-H₃PO₄ [●] and PC-H₃PO₄ [▼] gel electrolytes versus mass % of H₃PO₄. Samples containing 10.5 mass % of polymer matrix (GMA and MMA based respectively). (Reproduced with permission from ref. 18. Copyright ACS.)

cm^{-1} band decreases with increasing temperature indicating further dissociation of the phosphoric acid ions possibly producing "free" protons. This agrees with our observations from conduction studies of an increase in charge carrier concentration with increase in temperature.

For PC liquid or gel electrolytes a broadening of the band characteristic of the C=O stretching vibration ($\sim$1790 cm^{-1}) is also observed after the addition of H_3PO_4. However, the maxima of these vibrational bands do not shift down even for high concentrations of H_3PO_4. The maximum of the P-O(H) vibrational bands remains at 1000-1010 cm^{-1} over the entire concentration range of H_3PO_4 for both liquid and gel electrolytes.

Ambient temperature conductivities measured for the PC based electrolytes also exceeded 10^{-4} S/cm. For ambient to subambient temperatures $E_a \sim 0.52$ eV and protonic transport probably occurs via a vehicle type mechanism in which polyatomic molecules such as $H_2PO_4^-$ or $H_4PO_4^+$ are the dominant charge carriers (34,35) and play a different role than in the Grotthuss mechanism. In the higher temperature region used in Figure 7 the conduction activation energy (Figure 5) is $\sim$0.3 eV for the PC-D_3PO_4(23%) gel electrolyte (19). This is a little lower than the activation energy ($\sim$0.38 eV) measured for deuteron self diffusion by NMR indicating the possibility of some structure diffusion (Grotthuss). As seen in Figure 7 the diffusion coefficient for the PC-D_3PO_4 (45%) gel is slightly less than for the PC-D_3PO_4 (23%) gel from static field gradient NMR spectroscopy. Also the ^{31}P results indicate that the "vehicles" move more slowly than the proton and that as indicated in the IR and Raman studies there are some "free" protons present (36, 37).

The solvent (PC) is not involved in any protonation and lower ambient temperature conductivities result for gels with low H_3PO_4 concentrations (see Figure 6); the conductivity increases with increasing H_3PO_4 concentration. Although PC is a good electron pair donor solvent (38) it has considerable difficulty solvating free protons (39). The formation of charge carriers probably occurs via the self dissociation of H_3PO_4 as shown in a simplified form (35).

$$2\,H_3PO_4 \rightleftharpoons H_2PO_4^- + H_4PO_4^+$$

Here protonic transport probably occurs via a vehicle type mechanism in which polyatomic molecules such as $H_2PO_4^-$ or $H_4PO_4^+$ are the dominant charge carriers and play a different role than in the Grotthus mechanism. Here the solvent is not involved and there is a higher activation energy for conduction and lower ambient temperature conductivities result for gels with low H_3PO_4 concentrations. For high H_3PO_4 concentrations, acid-base pairs like $H_2PO_4^-/HPO_4^{2-}$ are probably also present leading to the possibility of a pseudo Grotthus type mechanism in which an exchange of protons between acid-base pairs of components occurs and therefore an increase in the conductivity is observed.

The Nernst-Einstein equation can be written as

Diffusion Coefficients

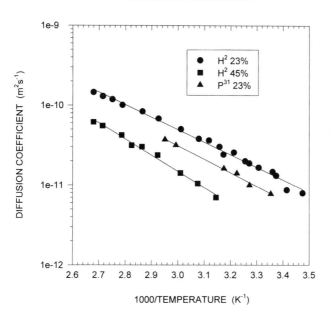

Figure 7.   Diffusion coefficient versus inverse temperature for the PC-D_3PO_4 (23%) gel electrolyte.

$$\sigma T = \frac{nq^2}{k} D$$

where n is the number density of charge carriers and q is the charge on charge carrier. If we consider the temperature region 285 K to 322 K in Figure 7 there is a linear relation between σT (from Figure 5) and D (from Figure 7) and the slope is ~2.1×10^5 S cm^{-3} s K. (See Figure 8.) This gives a value of n ~ 1.1×10^{20} cm^{-3} for the PC gel sample with 23 mass % of H_3PO_4. Assuming that $D_4PO_4^+$ and $D_2PO_4^-$ are the main charge carriers and using the dissociation strategy suggested (*34,35*) n ~ 1×10^{20} cm^{-3} for 0.69g D_3PO_4 and 1.98g PC in a 3g sample; in good agreement.

The higher viscosity of PC results in a lower ionic charge mobility, which can also be the reason for the lower ionic conductivities measured for PC systems. At temperatures close to but above T_g the viscosity of the electrolyte as reflected in the value of T_g has a dominant effect on the ionic conductivity plotted in Figures 5 and 6. The higher the H_3PO_4 concentration, the higher T_g and thus the lower the conductivity. Above ~-10°C the viscosity of the electrolytes containing PC is lower and comparable for electrolytes with various H_3PO_4 concentration and ionic transport depends mainly on the number of charge carriers participating in the vehicle mechanism. The possibility of proton transport via the vehicle mechanism in PC systems is also confirmed by a significant increase in conductivity over the entire temperature range (*31*). It can be noticed (compare Figures 4 and 5) that at temperatures higher than ~50°C conductivities obtained for PC based systems are comparable or even higher (see samples with 45 and 50 mass% of H_3PO_4 in Figure 5) than conductivities measured for the DMF based gel electrolytes.

Studies of the effect of changes in crosslink density on the ionic conductivity of gel protonic electrolytes were carried out using as crosslinking agent, triethylene glycol dimethacrylates. With this agent one can expect chemically that the crosslink density will increase as the concentration of crosslinking agent increases. This was confirmed by DSC measurements of T_g. For the nonaqueous PMMA/PC/H_3PO_4 system room temperature conductivities increase with a decrease in the crosslinking agent concentration as well as with a decrease in the polymer matrix concentration. For the gel electrolytes containing DMF-H_3PO_4 the crosslink density can be reduced by the utilization of GMA copolymerized with other acrylic monomers as polymer matrices instead of GMA homopolymers. A decrease in the density of junction points in the polymer matrix does not result in an increase in the ionic conductivity of these gel electrolytes. A small increase in conductivity (~3.1×10^{-4} to 3.8×10^{-4} S/cm) is only observed for gels based on a GMA-AN copolymer matrix (*18*). On the other hand the concentration of P-OH groups in matrices based on GMA copolymers is lower than in matrices based on GMA homopolymers which, according to previous assumptions, should lead to a decrease in conductivity. The final conductivity is a result of a competition between the effect of the density of matrix junction points and the effect of the concentration of P-OH groups. The above observations illustrate the role of P-OH polymer matrix groups in the conductivity of gel protonic electrolytes.

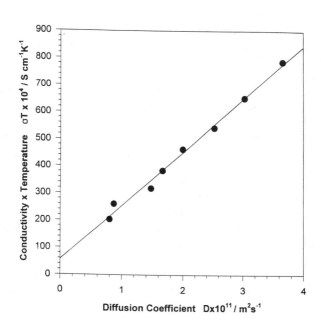

Figure 8.   The product of ionic conductivity and temperature (K) versus the diffusion coefficient for the PC-D$_3$PO$_4$ (23 mass %) gel.

68

Applications. Efforts are underway to transfer the knowledge gained in these studies to industrially viable electrochromic devices. These include "smart" windows, rear view mirrors and displays. The greatest progress has been made with EO/PO/PMMA/lithium salt systems. While the proton conducting gels have higher conductivities they react with electrochromic layers such as nickel and vanadium oxide after about 1000 cycles. WO_3 presents no problems.

Conclusions

Stable to atmospheric moisture, adhesive and transparent polymer blend electrolytes have been synthesized by the free radical polymerization of methyl methacrylate in a range of polyethers and doped with $LiCF_3SO_3$. The polyethers include PEG and EO/PO copolymers. Ionic conductivities range from 10^{-4} S/cm to 10^{-6} S/cm at room temperature. Electrolytes are viscoelastic ranging from soft elastomers to castable thin films. The most favourable blends are thin films based upon EO/PO ($= 50/50$ or $60/40$) copolymers with molecular weights in the range of 300-600 and with salt concentrations of 25:1, PMMA concentrations in the range 8 to 40 vol% and PC < 15 vol%. OMPEO/PAAM/$LiClO_4$ films are also quite satisfactory with room temperature conductivities up to 6×10^{-5} S/cm; a factor of two less than the EO/PO based system (13,14). These oxymethylene linked poly(ethylene oxide) (OMPEO) based polymers are of much higher molecular weight ($\sim 10^5$) and the best films were those with O:Li $= 10:1$ and a PAAM concentration of 20 vol%. Although these electrolytes are microphase separated those that are transparent exhibit only one T_g and have structural stability.

Nonaqueous gels swollen by the solvents DMF and PC are prospective materials for various electrochemical applications due to their high ambient and sub-ambient temperature conductivities. Electrolytes based on DMF exhibit higher ambient temperature conductivities (up to 8×10^{-4} S/cm) due to the possibility of solvent protonation and therefore the possibility of proton transport via a Grotthus type mechanism. FT-IR spectra confirm the possibility of the protonation of the carbonyl oxygens of DMF by H_3PO_4. Ambient temperature conductivities measured for the PC based electrolytes also exceed 10^{-4} S/cm. Conductivities of some of these gel electrolytes remain higher than 10^{-5} S/cm even at temperatures as low as -50°C. It has been suggested that at subambient temperatures the mobility of charge carriers, which is a function of the electrolyte viscosity, has a dominant effect on conductivity whereas at ambient and moderate temperatures conductivity depends mostly on charge carrier concentration.

Acknowledgements.

This work was financially supported by DSS-Energy Mines and Resources Canada 23440-2-94001/A-SQ Research contract and by the National Sciences and Engineering Research Council of Canada under an Industrial Oriented Research grant. This support is gratefully acknowledged.

Literature Cited.

(1) World Commission on Environment and Development Annual Report 1987.

(2) Solar Ion Energy Houses of IEA Task 13, Report of the International Energy Agency, Solar Heating and Cooling Programme, Robert Hastrup ed. (James and James (Science Publishers) Ltd. London, January 1995.

(3) Lampert, C.M.; Granqvist, G.G.; Eds. *Large-area Chromogenics; Materials and Devices for Transmittance Control* (SPIE Optical Engineering Press, Bellingham 1990).

(4) Granqvist, C.G., *Electrochromic Oxides: Thin Films and Devices*, Elsevier, Amsterdam **1995**, Chapter 26.

(5) Fenton, D.E.; Parker, J.M.; Wright, P.V., *Polymer*, **1973**, 14, 589.

(6) Wright, P.V., *Brit. Polym. J.*, **1975**, 7, 319.

(7) Armand, M.B.; Chabagno, J.M.; Duclot, M.J., in *Fast Ion Transport in Solids*, ed. Vashista, P.; Mundy, J.N.; Shenos, G.K., North Holland, New York, **1979**, p. 131.

(8) Bruce, P.G., ed. *Solid State Electrochemistry*, Cambridge University Press, Cambridge UK, **1995**.

(9) Mani, T,; Stevens, J.R., *Polymer*, **1992**, 33, 834.

(10) Mani, R.; Mani, T.; Stevens, J.R., *J. Polym. Sci. Polym. Chem. Ed.*, **1992**, 30, 2025.

(11) Mani, T,; Mani, R.; Stevens, J.R., *Solid State Ionics*, **1993**, 60, 113.

(12) Marco, C,; Fatou, J.G.; Gomex, M.A.; Tanaka, H.; Tonelli, A.E., *Macromolecules*, **1990**, 23, 2183.

(13) Such, K.; Stevens, J.R.; Wieczorek, W.; Siekierski, M.; Florjanczyk, Z., *J. Polym. Sci. Poly. Phys. Ed.*, **1994**, 32, 2221.

(14) Stevens, J.R.; Such, K.; Cho, N.; Wieczorek, W., *Solar Energy Materials and Solar Cells*, **1995**, 39, 223.

(15) Monikowska-Zygadlo, E.; Florjanczyk, Z.; Wieczorek, W., *J. Macromol. Sci. Pure Appl. Chem.*, **1994**, A31(9), 1121.

(16) Allcock, H.R.; Ravikiran, R.; O'Connor, J.M., *Macromolecules*, **1997**, 30, 3184.

(17) Fan, J.; Fedkiw, P.S., *J. Electrochem. Soc.*, **1997**, 144, 399.

(18) Raducha, D.; Wieczorek, W.; Florjanczyk, Z.; Stevens, J.R., *J. Phys. Chem.*, **1996**, 100, 20126.

(19) Stevens, J.R.; Wieczorek, W.; Raducha, D.; Jeffrey, K., *Solid State Ionics*, **1997**, 97, 347.

(20) K.-D. Kreuer, *Chem. of Matter*, **1996**, 8, 610.

(21) Lassegues, J.C., in *Proton Conductors: Solids, Membranes and Gels-Materials and Devices*, ed. Colomban, P. (Cambridge University Press, Cambridge 1992) chapter 20.

(22) Lassegues, J.C.; Desbat, B.; Trinquet, O.; Cruege, F.; Poinsignon, C., *Solid State Ionics*, **1989**, 35, 17.

(23) Donoso, P.; Gorecki, W.; Berthier, C.; Defendini, F.; Poinsignon, C.; Armand, M., *Solid State Ionics*, **1988**, 28-30, 969.

(24) Petty-Weeks, S.; Zupancic, J.J.; Swedo, J.R., *Solid State Ionics*, **1988**, 31, 177.

(25) Daniel, F.M.; Desbat, B.; Cruege, F.; Trinquet, O.; Lassegues, J.C., *Solid State Ionics*, **1988**, 28-30, 637.

(26) Tanaka, R.; Yamamoto, H.; Kawamura, S.; Iwase, T., *Electrochimica Acta*, **1995**, 40, 2421.

(27) Dabrowska, A.; Wieczorek, W., *Mater. Sci. and Eng.*, **1994**, B22, 107.

(28) Przyluski, J.; Wieczorek, W., *Synthetic Metals*, **1991**, 45, 323.

(29) Barthel, J.; Goves, H.-J.; Schmerr, G.; Wachter, R., in *Topics in Current Chemistry: Physical and Inorganic Chemistry*, ed. Boschke, F.L. (Springer-Verlag, Heilderberg **1983**, 33-145.

(30) See for instance Lengyel, J.; Conway, B.E., Chapter 4 in *Comprehensive Treatise of Electrochemistry* Vol. 5, eds. Conway, B.E.; Bockris, J.O.M.; Yeager, E. (Plenum Press: New York, London **1983**) and the references cited therein.

(31) Chandra, S., in *Proc. II Int. Symposium on Solid State Ionic Devices*, Singapore, July 18-23, 1988, Chowdari, B.V.R. and Radakrishna, S. Eds., World Scientific Publication, Singapore **1988**, p. 265.

(32) Bellamy, L.J. in *The Infra-red spectra of Complex Molecules*, Chapman and Hall, London **1975**, Chapter 18.

(33) Thomas, A.B.; Rochow, E.G., *J. Amer. Chem. Soc.*, **1957**, 79, 1843.

(34) Munson, R.A., *J. Phys. Chem.*, **1964**, 68, 3374.

(35) Dippel, Th.; Kreuer, K.-D.; Lassegues, J.C.; Rodriguez, D., *Solid State Ionics*, **1993**, 61, 41.

(36) Jeffrey, K.R.; Stevens, J.R.; Raducha, D.; Wieczorek, W., *Can. J. Chem.*, (to be published).

(37) Ferry, A.; Jeffrey, K.R.; Oradd, G.; Jacobsson, P.; Stevens, J.R., *J. Phys. Chem.* (to be published).

(38) Reichardt, C. in *Solvents and Solvent Effects in Organic Chemistry*, 2nd ed. VCH Verlagsgesellschaft mbH, Weinheim, Germany **1988**, p. 69.

(39) Marcus, Y. in *Ion Solvation*, John Wiley and Sons Ltd., New York **1985**, p. 166.

Chapter 5

Morphology and Luminescence Properties of Poly(phenylenevinylene) and Poly(N-vinylpyrrolidone) Polyblends

King-Fu Lin, Lu-Kuen Chang, and Horng-Long Cheng

Institute of Materials Science and Engineering, National Taiwan University, Taipei, Taiwan 10617, Republic of China

Poly(vinyl pyrrolidone) (PVP) was mixed with poly(phenylene vinylene) (PPV) precursor (III) aqueous solution to prepare the PPV/PVP polyblends and its dilution effect on the luminescence properties of PPV was investigated. PVP is miscible with PPV precursor but immiscible with the transformed PPV. In PPV/PVP polyblends, PPV conjugated segments preserved their configuration of C_{2h} symmetry but with less packing. The size of PPV phase is similar to the reported PPV crystallites and was decreased with the content of PVP. As to the photoluminescence (PL) properties, the energy gap to produce the excitons in the PPV conjugated chains was not changed by blending with PVP, whereas the PL intensity per mole of PPV conjugated units was increased. It was attributed to the dilution effect that decreased the non-radiative interchain's quenching of excitons. Similar results were also found for the electroluminescence (EL) properties of ITO/polyblend/Al light emitting diode (LED) device, except that the current to generate the excitons might leak to the PVP phase. As a result, the optimal content of PVP in polyblends to provide the best EL performance was 15~20 wt%.

Poly(p-phenylene vinylene) (PPV) was the first reported polymer having electroluminescence (EL) properties (1), which made this material attractive in view of the potential applications in large-area visible emitting diode (LED). In the LED devices incorporating PPV polymers, the injected positive and negative charges move through the conjugated chains under the influence of the applied electric field. Some of the charges annihilate one another in pairs and form a singlet exciton that decays to the ground state with a fluorescence emission, the process of which is called EL. The emitted EL spectrum is similar to that of the photoluminescence (PL) excited by ultraviolet (UV) light. Thus, the π^{*} to π interband transition of excitons during emission has no different between EL and PL.

However, the EL efficiency of PPV is rather low. Only up to 0.05% photons/electron was reported (1). Many efforts were aimed to improve the EL efficiency, such as (i) using multilayer LED devices to enhance the EL through charge carrier confinement (2,3); (ii) minimizing the energy barriers between polymer and the LED injection electrods (4,5); (iii) introducing side groups to the PPV units (6,7); (iv) copolymerizing PPV with other monomers (8,9); and (v) blending PPV with other polymers (10,11). The first two efforts were to increase the concentration of excitons, whereas the last two were to decrease the non-radiation decay of generated excitons. It has been indicated that one of the major non-radiation decay was through the migration of excitons to the quenching sites, which might be in intrachains or interchains of conjugated polymers (12). Several evidences have been reported for the intrachain's quenching. For examples, the photoluminescence was longer lived for less-conjugated PPV in time-resolved measurements of PL and it accounted for the higher EL efficiency of light-emitting devices made from less-conjugated materials (13). Wong et al. (14) also reported that the decay of PL was faster in more conjugated materials. Due to the fact that more-conjugated polymer chains contain more intrachain's quenching sites for excitons, blockcopolymers of PPV with non-conjugated segments showed much higher quantum yield than pure PPV in PL measurements (12).

By the same token, the non-radiation interchain's quenching of PPV might be reduced by dilution of the conjugated polymer chains with non-conjugated polymers. It has been indicated that the dilution of poly(phenyl-p-phenylenevinylene) in a blend with polycarbonate leaded to the increase of radiation recombination (15). Significant enhancement of PL by isolation of extended conjugated polymer chains in the PPV-incorporated nanocomposites with a well-defined hexagonal architecture was also reported and has been attributed to the reduction of non-radiative interchain's quenching (16).

In this study, we blended PPV (I) with poly(N-vinylpyrrolidone) (PVP) (II), a water-soluble non-conjugated polymer. The polyblends were prepared from the PPV precursor (III) aqueous solution mixed with various amounts of PVP. When their mixture was spin-coated on a glass plate and heated in vacuum to transform the PPV precursor into PPV, the converted PPV became immiscible with PVP. Since the non-radiative interchain's quenching of excitons depends on the characteristics of polyblends, we investigated the chemical structure and morphology of PPV/PVP polyblends first by Fourier transform infrared spectroscopy (FTIR), Raman spectroscopy, transmission electron microscopy (TEM) and wide angle X-ray scattering (WAXS). Then, the effects of morphology on the PL and EL properties of PPV/PVP polyblends will be discussed.

Experimental

Sample Preparation. The preparation of PPV precursor is briefly shown in Figure 1. The synthesis of p-xylylene-bis(tetrahydrothiophenium chloride) monomer IV was according to the method of Lenz et al (17): 10 g dichloro-p-xylene (V, Tokyo Kasei) was dissolved in 150 mL methanol and then added with 10 mL tetrahydrothiophene (VI, Janssen Chimica) for reaction. After reacted at 50 °C for 20 h, the solution was concentrated and then added with 250 mL cold acetone (0 °C) to precipitate the monomer. A white monomer crystalline with m.p.=149~151 °C was obtained after filtrated, washed with cold acetone several times, and dried. The PPV precursor III was polymerized by the following method (18-20): 0.4 M monomer IV aqueous solution (20 mL) was mixed with 80 mL pentane and then cooled to 0~5 °C. The polymerization was carried out by further addition of 0.4 M sodium hydroxide (20 mL, already cooled to 0~5 °C) under nitrogen and proceeded for 1 h. The reaction was terminated with 0.1 M HCl to pH 7.

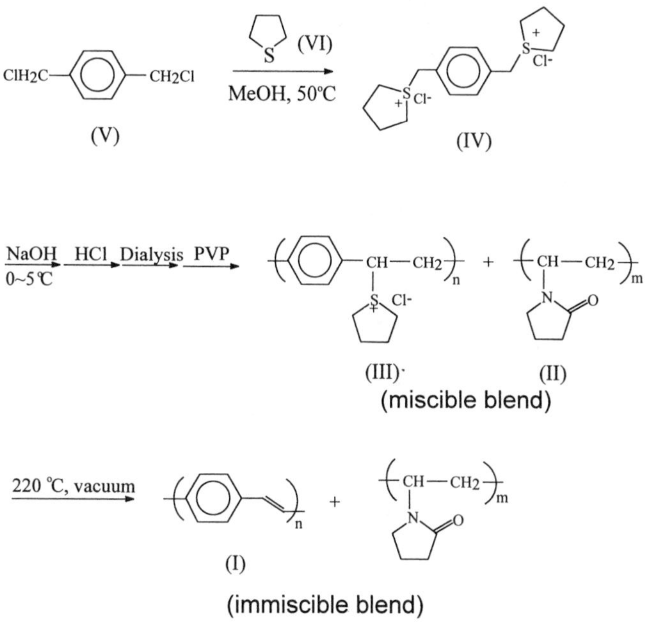

Figure 1. The preparation procedure of PPV/PVP polyblends.

The prepared PPV precursor solution after removed pentane was dialyzed in membrane (molecular weight cut-off: 3,500) against deionized water for at least a week. At this stage, the sodium ion content remained 3.57 ppm, measured by using a GBC 902 model atomic analyzer. Its PPV content measured from the thermal gravimetric analysis (TGA) was ~0.25 wt% (21). Then, the aqueous solution was added with various amount of PVP II (Sigma, molecular weight=40,000) and mixed until homogeneous. The aqueous mixture to prepare the PPV/PVP blended samples for EL tests were spin-coated on an indium-tin oxide (ITO)-deposited glass plate, whereas those for PL, infrared and Raman spectroscopies, and WAXS tests were spin-coated on a regular glass plate. All the samples were vacuumed overnight at room temperature to remove water and then heat-treated in a high vacuum oven (<10^{-5} torr) at 220 °C for 2 h to convert the PPV precursor into PPV. The thickness of coated films after heat-treated was measured ranging 1000~1300 Å by using a Dektek 3030 model surface texture analysis system. We designated the prepared PPV/PVP polyblend containing 20 wt% PVP as PPV-20PVP, and so on. To prepare the LED devices, the samples on ITO glass were coated with aluminum (Al) by thermal evaporation at high vacuum (<$5x10^{-6}$ torr) to give an ITO/Polymer/ Al device. The thickness of coated Al was 1500~2000 Å.

Characterization. PL spectra of the PPV and PPV/PVP polyblends excited by ultraviolet light with a wavelength of 365 nm were recorded on a Jasco FR-777 spectrofluorometer. The EL spectra of their prepared LED devices at a drive voltage of 8 volts (V) were also recorded on the same spectrofluorometer. Their current-voltage curves were recorded on a keithley 2400 model electrometer, where the concurrent EL intensities were recorded using a photodiode detector. The IR spectra of PPV and PPV/PVP films (removed from the coated glass plates) were recorded on a Jasco 300E model FTIR spectrometer. Their Raman spectra excited by a He-Ne laser (632.8 nm) were recorded on a Renishaw 127 model Raman spectrometer. WAXS of the PPV and PPV/PVP polyblends was performed on a Philips model PW1710 X-ray diffractometer using Cu K $_α$ radiation and a graphite monochromator. Their ultraviolet/visible (UV/vis) absorption spectra were recorded on a Jasco 7800 model spectrometer. The samples for TEM were prepared by the following method: A 300-mesh carbon-coated copper grid was dipt into the PPV precursor/PVP aqueous solution and quickly withdrawn. After dried under vacuum at room temperature, the sample-coated grid was heat-treated under high vacuum at 220 °C for 2 h to transform the PPV precursor into PPV. Then the specimens were stained by exposing to the vapor from an Osmium tetraoxide (OsO_4)/acetone/H_2O (2/49/49 wt%) solution at room temperature for 72 h. A Hitachi H-600 model TEM was used to investigate the microstructure; the stained PPV-rich phase appeared as darkened areas, whereas the PVP-rich phase appeared as white areas.

Results and Discussion

Morphology of PPV/PVP Polyblends. Figure 2 shows the FTIR spectra of PPV, PVP and PPV/PVP polyblends. No significant chemical reaction was found between PPV and PVP during the preparation of polyblends. However, the absorption peak at 959 cm^{-1} contributed by the CH out-of plane bending of trans-vinyl goups of PPV was shifted to 967 cm^{-1} whereas that at 1660 cm^{-1} contributed by C=O stretching of PVP became broader, especially for PPV-80PVP. Apparently,

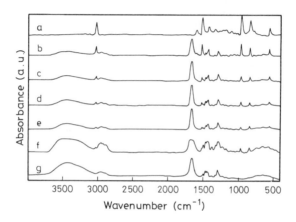

Figure 2. FTIR spectra of (a) PPV, (b) PPV-20PVP, (c) PPV-40PVP, (d) PPV-50PVP, (e) PPV-60PVP, (f) PPV-80PVP, and (g) PVP.

certain interaction between PPV and PVP was expected. Figure 3 shows the Raman spectra of PPV and PPV-50PVP polyblend excited by a wavelength of 632.8 nm. The Raman spectrum of PPV has been well studied (22-25). The main bands peaked at 1330 and 1628 cm^{-1} were contributed by the vinylene C=C stretching, those peaked at 1550 and 1586 cm^{-1} were contributed by the phenylene C-C ring stretching, whereas the peak at 1174 cm^{-1} was contributed by the phenylene C-H ring in-plane bending. The appearance of very weak 963 cm^{-1} band contributed by the vinylene C-H out-off-plane bending was indicated as an evidence of slight distortion of the vinylene group from a planar trans form of PPV polymer chains (22). According to their studies, PPV conjugated segments in solid state have approximate C_{2h} symmetry. When PPV was blended with PVP, no significant change in Raman spectrum was observed as seen in Figure 3. Apparently, the chemical structure and C_{2h} symmetry of PPV conjugated segments was not affected by blending with PVP.

Figure 4 shows the WAXS patterns of PPV, PVP and PPV-50PVP polyblend. The thin film of PPV was reported to possess measurable equatorial anisotropy with [110] and [200] lattice planes preferentially oriented parallel to the film surface (26). According to their calculation based on a monoclinic unit cell with nominal lattice parameters of a=0.80 nm, b=0.60 nm, c(chain axis)=0.66 nm and α=123°, the d-spacings of [110], [200] and [210] are 4.29 Å (2 Θ=20.7°), 3.95 Å (2 Θ=22.5°) and 3.19 Å (2 Θ=28°) respectively. Apparently, the d-spacing values of our PPV films observed in Figure 3 are in agreement with the calculated values. When PPV was blended with 50 wt% PVP which is amorphous in nature, the [200] diffraction peak reduced significantly. The WAXS pattern of PPV-50PVP shown in Figure 4 is similar to the reported patterns of incompletely-transformed PPV, which was indicated as the imperfect PPV crystalline co-existed with the amouphous PPV precursor (27). When the latter was transformed more, the crystalline domains had better packing and their size grew bigger. The reported maximum size of pristine PPV crystallites was 4.5 nm, slightly smaller than those of oriented samples (27~29). The oriented films of PPV have been shown to possess crystallites approximately 5.0 nm in size with aspect ratios near 1.0 (29).

We did not observe the individual PPV crystallites in the TEM investigation of OsO$_4$-stained PPV films. Instead, the PPV image showed a variety of rectangular, square and triangular shapes with a size ranging from 30 to 300 nm. It might be due to the fact that heavy electron density of the staining osmium atoms blurred the density difference between the disordered grain-boundary regions and the PPV crystallites. When PPV was blended with 40 wt% PVP, most of the PPV broke into small particles with a size in the same order as that of the crystallites although it still had a small portion of PPV remained the original shape as illustrated in Figure 5a. When PPV was blended with 60 wt% PVP, the PPV particles became smaller and dispersed in PVP matrix as illustrated in Figure 5b. Although some of them coagulated together, they did not form larger particles. When PPV was blended with 80 wt% PVP, the small particles coagulated into larger particles with a size ranging 20~150 nm as illustrated in Figure 5c. Because the size and behavior of small particles are similar to those of the PPV crystallites, it implies that they were identical.

Larger particles in the PPV-80PVP polyblend were formed due to the fact that small particles have a higher tendency to coagulate with each other in order to reduce the surface energy. The boundary between small particles in a coagulated large particle is still barely visible, in which we believe that the PVP was inserted

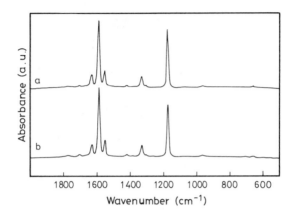

Figure 3. Raman spectra of (a) PPV and (b) PPV-50PVP.

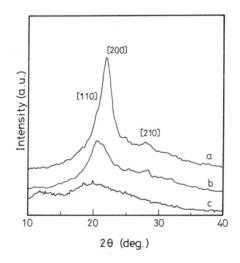

Figure 4. WAXS patterns of (a) PPV, (b) PPV-50PVP, and(c)PVP.

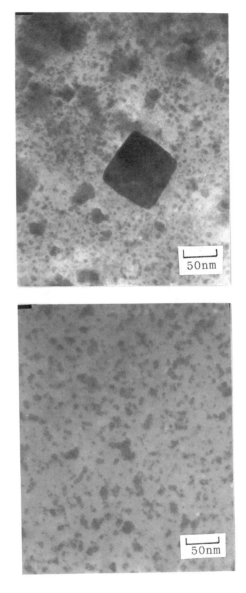

Figure 5. TEM micrographs of (a) PPV-40PVP, (b) PPV-60PVP, and (c) PPV-80PVP stained with OsO_4.

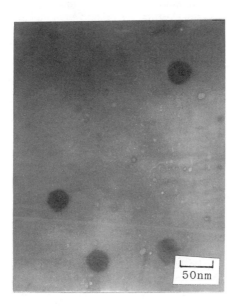

Figure 5. *Continued.*

because higher interactions between two polymers in PPV-80PVP polyblend than other compositions were found in their FTIR spectra (Figure 2).

Luminescence Properties. Figure 6 shows the UV/vis spectra of PPV and its blends with PVP. The energy gap, E, of π - π^* interband onset transition to yield the exciton can be calculated by the following equation,

$$E=1240/\lambda \qquad (1)$$

where λ is the onset UV/vis absorption wavelengh. From equation 1, the energy gap of PPV to produce the excitons was calculated as 2.34 eV. Because no significant change of UV/vis absorption spectrum was found when PPV was blended with PVP, the similar excitation energy to produce the excitons was expected. The same energy gap also indicates that the effective conjugation length of PPV was not changed by blending with PVP. Figure 7 shows the PL spectra of PPV and its blends with PVP. The PPV has a major emission peak appeared at 550 nm wavelength and a shoulder at 520 nm. Because only PPV conjugated chains emit fluorescence light, to evaluate the effect of blending on the PL intensity of PPV we calculated their relative intensities, RI, of emission peaks both at ~520 and 550 nm wavelength per mole of PPV conjugated unit by the following equation,

$$RI= I_b/(x_p I_p) \qquad (2)$$

where I_b is the peak intensity of polyblends, x_p is the molar fraction of PPV in polyblends, and I_p is the peak intensity of pristine PPV. The results were plotted in Figure 8, indicating the increase of PPV emission by blending with PVP and that the emission intensity at ~520 nm was increased more. Apparently, the incorporated PVP has a dilute effect and thus decreased the interchain's quenching of excitons. A slight blue shift of the peak at 520 nm to 515 nm was also found when PPV was blended with 80 wt% PVP.

Figure 9 shows the EL spectra of ITO/polymer/Al LED devices incorporating PPV and its blends with PVP respectively, at a drive voltage of 8 V. ITO was used as a positive electrode, whereas Al as a negative electrode. The EL spectrum of PPV LED is basically the same as its PL spectrum except that the peak intensity at 520 nm is relatively higher. When PPV was blended with 20 wt% PVP, the EL spectrum became broader and had much higher intensity compared to the pristine PPV. The peak at 520 nm also had a blue shift to 515 nm. However, with further increase of PVP content, the peak intensities decreased. The peak at 550 nm reduced its intensity at much higher rate than the left peak, while the latter further shifted to 502 nm when the content of PVP was increased to 50 wt%.

Figure 10 illustrates the current-voltage characteristic of LED devices incorporating PPV and its blends with PVP respectively. When the forward bias exceeded the threshold voltage, a yellow-green light emitted. It can be seen from the figure that the threshold voltage of PPV LED was decreased from 7 to 5 V by blending with 20 wt% PVP. Further increasing the content of PVP to 60 wt% in PPV/PVP polyblends increased the threshold voltage of LED to 6 V. Figure 11 illustrates the concurrent EL intensities emitted by LED as a function of the applied voltage swept in the same speed as that of the current-voltage measurements. Since EL is emitted by the singlet excitons generated by way of the annihilation

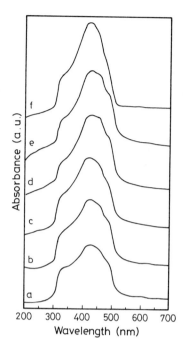

Figure 6. UV/vis spectra of (a) PPV, (b) PPV-20PVP,(c)PPV-40PVP, (d) PPV-50PVP, (e) PPV-60PVP, and (f) PPV-80PVP.

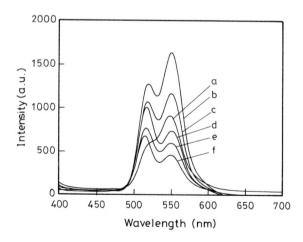

Figure 7. PL spectra of (a) PPV, (b) PPV-20PVP,(c)PPV-40PVP, (d) PPV-50PVP, (e) PPV-60PVP, and (f) PPV-80PVP.

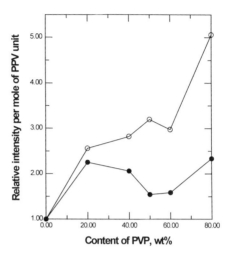

Figure 8. Relative intensities of emission peaks at (○) 520 nm and (●) 550 nm wavelength per mole of PPV conjugated units as a function of PVP content in PPV/PVP polyblends.

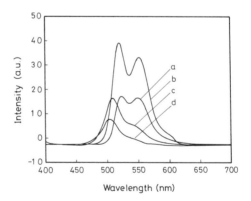

Figure 9. EL spectra of ITO/polymer/Al LED devices incorporating (a) PPV, (b) PPV-20PVP, (c)PPV-40PVP, and (d) PPV-50PVP.

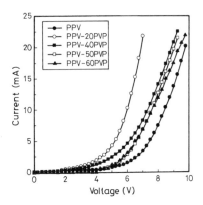

Figure 10. Current-voltage characteristics of LED devices incorporating (●) PPV, (○) PPV-20PVP, (■) PPV-40PVP, (□) PPV-50PVP, and (▲) PPV-60PVP.

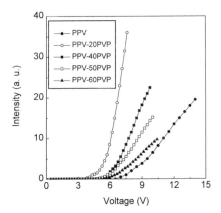

Figure 11. EL intensity-voltage (I-V) curves of LED devices incorporating (●) PPV, (○) PPV-20PVP, (■) PPV-40PVP, (□) PPV-50PVP, and (▲) PPV-60PVP.

between the injected holes and electrons from the electrodes, higher current produces more excitons and hence gives off higher EL intensity. This phenomenon can be seen in Figure 12 illustrating the EL intensities per mole of PPV conjugated units as a function of the current in the LED devices. The figure was plotted by rearrangement of the data from Figures 10 and 11 based on the uniform current across the specimen. As seen in the figure, the PPV phase in PPV/PVP polyblends emitted higher EL intensity than pristine PPV based on the same current. Again, it was attributed to the dilution effect that the generated excitons had less tendency to undergo the non-radiation decay through the interchain's quenching. However, higher content of PVP in the PPV/PVP polyblends decreased the EL intensity. This contradictory phenomenon might be due to the leaking of current through the PVP phase more than through the PPV phase, because the PVP polymers turned to form ionic clusters with the sodium ions remained during the sample preparation leading to higher dielectric constant of the PPV/PVP polyblends than the pristine PPV (21). In that case, there should have an optimum content of PVP in the blends to give the highest EL efficiency.

Figure 13 shows the current-voltage characteristic of PPV/PVP LED with low content of PVP. It is noteworthy that gradual decrease of the threshold voltage was found as the content of PVP was increased from 5 to 20 wt%. Thus, based on our studies, PPV-20PVP LED possessed the lowest threshold voltage. However, when the emitted EL intensity per mole of PPV conjugated units was plotted versus current, the gradual increase of intensity with the content of PVP was found (based on the same current) until 15 wt% was reached as illustrated in Figure 14. Therefore, the PPV-15PVP LED possesses the highest EL intensity. In summary, the optimam content of PVP in polyblends to provide the best performance of EL is 15~20 wt%.

Acknowledgments

Financial support of this work by the National Science Council in Taiwan, ROC, through Grant NSC 87-2216-E-002-001 is gratefully acknowledged.

Literature Cited

1. Burroughes, J. H.; Bradley, D. D. C.; Brown, A. R.; Marks, R. N.; Mackay, K.; Friend, R. H.; Burn, P. L.; Holmes, A. B. *Nature* 1990, *347*, 539.
2. Brown, A. R.; Greenham, N. C.; Burroughes, J. H.; Bradley, D. D. C.; Friend, R. H.; Burn, P. L.; Kraft, A.; Holmes, A. B. *Chem. Phs. Letters* 1992, *200*, 46.
3. Brown, A. R.; Bradley, D. D. C.; Burroughes, J. H.; Friend, R. H.; Greenham, N. C.; Burn, P. L.; Holmes, A. B.; Kraft, A. *Appl. Phys. Lett.* 1992, *61*, 2793.
4. Greenham, N. C.; Moratti, S. C.; Bradley, D. D. C.; Friend, R. H.; Holmes, A. B. *Nature* 1993, *365*, 628.
5. Heeger, A. J.; Parker, I. D.; Yang, Y. *Synth. Met.* 1994, *67*, 23.
6. Gettinger, C. L.; Heeger, A. J.; Drake, J. M.; Pine, D. J. *J. Chem. Phys.* 1994, *101*, 1673.
7. Jin, J.-I.; Lee, Y.-H.; Shim, H.-K. *Macromolecules* 1993, *26*, 1805.
8. Jin, J.-I.; Kim, J.-C.; Shim, H.-K. *Macromolecules* 1992, *25*, 5519.
9. Yang, Z.; Hu, B.; Karasz, F. E. *Macromolecules* 1995, *28*, 6151.
10. Kang, I.-N.; Hwang, D.-H.; Shim, H.-K.; Zyung, T.; Kim, J.-J. *Macromolecules* 1996, *29*, 165.

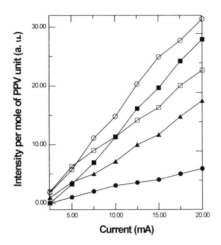

Figure 12. EL intensity per mole of PPV units as a function of current in LED devices incorporating (●) PPV, (○) PPV-20PVP, (■) PPV-40PVP, (□) PPV-50PVP, and (▲) PPV-60PVP.

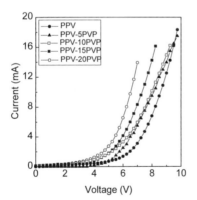

Figure 13. Current-voltage characteristics of LED devices incorporating (●) PPV, (▲) PPV-5PVP, (□) PPV-10PVP, (■) PPV-15PVP, and (○) PPV-20PVP.

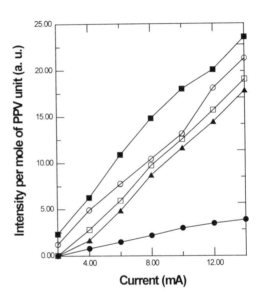

Figure 14. EL intensity per mole of PPV units as a function of current in LED devices incorporating (●) PPV, (▲) PPV-5PVP, (□) PPV-10PVP, (■) PPV-15PVP, and (○) PPV-20PVP.

11. Chang, W.-P.; Whang, W.-T. *Polymer* 1996, *37*, 3493.
12. Bazan, G. C.; Miao, Y.-J.; Renak, M. L.; Sun, B. J. *J. Am. Chem. Soc.* 1996, *118*, 2618.
13. Samuel, I. D. W.; Crystall, B.; Rumbles, G.; Burn, P. L.; Holmes, A. B.; Friend, R. H. *Synth. Met.* 1993, *54*, 281.
14. Wang, K. S.; Bradley, D. D. C.; Hayes, W.; Ryan, J. F.; Friend, R. H.; Lindenberger, H.; Roth, S. *J. Phys. C* 1987, *20*, L187.
15. Lemmer, U.; Mahrt, R. F.; Wada, Y.; Greiner, A.; Bassler, H.; Gobel, E. O. *Appl. Phys. Lett.* 1993, *62*, 2827.
16. Smith, R. C.; Fischer, W. M.; Gin, D. L. *J. Am. Chem. Soc.* 1997, *119*, 4092.
17. Lenz, R. W.; Han, C. C.; Smith, J. S.; Kazasz, F. E. *J. Polym. Sci., Part A. Polym. Chem.* 1988, *26*, 3241.
18. Wessling, R. A.; Zimmerman, R. G. US Patent 3 401 152, 1968.
19. Beerden, A.; Venderzande, D.; Gelan, J. *Synth. Met.* 1992, *52*, 387.
20. Massardier, V.; Guyot, A.; Tran, V. H. *Polymer* 1994, *35*, 1561.
21. Chang, L.-K. Master's Thesis, National Taiwan University, Taipei, Taiwan, 1997.
22. Sakamoto, A.; Furukawa, Y.; Tasumi, M. *J. Phys. Chem.* 1992, *96*, 1490.
23. Rakovic, D.; Kostic. R.; Gribov, L. A.; Davidova, I. E. *Phys. Rev. B* 1990, *41*, 10745.
24. Tian, B.; Zerbi, G.; Schenk, R.; Mullen, K. *J. Chem. Phys.* 1991, *95*, 3191.
25. Lefrant, S.; Perrin, E.; Buisson, J. P.; Eckhardt, H.; Han, C. C. *Synth. Met.* 1989, *29*, E91.
26. Chen, D.; Winokur, M. J.; Masse, M. A.; Karasz, F. E. *Polymer* 1992, *33*, 3116.
27. Ezquerra, T. A.; Lopez-Cabarcos, E.; Balta-Calleja, F. J.; Stenger-Smith, J. D.; Lenz, R. W. *Polymer* 1991, *32*, 781.
28. Bradley, D. D. C. *J. Phys. (D) Appl. Phys.* 1987, *20*, 1389.
29. Masse, M. A.; Martin, D. C.; Thomas, E. L.; Karasz, F. E.; Petermann, J. H. *J. Mater. Sci.* 1990, *25*, 311.

Chapter 6

An Overview of the Piezoelectric Phenomenon in Amorphous Polymers

Zoubeida Ounaies[1], Jennifer A. Young[2], and Joycelyn S. Harrison[3]

[1]National Research Council and [3]Composites and Polymers Branch, NASA Langley Research Center, Hampton, VA 23681
[2]Department of Materials Science and Engineering, University of Virginia, Charlottesville, VA 22903

An overview of the piezoelectric activity in amorphous piezoelectric polymers is presented. The criteria required to render a polymer piezoelectric are discussed. Although piezoelectricity is a coupling between mechanical and electrical properties, most research has concentrated on the electrical properties of potentially piezoelectric polymers. In this work, we present comparative mechanical data as a function of temperature and offer a summary of polarization and electromechanical properties for each of the polymers considered.

Kawai's (1) pioneering work almost thirty years ago in the area of piezoelectric polymers has led to the development of strong piezoelectric activity in polyvinylidene fluoride (PVDF) and its copolymers with trifluoroethylene and tetrafluoroethylene. These semicrystalline fluoropolymers represent the state of the art in piezoelectric polymers. Research on the morphology (2-5), piezoelectric and pyroelectric properties (6-10), and applications of polyvinylidene fluoride (11-14) are widespread in the literature. More recently Scheinbeim et al. have demonstrated piezoelectric activity in a series of semicrystalline, odd numbered nylons (15-17). When examined relative to their glass transition temperature, these nylons exhibit good piezoelectric properties (d_{31} = 17 pC/N for Nylon 7) but have not been used commercially primarily due to the serious problem of moisture uptake. In order to render them piezoelectric, semicrystalline polymers must have a noncentrosymmetric crystalline phase. In the case of PVDF and nylon, these polar crystals cannot be grown from the melt. The polymer must be mechanically oriented to induce noncentrosymmetric crystals which are subsequently polarized by an electric field. In such systems the amorphous phase supports the crystalline orientation and polarization is stable up to the Curie temperature.

Nalwa et al. have also examined piezoelectricity in a series of polythioureas (*18,19*). Though not highly crystalline, these thiourea polymers have a very high degree of hydrogen bonding which stabilizes the remanent polarization in such systems after poling.

The literature on amorphous piezoelectric polymers is much more limited than that for semicrystalline systems. This is in part due to the fact that no amorphous piezoelectric polymers have exhibited responses high enough to attract commercial interest. Much of the previous work resides in the area of nitrile substituted polymers including polyacrylonitrile (PAN) (*20-22*), poly(vinylidenecyanide vinylacetate) (PVDCN/VAc) (*23-26*), polyphenylethernitrile (PPEN) (*27,28*) and poly(1-bicyclobutanecarbonitrile) (*29*). The most promising of these materials are the vinylidene cyanide copolymers which exhibit large dielectric relaxation strengths and strong piezoelectricity. The carbon-chlorine dipole in polyvinylchloride (PVC) has also been oriented to produce a low level of piezoelectricity (*30,31*). Motivated by a need for high temperature piezoelectric sensor materials, NASA has recently begun research in the development of amorphous piezoelectric polymers. In this paper an amorphous, aromatic piezoelectric polyimide developed at NASA (*32*) is presented along with other amorphous and paracrystalline piezoelectric polymers shown in Table I. The purpose of this overview is to explain the mechanism and key components required for developing piezoelectricity in amorphous polymers and to present a summary of polarization and electromechanical properties of currently researched amorphous systems.

Background

The piezoelectricity in amorphous polymers differs from that in semi-crystalline polymers and inorganic crystals in that the polarization is not in a state of thermal equilibrium, but rather a quasi-stable state due to the freezing-in of molecular dipoles. As mentioned by Broadhurst and Davis (*33*), four criteria are essential to make an amorphous polymer exhibit piezoelectric behavior. First, molecular dipoles must be present. As seen in Table 1, these dipoles are typically pendant to the polymer backbone as are the nitrile groups in PAN, PVDCN-VAC, and (β-CN) APB/ODPA. However, the dipoles may also reside within the main chain of the polymer such as the anhydride units in the (β-CN) APB/ODPA polyimide. In addition to a dipole moment μ, the dipole concentration N (number of dipoles per unit volume) is also important in determining the ultimate polarization, P_u, of a polymer,

$$P_u = N\mu \qquad (1)$$

Equation (1) is for a rigid dipole model and gives a maximum value for the polarization which assumes all dipoles are perfectly aligned with the poling field. Table II lists some amorphous piezoelectric polymers along with ultimate polarizations, remanent polarizations, and calculated and measured dielectric relaxation strengths, $\Delta\varepsilon$. The dielectric relaxation strength is defined as the change in dielectric constant as the polymer traverses the glass transition temperature. Semicrystalline PVDF is added for comparison.

Table I. Structure, morphology and T_g for piezoelectric polymers.

Polymer	Repeat Unit	Morphology	T_g (°C)
PVC	$-(CH_2-CH)_n$ with Cl	Amorphous	80
PAN	$-(CH_2-CH)_n$ with C≡N	Paracrystalline	90
PVAc	$-(CH-CH_2)_n$ with O, $O=C-CH_3$	Amorphous	30
P(VDCN-VAc)	$-(CH_2-C-CH_2-CH)_n$ with C≡N, C≡N, O, $O=C-CH_3$	Paracrystalline	170
PPEN	aromatic ether with C≡N	Amorphous	145
(β-CN) APB/ODPA	aromatic imide with C≡N	Amorphous	220
PVDF	$-(CH_2-C)_n$ with F, F	Semicrystalline	-35

Table II. Polarization data for some amorphous piezoelectric polymers and PVDF.

Polymer	μ (10^{-30} Cm)	N (10^{28} m^{-3})	P_u^a (mC/m^2)	$\Delta\varepsilon^b$ calculated	$\Delta\varepsilon^c$ measured	$P_{remanent}^d$ (mC/m^2)	P_r/P_u (%)
PVC	3.7	1.33	50	7.0	10.0	16.0 (E_p = 32 MV/m)	32
PAN	11.3	1.34	152	35.0	38.0	25.0	17
PVAc	6.0	0.83	50	6.6	6.5	5.0	10
PVDCN/VAc	19.0	0.44	84	30.0	125.0	50.0	60
PPEN	14.0	0.37	52	12.0	12 .0	10.0	19
(β-CN) APB/ODPA	29.5	0.14	40	23.0	17.6	14.0	35
PVDF	7.0	18.40	130	-	-	40.0-55.0	30-42

[a]Calculated using equation 1.
[b] $\Delta\varepsilon$ is calculated by equation 2.
[c] $\Delta\varepsilon$ is measured as ($\varepsilon_{above\ Tg} - \varepsilon_{below\ Tg}$).
[d] P_r is the actual polarization in the polymer, measured by the thermally stimulated current method, or hysteresis measurements in the case of PVDF.

The importance of dipole concentration on ultimate polarization is evident from a comparison of polyacrilonitrile (PAN) and the polyimide (β-CN) APB/ODPA. PAN has a single nitrile dipole per repeat unit ($\mu = 3.5D$) resulting in a dipole concentration of 1.34×10^{28} m^{-3}. This translates into an ultimate polarization of 152 nC/m^2 (20). The (β-CN) APB/ODPA polyimide, on the other hand, has a single nitrile dipole pendant to a phenyl ring ($\mu = 4.2$ D), as well as two anhydride dipoles ($\mu = 2.34$ D) resulting in a total dipole moment per monomer of 8.8 D (Young, J. A.; Farmer, B. L. *Polymer*, in press). However, the dipole concentration of (β-CN) APB/ODPA is only 0.136×10^{28} m^{-3}, resulting in an ultimate polarization of 40 mC/m^2, which is less than a fourth of that of PAN.

The second criterion for piezoelectricity is the ability to align the dipoles. Orientation polarization of molecular dipoles is responsible for piezoelectricity in amorphous polymers. It is induced by applying an electric field (E_p) at an elevated temperature ($T_p \geq T_g$) where the molecular chains are sufficiently mobile to allow dipole alignment with the electric field. Partial retention of this orientation is achieved by lowering the temperature below T_g in the presence of E_p as shown in Figure 1. The resulting remanent polarization (P_r) is directly proportional to E_p and the piezoelectric response. The procedure used to prepare a piezoelectric amorphous polymer clearly results in both oriented dipoles and space or real charge injection. The real charges are usually concentrated near the surface of the polymer as schematically shown in Figure 2, and they are introduced due to the presence of the electrodes. However, Broadhurst et.al. (34) have shown that the presence of space charges does not have a significant effect on the piezoelectric behavior. The reason for this is two fold. The magnitude of the space charges is usually not significant with respect to the polarization charges. Secondly, space charges are essentially symmetrical with respect to the thickness of the polymer therefore when the material is strained uniformly their contribution to the piezoelectric effect is negligible. A number of authors have demonstrated this by use of phenomenological models (33,34).

A study of the relationship between relaxation times, poling temperatures and poling fields is crucial to achieve optimal dipole alignment. Theoretically, the higher the electric field, the better the dipole alignment. However, the value of the electric field is limited by the dielectric breakdown of the polymeric material. In practice, 100 MV/m is the maximum field that can be applied to these materials. Poling times need to be of the order of the relaxation time of the polymer at the poling temperature. It is unlikely that a high degree of alignment is achievable in amorphous polymers as evidenced by P_r/P_u data in Table II. Using computational chemistry techniques the orientation polarization of the (β-CN) APB/ODPA polymer has been assessed by monitoring the angle, θ, that the dipoles make with the applied electric field (Young, J. A.; Farmer, B. L. *Polymer*, in press). The unpoled state is found to exhibit random orientation of the dipoles, $\theta = 90°$, as shown in Figure 3a. Upon poling, the nitrile and anhydride dipoles are perturbed by the electric field to form average angles of $\theta = 50°$ and $\theta = 63°$, respectively, Figure 3b. As shown in Table II, for most polymers the ratio of measured remanent polarization to the calculated ultimate polarization is 30%.

When local ordering or paracrystallinity is inherent in the polymer or is induced by mechanical stretching, an increase in the value of the remanent polarization

92

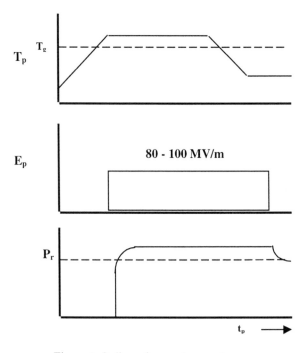

Figure 1. Poling of amorphous polymer.

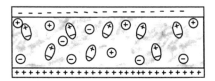

Figure 2. A model of a polymer with real charges and oriented dipolar charges after poling.

Before Poling

After Poling

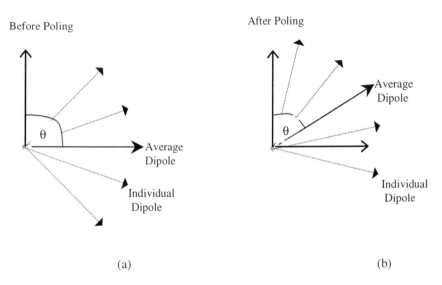

θ

Average
Dipole

Individual
Dipole

θ

Average
Dipole

Individual
Dipole

(a)

(b)

Figure 3. (a) Before poling, dipoles form an average angle of 90° with respect to the vertical axis.
(b) After poling, the angle θ between the vertical axis and the average dipoles is decreased.

is observed. For example, some researchers (*23, 35, 36*) assert that the large discrepancy between the measured and calculated $\Delta\varepsilon$ for PVDCN-VAc (Table II) may be attributed to locally ordered regions in the polymer. A number of authors have suggested that PVDCN-VAc also exhibits ferroelectric-like behavior (*37*) due to switching of the nitrile dipoles under AC-field. Several investigators (*20, 33, 38*) have proposed that the difficulty of poling PAN in the unstretched state is related to the strong dipole-dipole interaction of nitrile groups of the same molecule which repel each other, thus preventing normal polarization. Upon stretching, the intermolecular dipole interactions facilitate the packing of the individual chains and give rise to ordered zones (*38*). Comstock et al. (*39*) measured the remanent polarization of both unstretched and stretched PAN using the thermally stimulated current method (TSC) and observed a two-fold increase in the remanent polarization (TSC peak at 90°C) for PAN that was stretched four times its original length.

The third criterion for making an amorphous piezoelectric polymer is the locking-in of dipole alignment and its subsequent stability. As explained earlier, the temperature is lowered to room temperature while the field is still on, to freeze in the dipole alignment. In a semi-crystalline material, however, the locking-in of the polarization is supported by the crystalline structure of the polymer, and is therefore stable above the glass transition temperature of the polymer. It is for that reason that PVDF (T_g = -35° C) can be used from room temperature to about 100° C. In semi-crystalline materials, piezoelectricity remains until the Curie temperature is reached. Although there is little data addressing the stability of piezoelectric activity in amorphous polymers, the general effect of time, temperature and pressure has been noted. Broadhurst and Davis (*33*) state that as temperature decreases the structural relaxation time of PVC increases rapidly to the order of years at room temperature. This is probably the case for most of the polymers mentioned in this discussion. It has been shown by TSC measurements that the remanent polarization of (β-CN) APB/ODPA is stable when heated at 1°C/min up to 200°C, where over 80% of the P_r is retained (*40*). It is clear that time, pressure and temperature can all contribute to dipole relaxation in these polymers. For a given application and use temperature, the effect of these parameters on the stability of the frozen-in dipole alignment should be determined.

The final determining factor for a material's degree of piezoelectric response is the ability of the polymer to strain with applied stress. Since the remanent polarization in amorphous polymers is lost in the vicinity of T_g, the use of these piezoelectric polymers is limited to temperatures well below T_g. This means that the polymers are in their glassy state, and the further away from T_g the use temperature is, the stiffer the polymer. This also means that measurement of the bulk physical properties is crucial both for identifying practical applications and for comparing polymers. The electromechanical coupling coefficient, k_{31}, is a measure of the combination of piezoelectric and mechanical properties of a material (refer to Table III). It can be calculated using the equation below:

$$k_{31} = d_{31}\sqrt{\frac{Y_{11}}{\varepsilon\varepsilon_0}} \qquad (2)$$

Table III. Piezoelectric and mechanical properties (at 25°C).

	$E_{11}{}^a$ (10^9 Pa)	d_{31} (pC/N)	$k_{31}{}^f$
PVC	0.34	0.7^b	0.001
PAN	1.70	1.7^c	0.010
PVDCN-VAc	-	7.0^d	0.050
(β-CN)APB/ODPA	2.80	0.3^e	0.002
PVDF	2.60	27.0^e	0.120

[a]Measured in our laboratories using a Rheovibron.
[b][from Ref. 30].
[c][from Ref. 21].
[d][from Ref.27].
[e]Measured in our laboratories.
[f]Calculated using equation 2.

where d_{31} is the piezoelectric strain coefficient, Y_{11} is Young's modulus, ε is the dielectric constant of the polymer and ε_0 is the dielectric constant of free space. The piezoelectric amorphous polymer may be used at temperatures near its T_g to optimize the mechanical properties, but not too close so as not to lose the remanent polarization.

Dielectric properties as predictors of piezoelectric behavior

This section addresses the origins of the dielectric contribution to the piezoelectric response of amorphous polymers. The potential energy U of a dipole μ at an angle θ with the applied electric field is U = μ **E** cos θ. Using statistical mechanics and assuming a Boltzman's distribution of the dipole energies, the mean projection of the dipole moment, $<\mu>_E$, in the direction of the applied electric field is obtained.

$$\frac{<\mu_E>}{\mu} = \coth\frac{\mu E_p}{kT} - \frac{kT}{\mu E} \tag{3}$$

This is the Langevin equation which describes the degree of polarization in a sample when an electric field, E, is applied at temperature T. Experimentally, a poling temperature in the vicinity of T_g is used to maximize dipole motion. The maximum electric field which may be applied, typically 100 MV/m, is determined by the dielectric breakdown strength of the polymer. For amorphous polymers μ E / kT <<1, which places these systems well within the linear region of the Langevin function. The following linear equation for the remanent polarization results when the Clausius Mossotti equation is used to relate the dielectric constant to the dipole moment (41).

$$P_r = \Delta\varepsilon \ \varepsilon_0 E_p \tag{4}$$

It can be concluded that remanent polarization and hence piezoelectric response of a material is determined by $\Delta\varepsilon$, making it a practical criterion to use when designing piezoelectric amorphous polymers. The dielectric relaxation strength, $\Delta\varepsilon$ may be the result of either free or cooperative dipole motion. Dielectric theory yields a mathematical way of examining the dielectric relaxation due to free rotation of the dipoles, $\Delta\varepsilon$. The equation incorporates Debye's work based on statistical mechanics, the Clausius Mossotti equation, and the Onsager local field and neglects short range interactions (42)

$$\Delta\varepsilon_{calculated} = \frac{N\mu^2}{3kT\varepsilon_0}(\frac{n^2+2}{3})^2(\frac{3\varepsilon(0)}{2\varepsilon(0)+n^2})^2 \tag{5}$$

N is the number of dipoles per unit volume, k is the Boltzmann constant, $\varepsilon(0)$ is the static dielectric constant and n is the refractive index. If the experimental value of $\Delta\varepsilon$ ($\Delta\varepsilon_{measured}$) agrees with the theoretical value of $\Delta\varepsilon$ ($\Delta\varepsilon_{calculated}$), then the material exhibits free dipolar motion. Table II shows that in polymers such as PAN, VAc, PVC, PPEN, and (β-CN) APB/ODPA the dielectric relaxation strength corresponds to free dipolar motion since $\Delta\varepsilon_{calculated}$ and $\Delta\varepsilon_{measured}$ are in agreement. This table also shows that for the copolymer PVDCN/VAc, $\Delta\varepsilon_{calculated} = 30$ while $\Delta\varepsilon_{measured} = 125$ (25, 36). This large discrepancy in the values of $\Delta\varepsilon$ is indicative of cooperative motion of several CN dipoles within the locally ordered regions of the polymer. Cooperativity means that instead of each dipole acting independently, multiple CN dipoles respond to the applied electric field in a unified manner. When x dipoles act cooperatively, the number density of dipoles decreases by 1/x yet the effective dipole moment increases by x^2 to yield a large dielectric relaxation strength. Intramolecular and/or intermolecular interactions between individual dipoles may be responsible for this particular phenomenon (25). Such interactions are manifested in the existence of paracrystalline regions within the PVDCN/VAc polymer (27). The large relaxation strength exhibited by PVDCN/VAc gives it the largest value of P_r (Table II) and hence d_{31} (Table III) of all the amorphous polymers. Although the existence of cooperative dipole motion clearly increases the piezoelectric response of amorphous polymers, the mechanisms by which cooperativity can be systematically incorporated into the polymer structure remain unclear at this time. Finally, Table II demonstrates the efficiency with which polarization may be imparted into the sample. The polymers which exhibit free dipolar motion show relatively low ratios of P_r / P_u (10-30%) which indicates low dipole orientation, while the ratio of P_r / P_u for PVDCN/VAc polymer is as high as 60%. It is noted that in Table II, P_r is measured using the previously mentioned TSC method. As a result, P_r could include space charge effects as well as dipolar reorientation. However, for all the amorphous polymers of Table II except PVDCN-VAc, P_r measured was seen to vary linearly with E_p, which is an indication that space charge effect is negligible since space charge polarization varies nonlinearly with the poling field. Also for these polymers, P_r measured is of the order of P_r given by equation (4), again indicating a linear relationship.

Designing an amorphous polymer with a large dielectric relaxation strength and hence piezoelectric response would require the ability to incorporate highly polar groups at high concentrations and cooperative dipole motion.

Mechanical and electromechanical properties

Mechanical properties are often overlooked when investigating piezoelectric polymers. It is important to note that the piezoelectric response is a result of the coupling between the mechanical and dielectric properties in an amorphous polymer. The piezoelectric coefficient, d, is defined as

$$d = \frac{1}{A}\frac{\partial Q}{\partial T}\Big|_{E=0,T} \tag{6}$$

where Q is the charge per unit area displaced through a closed circuit between two electrodes, T is the applied stress, A is the area of the electrodes, and E is the applied field. This equation is used for the direct measurement of the d_{31} piezoelectric coefficient by stressing the polymer in the plane of the film, and measuring the charge that forms on the electrodes under zero field. Figure 4 shows the Young's modulus, Y'_{11} as a function of temperature. A decrease in the modulus of the four amorphous

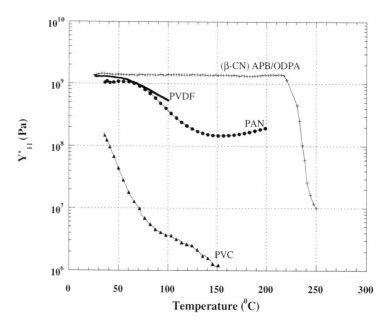

Figure 4. Comparison of Young's modulus Y'_{11} (measured in our laboratories).

polymers, PVDCN-VAc, unstretched PAN, PVC, and (β-CN) APB/ODPA, occurs in the region of the glass transition. Figure 5 presents d_{31} as a function of temperature for several piezoelectric polymers. As the polymers approach their respective glass transition temperatures, d_{31} increases due to the decrease in the modulus. The trend continues until the thermal energy randomizes the molecular dipoles to yield a decrease in d_{31} as shown for PAN which has a $T_g = 90\ °C$. The effect of the mechanical properties on the piezoelectric response is also evident by comparing two polymers with comparable remanent polarizations, PVC, and (β-CN) APB/ODPA. The lower modulus (higher compressibility) of PVC results in a larger piezoelectric response relative to (β-CN) APB/ODPA. It is important to note that data for identical processing conditions for the various polymers (E_p, t_p and T_p) is not readily available in the literature. This type of data would be very useful for a comparative analysis of the relative effects of mechanical and dielectric properties on the piezoelectric response.

Stretching can also have an effect on the piezoelectric coefficient of a polymer as shown for PAN in Figure 5. The increase in d_{31} with stretching has both mechanical and polarization contributions. Stretching in the 1-direction aligns the chains in the plane of the film, which results in an increase in the compressibility in the 3-direction. This chain alignment also facilitates dipole orientation in response to an applied electric field (38,39) which results in a higher P_r than is achievable in unstretched PAN.

Figure 6 stresses the importance of the relative effect of the temperature on the piezoelectric activity. Although from Figure 5, PAN (stretched) looked like it outperforms (β-CN) APB/ODPA, a closer look of Figure 6 shows that at 70°C below T_g, d_{31} of PAN is 1.7 pC/N whereas that of (β-CN) APB/ODPA is 5 pC/N. It is clear that the amorphous piezoelectric polymers have to be used below their T_g as PAN quickly depolarizes as T_g is approached. This is in contrast to PVDF (and other semicrystalline polymers) which are used well above their T_g's.

The coupling between mechanical and dielectric properties is also evident in the hydrostatic piezoelectric coefficient, d_h which is given by (33):

$$d_h = -\beta\ \Delta\varepsilon\ \varepsilon_0\ \varepsilon_\infty\ E_p /3 \qquad (7)$$

As seen in equation 7, both the mechanical properties (through the compressibility, β) and the dielectric properties (represented by $\Delta\varepsilon$) affect the piezoelectric coefficient. Figure 7a presents the mechanical and electrical properties as a function of temperature for (β-CN) APB/ODPA. The compressibility of (β-CN) APB/ODPA increases slightly with temperature until T_g is reached. The remanent polarization is relatively stable until about 50 degrees below the glass transition at which point it decreases due to dipole randomization. Consequently as shown in Figure 7b, d_h increases slightly with temperature prior to the onset of depolarization.

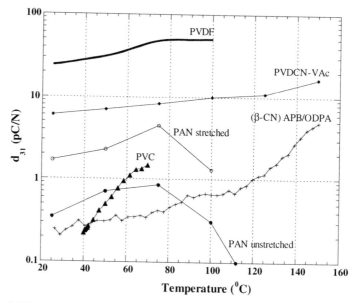

Figure 5. Piezoelectric coefficient d_{31} for various amorphous polymers and PVDF. (PVC,CN-APB/ODPA and PVDF measured in our laboratories, PAN from Ref. 21, PVDCN from Ref. 25).

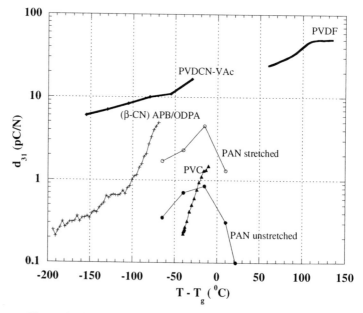

Figure 6. Piezoelectric coefficient d_{31} as a function of $(T-T_g)$.

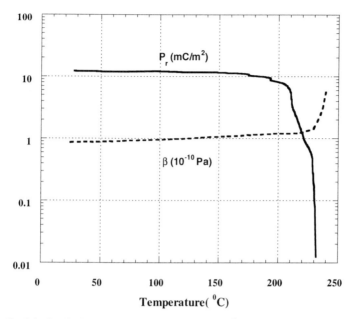

Figure 7a. Mechanical and Piezoelectric properties (β-CN)APB/ODPA(measured in our laboratories).

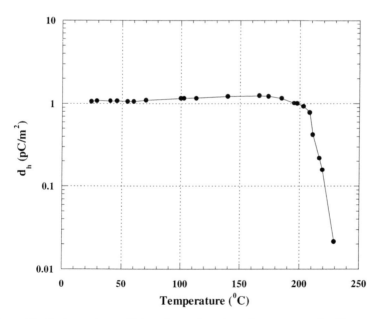

Figure 7b. Hydrostatic coefficient d_h as a function of temperature for (β-CN)APB/ODPA(calculated using equation 7).

102

Summary

This review has brought together the dielectric theory and the mechanical properties which define the piezoelectric response in amorphous polymers. The basic requirements for designing an amorphous piezoelectric polymer are the presence and concentration of dipoles, the ability to orient these dipoles and to lock them in this alignment, and the ability to sufficiently strain the polymer. Calculating the ultimate polarization P_u is a good starting point when designing piezoelectric amorphous polymers. A primary weakness of amorphous polymers is poor dipole alignment during poling (low P_r/P_u value). This would be overcome by incorporating cooperativity such as the case of PVDCN-VAc. Until this phenomenon is understood where the polymer structure may be engineered to include cooperativity, incremental steps are being made to improve the response by incorporating large dipoles (primarily CN) at high concentrations.

References

1. Kawai, H. *Jpn, J. Appl. Phys.* **1969**, *8*, p. 975
2. Lovinger, A.J. In *Developments in Crystalline Polymers*; Basset, D.C., Ed.; Applied Science Publishers: London, UK, 1982, Vol. 1; p. 242
3. Carbeck, J.D.; Lacks, D.J.; Rutledge, G.C. *J. Chem. Phys.* **1995**, *103*, p. 10347
4. Kepler, R.G.; Anderson, R.A. *J. Appl. Phys.* **1978**, *49*, p. 1232
5. Davis, G.T.; McKinney, J.E.; Broadhurst, M.G.; Roth, S.C. *J. Appl. Phys.* **1978**, *49*, p. 4998
6. Lovinger, A.J. *Science* **1983**, *220*, p. 1115
7. Davis, G.T. In *Polymers for Electronic and Photonic Applications*, Wong, C. P., Ed., Academic Press, Inc.: Boston, MA, 1993; p. 435
8. Sessler, G.M. *J. Acoust. Soc. Am.* **1981**, *70*, p. 1596
9. Kepler, R. G.; Anderson, R. A. *Adv. Phys.* **1992**, *41*, p. 1
10. Wada, Y.; Hayakawa, R. *Jpn. J. Appl. Phys.* **1976**, *15*, p. 2041
11. Furukawa, T.; Wang, T.T. In *The Applications of Ferroelectric Polymers*, Wang, T.; Herbert, J.; Glass, A., Ed.; Blackie: London, U.K., 1988; p. 66
12. Garner, G. M. In *The Applications of Ferroelectric Polymers*, Wang, T.; Herbert, J.; Glass, A., Ed.; Blackie: London, U.K., 1988; p. 190
13. Meeker, T. R. In *The Applications of Ferroelectric Polymers*, Wang, T.; Herbert, J.; Glass, A., Ed.; Blackie: London, U.K., 1988; p. 305
14. Yamaka, E. In *The Applications of Ferroelectric Polymers*, Wang, T.; Herbert, J.; Glass, A., Ed.; Blackie: London, U.K., 1988; p. 329
15. Newman, B.A., Chen, P., Pae, K.D., and Scheinbeim, J.I. *J. Appl. Phys.* **1980**, *51*, p. 5161
16. Scheinbeim, J.I. *J. Appl. Phys.* **1981**, *52*, p. 5939
17. Mathur, S.C.; Scheinbeim, J.I.; Newman, B.A. *J. Appl. Phys.* **1984**, *56*, p. 2419
18. Nalwa, H.S.; Taneja, K. L.; Tewari, U.; Vasudevan, P. *Proc. Nucl. Phys. Sol. St. Phys.Symp.* **1978**, *21C*, p. 712

19. Fukada, E.; Tasaka, S.; Nalwa, H. S. In *Ferroelectric Polymers, Chemistry, Physics and Application*, Nalwa, H. S., Ed.; Plastics Engineering, 28; Marcel Dekker, Inc.:New York, NY, 1995, p. 353

20. Ueda, H.; Carr, S.H. *Polymer J.* **1984**, *16*, p. 661

21. von Berlepsch, H.; Pinnow, M.; Stark, W. *J. Phys. D: Appl. Phys.* **1989**, *22*, p. 1143

22. von Berlepsch., H.; Kunstler, W.; Wedel, A.; Danz, R.; Geiss, D. *IEEE Trans. Elec. Ins.* **1989**, *24*, p. 357

23. Jo, Y. S.; Sakurai, M.; Inoue, Y.; Chujo, R.; Tasaka, S.; Miyata, S. *Polym.* **1987**, *28*, p. 1583

24. Miyata, S.; Yoshikawa, M.; Tasaka, S.; Ko, M. *Polymer J.* **1980**, *12*, p. 857

25. Furukawa, T.; Tada, M.; Nakajima, K.; Seo, I. *Jpn. J. Appl. Phys.* **1988**, *27*, p. 200

26. Sakurai, M.; Ohta, Y.; Inoue, Y.; Chujo, R. *Polym. Comm.* **1991**, *32*, p. 397

27. Tasaka, S.; Inagaki, N.; Okutani, T.; Miyata, S. *Polymer* **1989**, *30*, p. 1639

28. Tasaka, S.; Toyama, T.; Inagaki, N. *Jpn. J. Appl. Phys.* **1994**, *33*, p. 5838

29. Hall, H.K.; Chan, R.; Oku, J.; Huges, O.R.; Scheinbeim, J.; Newman, B. *Polym. Bulletin* **1987**, *17*, p. 135

30. Broadhurst, M. G.; Malmberg, C. G.; F. I. Mopsik; Harris, W. P., In Electrets, Charge Storage and Transport in Dielectrics; Perlman, M. M., Ed.; The Electrochemical Society: New York, NY; p. 492

31. Mopsik, F. I.; Broadhurst, M. G. *J. Appl. Phys.* **1975**, *46*, p. 4204

32. Ounaies, Z.; Young. J. A.; Simpson, J.O.; Farmer, B. L. In *Materials Research Society Proceedings: Materials for Smart Systems II*, George, E., P.; Gotthardt, R.; Otsuka, K.; Trolier-McKinstry, S.; Wun-Fogle, M., Ed.; Materials Research Society: Pittsburgh, PA, 1997, Vol. 459; p. 59

33. Broadhurst, M. G. and Davis, G. T., In *Electrets*; Sessler, G. M., Ed.; Springer-Verlag: New York, NY, 1980; Vol 33; p. 283

34. Broadhurst, M. G.; Harris, W. P.; Mopsik, F. I.; Malmberg, C. G.*Polym. Prep.* **1973**, *14*, p. 820

35. Furukawa, T. *IEEE Trans. Elect. Insul.* **1989**, *21*, p. 375

36. Furukawa, T.; Date, M.; Nakajima, K.; Kosaka, T.; Seo, I. *Jpn. J. Appl. Phys.* **1986**, *25*, p. 1178

37. Wang, T. T.; Takase, Y., *J.Appl.Phys.* **1987**, *62*, p. 3466

38. Tasaka, S. In *Ferroelectric Polymers;* Nalwa, H. S., Ed.; Plastics Engineering 28; Marcel Dekker Inc.: New York, NY, 1995; p. 325

39. Comstock, R. J.; Stupp, S. I.; Carr, S. H. *J. Macrom. Sci. Phys.* **1977**, *B13*, p. 101

40. Simpson, J.O., Ounaies, Z., and Fay, C., In *Materials Research Society Proceedings: Materials for Smart Systems II*, George, E., P., Gotthardt, R., Otsuka, K., Trolier-McKinstry, S., and Wun-Fogle, M., Ed., Materials Research Society: Pittsburgh, PA, 1997, Vol. 459; p. 53

41. Hilczer, B.; Malecki, J. In *Electrets*; Studies in Electrical and Electronic Engineering 14; Elsevier: New York, NY, 1986; p. 19

42. Frohlick, H. In *Theory of Dielectrics*; Monographs on the Physics and Chemistry of Materials 42; Oxford University Press: Oxford, U. K., 1958; p. 15

Chapter 7

The Electrorheological Properties of Chitosan Sulfate Suspensions

Shuizhu Wu and Jiarui Shen

Department of Polymer Science and Engineering, South China University of Technology, Guangzhou 510641, Peoples Republic of China

The activator-free electrorheological suspensions based on chitosan sulfate particles exhibit significant electrorheological effect under the applied electric field, and have good thermal stability and low conductivity. The suspension's electrorheological effect increases with the increasing field strength and suspension concentration but decreases with the increasing shear rate, and the dynamic yield stress of the suspension increases with the increasing concentration.

Electrorheology is the term applied to the phenomenon in which the fluidity of suspensions is modified by the application of electric fields(*1*). This phenomenon concerns the formation of a fibrilated microstructure in dense suspension due to dipole interactions. These interparticle forces result in a fluid with an enhanced viscosity and that is capable of sustaining a large yield stress(*2*). The magnitude of these stress, and the rapid time scales of the structure formation make these systems ideal working fluids in electromechanical applications(*3*). ER devices currently being developed(*4,5*), including engine mounts and shock absorbers, require large field-induced viscosities as well as rapid responses, with times scales on the order of milliseconds. However, there are still problems to be solved before ER fluids find extensive commercial applications.

It has been long observed that wet particulates are most ER active. But these moist fluids are limited to a narrow temperature range (< 70°C) and show undesirable levels of conductance arising from mobile ions(*4,6*). Solution of these problems and the development of better ER fluids depend on improving our understanding of how the phenomenon depends upon the properties of the materials which make up ER fluids. Recent development of anhydrous suspensions

104

based on conducting materials seems to have overcome some of these problems, however they show undesirable levels of conductance as well(*3,4,6*).

Natural polymers such as cellulose have been used in ER fluid preparation, but, these fluids usually require water or other polar liquids as activator(*7,8*). In this article, we made use of the unique features of chitosan, such as it contains lots of polar groups along its molecular chain, thus it has relatively higher dielectric constant, and is easy to go through modification reactions including sulfation. In present study, the activator-free chitosan sulfate--silicone oil suspensions are prepared, these suspensions appear to be able to avoid the disadvantages of the moist ER suspensions. On the other hand, the biodegradable feature of chitosan might be good for their future applications in terms of environmental protection. The electrorheological properties of these suspensions are investigated for a range of field strengths, particle concentrations and shear rates. The conductance of ER fluid is believed to be an important parameter in ER effect especially when it comes to actual applications. Large increase in the conductance of the fluid would result in excessive power demands with possible serious implications in terms of power supply and energy dissipation in the ER devices, it could even cause dielectric breakdown(*1,9*). For this reason, low conductance is an important goal for future ER fluids. The chitosan sulfate --- silicon oil suspensions prepared in the present study have extremely low conductance. The relationship between the suspension's dynamic yield strength and particle concentration is determined experimentally as well.

Experimental

Synthesis of Chitosan Sulfate. The deacetylation of chitin (Katakura Chikkarin Co., Japan) was carried out according to literature(*10*), namely, the chitin samples were treated with 50wt% NaOH at 100°C for 1 hour to produce chitosan. The sulfation of chitosan was then conducted with the method of Wolfrom et al(*11*): First, chitosan sample was dissolved in dilute acetic acid solution; after the undissovled part being filtered out, chitosan was reprecipitated with NaOH solution, then washed with distilled water till the washings tested neutral with pH paper. After that, it was washed with ethanol, absolute ethanol, diethyl ether and dimethylformide (DMF). Then chitosan sample was suspended in DMF and treated with a SO_3 (fuming sulfuric acid) -- DMF mixture for 12h at 20°C. After neutralization, the reaction mixture then underwent the purification process by using the dialysis membrane. After drying, the sulfur content of the final product was determined by elemental analysis, the product has 13.68% S. The degree of sulfation (DS) is calculated as follows(*11*):

$$DS = \frac{S(\%) \times 161}{32 \times 100 - S(\%) \times 103}$$

where S is the sulfur content of the sample, 161 is the mole molecular weight of the repeating unit of chitosan, 103 is the mole molecular weight of SO_3Na, and 32 is the molecular weight of S. Therefore, the DS of the product is 1.23. The sulfation of chitosan is shown schematically as follows:

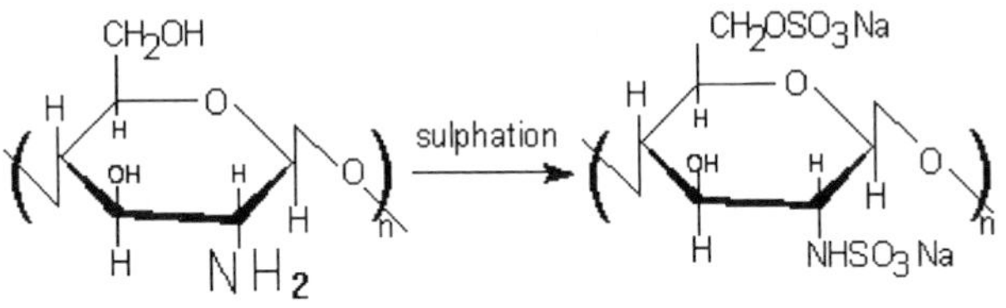

Suspension Preparation. The preparation for the activator-free suspension: After drying, chitosan sulfate samples were dispersed in a certain amount of silicone oil and ball-milled until microscopic examination indicated a mean particle size of 10 μm and the absence of particles > 20 μm. Particles were irregular in shape but without any tendency to anisometry. The silicone oil used is a colorless oil with the following physical properties: density 0.97 g/cm^3, viscosity 100 mPa·s at 20°C, dielectric constant 2.8, and boiling temperature 300°C. There was little tendency for these dispersions to separate in the short term, and such dispersions that had separated after lengthy standing readily redispersed on agitation.

The preparation for the activator-containing suspension: The chitosan sulfate samples were immersed into the glycerin-methanol solution for 72 hours, to make sure the samples adsorb a certain amount of glycerin. The fraction of the adsorbed glycerin is determined by the weight method, and the glycerin content is 5wt%. After the removal of methanol, the chitosan sulfate suspension was made with the same procedures as the above-mentioned.

Methods of Measurement. For electrorheological measurement, a concentric cylinder rheometer was used. To apply large electric field strength across the concentric cylinders, each cylinder was insulated from the rest of the rheometer. The inner cylinder has an outer diameter of 14.6 mm and height of 30mm. The outer cylinder has an inner diameter of 20 mm and height of 35 mm. The annular gap is 2.7 mm. The electric field strength was applied to the gap by grounding the outer cylinder and connecting the inner cylinder to a high-voltage source. The DC voltage of 100 ~ 2500 V were used in the experiments. The current passing the suspension was monitored by using a multimeter attached in series to the ground wire of the circuit. All measurements were carried out at room temperature.

Results and Discussion

Effects of Field Strength and Particle Concentration. The shear stress of the activator-free suspensions containing chitosan sulfate particles at different

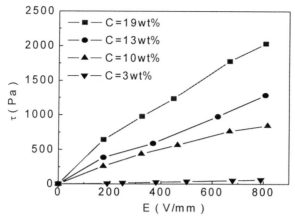

Figure 1 The suspension's shear stress versus field strength curve. ($\dot{\gamma} = 1.441s^{-1}$)

concentrations is shown in Figure 1, and the ratio of the suspension's shear stress under the applied electric field over the zero-field shear stress is plotted as a function of particle concentration in Figure 2. It can be seen that, the suspension's shear stress increases with the increasing field strength, i.e., the suspension's ER effect increases with the field strength. There exists a critical concentration (about 3wt%), under which the suspension barely displays any ER effect. While over the critical concentration, the suspension's ER effect increases with the increasing concentration.

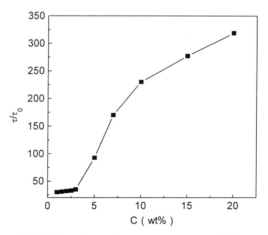

Figure 2 Relative shear stress versus concentration curve.
($E = 800$ V/mm, $\dot{\gamma} = 17s^{-1}$)

These experimental phenomena can be understood by considering the polarization forces between the suspended particles. Since the polarization forces scale as $\pi a^2 \varepsilon_0 \varepsilon_1 \beta E^2$, where a is the radius of particle, ε_0 is the permittivity of free space, ε_1 is the dielectric constant of the dispersing medium, and β is the polarization coefficient of particle: $\beta = (\varepsilon_2 - \varepsilon_1)/(\varepsilon_2 + 2\varepsilon_1)$, where ε_2 is the dielectric constant of particle(*12*). Therefore, with the enhancing field strength, the polarization forces between particles increases, as a result, more and stronger particle chains or strands form, hence the suspension exhibits more obvious ER effect.

On the other hand, the formation of particle chains is a percolation process(*13*). Only when there are enough amount of particles in the fluid, could the particle chains or strands span the gap of the electrodes, this is why there exists a critical concentration. Under the critical concentration, particles are simply dispersing in the continuous medium or form short chains, which couldn't cause significant ER effect. Over the critical concentration, the ER effect of the suspension increases with the concentration. It is because, with the increasing concentration, there are more particles dispersed in the continuous medium, therefore, more particle chains can form under the applied electric field, and the suspension exhibits much stronger ER effect.

At the same time, the suspension's thermal stability was examined as well, the results are shown in Table I.

Table I. The shear stress of the suspension before and after thermal treatment (C= 19wt%, E= 800V/mm, $\dot{\gamma} = 1.441s^{-1}$).

$T\ (^{\circ}C)$	$\tau\ (Pa)$
20	2031
50	2058
80	2064
110	2066
130	2071

The suspension was placed in an oven at 130°C for 72 hours; then its ER effect was examined again, the suspension's shear stress didn't decrease but increase a little, as shown in Table I. This indicates that activator-free suspension has quite good thermal stability.

Effects of Shear Rate. The suspension viscosity at different shear rates are shown in Figure 3. As shown in this Figure, the suspension viscosity decreases with the increasing shear rate.

According to the experimental facts, it is considered that, under the applied electric field, the interparticle polarization forces lead to the aggregation of

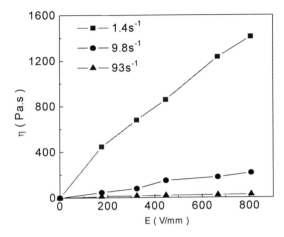

Figure 3 Suspension viscosity versus field strength
at different shear rates. (C = 19wt%)

particles or even fibril formation between the electrodes. Such a structural skeleton is across the direction of the shear field and leads to an increased suspension viscosity. In the presence of a shear field simultaneously, the particles are also acted on by the viscous forces, which is modulated by hydrodynamic interactions with other particles in the suspension. These viscous forces are in proportion with the shear rate $\dot{\gamma}$, and intend to disrupt the suspension structure(12). As $\dot{\gamma}$ is increased, the viscous forces increase, so that the tendency to break down the structural skeleton of the suspension is increased; Therefore, the suspension structure is much easier to damage and the increment of the viscosity is much smaller, while at high enough shear rate, the suspension viscosity becomes almost independent of the electric field. This suggests that, at high enough shear rate, the viscous forces are dominant, and the suspension structure does not vary appreciably with the field strength.

The characteristic parameter describing the interplay between dipole forces and flow, the Mason number, $Mn = 6\eta_1 \dot{\gamma} /[\varepsilon_0\varepsilon_1(\beta E)^2]$, is the ratio of the viscous forces tending to disrupt the structure and the polarization forces responsible for the structure, where η_1 is the viscosity of the dispersing medium(12). Based on the above analysis, it is expected that the dimensionless suspension viscosity η/η_1 depends on two parameters, suspension concentration C and Mn. At a given concentration and temperature, a plot of suspension viscosity against Mn should reduce the data at different field strengths and shear rates to a single curve. Figure 4 shows the dimensionless suspension viscosity η/η_1 as a function of the Mason number for different concentrations studied. All the results show an identical form; at low Mn values, the curve is almost linear; while at high Mn values, the curve approaches a constant viscosity. This behavior suggests that, at a given

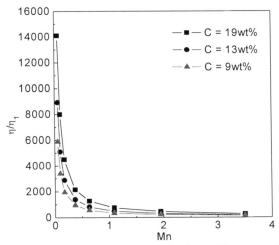

Figure 4 Relative viscosity as a function of Mason number

concentration, the shear and electric field dependence of ER suspensions can be expressed as a single function of Mn. That the data collapse onto a single curve in the transition from polarization-controlled structures to the domination of the viscous forces suggests that the suspension structure does not vary appreciable with field strength.

Suspension Current Density. In order to assess the suspension's conductivity, we measure the current passing the suspension, and the current density (j) is calculated by dividing the current over the surface area of the electrode. The current densities of the suspensions with or without activator are shown in Table II.

Table II. Relationship between the suspension current density and electric field strength. (C = 19wt%, $\dot{\gamma} = 0$ s^{-1})

E (V/mm)	j ($\mu A/cm^2$) (suspension with activator)	j ($\mu A/cm^2$) (suspension without activator)
500	1.84	----
1000	2.57	----
1500	3.18	----
2000	4.78	0.07

As shown in Table II, the current density of the suspension with activator increases with the increasing electric field strength. The current density of the suspension with activator is much larger than that of the suspension without activator. These phenomena can be explained as follows: The adsorption of glycerin will enhance the surface conductance of chitosan sulfate particles; on the other hand, glycerin as an impurity, will also enhance the system's volume conductance; and both these conductances increases with the increasing field strength(*14*). So the suspension's current density increases with the field strength. In addition, The field-induced particle chains or strands bridging the electrodes could provide conducting pathways as well.

Furthermore, under higher field strength, the space charge current could occur due to the electrons discharged into the fluid from the electrode. The space charge current is non-ohmic, and proportional to the square of the field strength(*14*). Under higher field strength, the space charge current could contribute to the current density of the suspension.

Due to the instrumental limitations, the current density of the suspension without activator under the field strength less than 2000 V/mm couldn't be obtained. These experimental results indicate that the activator-free chitosan sulfate suspension has very low conductivity, compared with other fluids (*6, 15*).

Rheograms. The suspensions' shear stress (τ) as a function of shear rate ($\dot{\gamma}$) is shown in Figure 5 and 6. In the absence of an electric field, the dependence of τ on

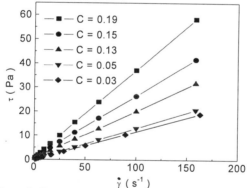

Figure 5 Shear stress versus shear rate
under zero field strength.

$\dot{\gamma}$ is almost linear. Under the applied electric field, at lower concentration, the $\tau \sim \dot{\gamma}$ curves are similar to the zero-field curves, and the suspension almost behaves as a Newtonian fluid; and the higher the concentration is, the greater the anomaly of the viscoplastic behavior is. In the presence of the electric field, at higher concentration range, there appears a yield limit (τ_d), which represents the limiting value of the shear stress as the shear rate approaches zero. The value of the dynamic yield stress is a function of concentration and increases with the increasing concentration (see Figure 6). This supports the assumption about a

112

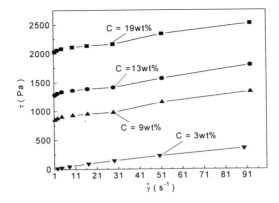

Figure 6 Shear stress versus shear rate curves.
(E = 800 V/mm)

growth of the interpaticle interaction forces and, therefore, strengthening of the system skeleton with the increasing concentration.

Acknowledgments

The authors gratefully acknowledge the support of Guangdong Natural Science Foundation and Guangdong High Education Bureau of China.

Literature Cited

(1). Block, H.; Kelly, J. P.; Qin, A.; Watson, T. *Langmuir* **1990**, *6*, 6.
(2). Winslow, W. M. *J. Appl. Phys.* **1949**, *20*, 1137.
(3). Block, H.; Kelly, J. P. *J. Phys. D* **1988**, *21*, 1661.
(4). Scott, D.; Yamaguchi, J. *Automotive Engineering* **1983**, *91*, 61.
(5). Jordan, T. C.; Shaw, M. T. *IEEE Transactions on Electrical Insulation* **1989**, *24*, 849.
(6). Block, H.; Kelly, J. P. U.S. Patent 4,687,589   1987
(7). Yoshimura, R. JP 04,25,596   1992
(8). Marakami, K. JP 02,255,798   1990
(9). Klass, D. L.; Martinek, T. W.; *J. Appl. Phys.* **1967**, *38*, 75.
(10). Roberts, G. A. F. *Chitin Chemistry;* Macmillan Press: Hampshire, 1992; pp121-130
(11). Wolfrom, M. L.; Han, Shen T. M. *J. Amer. Chem. Soc.* **1959**, *81*, 1764.
(12). Gast, A. P.; Zukoski, C. F. *Adv. Colloid Interface Sci.* **1989**, *30*, 153.
(13). Klingenberg, D. J.; Zukoski, C. F. *J Chem. Phys.* **1989**, *91*,7888.
(14). Coelho, R. *Physics of Dielectrics for the Engineer;* Elsevier: London, 1979; pp153-200.
(15). Gow, C. J.; Zukoski, C. F. *J. Colloid Interface Sci.* **1990**, *36*, 175.

Chapter 8

Field-Responsive Conjugated Polymers

Karim Faïd and Mario Leclerc

Département de Chimie, Université de Montréal, C.P. 6128, Succ. Centre Ville, Montréal, Québec H3C 3J7, Canada

Neutral, highly regioregular polythiophene derivatives undergo striking optical changes upon exposure to various external stimuli. These optical changes are believed to be related to a conformational transition of the polymer backbone, from a planar to non-planar form, triggered by adequately functionalized side-chains. In addition to the well-known chromic transitions induced by heating (thermochromism) or solvent quality changes (solvatochromism), novel phenomena have been generated including the detection of alkali metal cations (ionochromism), UV-induced dual photochromism and molecular recognition of chemical or biological moieties (affinitychromism).

In addition to the use of colorimetric detection, due to the change in the absorption characteristics of the polymer backbone, electrochemical techniques can be also advantageously employed. The recognition or binding events, between the functionalized side chains and the external stimuli, could be detected and measured by taking advantage of the large difference in the electronic structure between a planar and a nonplanar form of the polymer backbone. This results in a very significant shift of the oxidation potentials, allowing the design of highly selective and efficent electrochemical sensors.

The search for smart materials is an exploding research field due to the high demand for materials capable of carrying out increasingly complex tasks. One of the main requirements is the obtention of materials that can perform various functions while shrinking cost requirements. Field-responsive materials are one of these fast developing areas and can be defined as materials in which a given property might be changed in a

114

measurable way through its interaction with some external stimuli. The detectable characteristic can be any optical, electrical or magnetic properties while the external stimuli can be any form of energy or matter.

Among field-responsive materials, functionalized regioregular conjugated polythiophene derivatives have been found to be a very promising family, with impressive conformational changes upon exposure to specific stimuli. These conformational changes, which are believed to be related to a planar to non-planar transition of the conjugated backbone, result in very pronounced chromic effects, from deep violet to bright yellow.

The utilization of such field-induced chromic effects will be the subject of this chapter and we will review some examples in which the external stimuli can be varied from heat, light, ions or biological moieties while the side-chain moieties are tuned accordingly. A brief presentation of conjugated polymers will be also provided as well as a presentation of different chromic polymers and finally some possible applications will be discussed.

Conjugated Polymers

During the last twenty years, conjugated polymers (Figure 1), such as polyacetylenes, polyanilines, polypyrroles, polythiophenes, etc., have attracted tremendous attention, mainly because of their interesting optical, electrochemical and electrical properties. These properties may lead to a variety of applications such as information storage, electroluminescent devices, optical signal processing, solar energy conversion materials, electrochemical cells, EMI shieldings, antistatic coatings, bioelectronic devices, etc.[1-4].

For instance, these materials are well known for their high electrical conductivity arising upon doping (oxidation, reduction, protonation). The delocalized electronic structure of these polymers is partly responsible for the stabilization of the charge carriers created upon doping and electrical conductivities in the range of 1-1000 S/cm can be reached in most cases. Moreover, processability and a high level of conjugation have been obtained through the incorporation of alkyl side chains on polythiophenes[5-10]. However, the asymmetric nature of the starting monomers usually leads to the

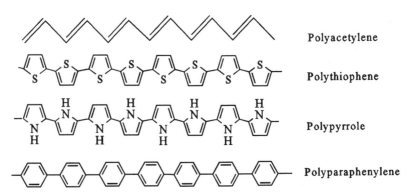

Figure 1: Structure of Different conjugated polymers

occurrence of head-to-tail, head-to-head or tail-to-tail couplings upon polymerization, which can yield up to four different triads along the backbone[*11-14*] (Figure 2).

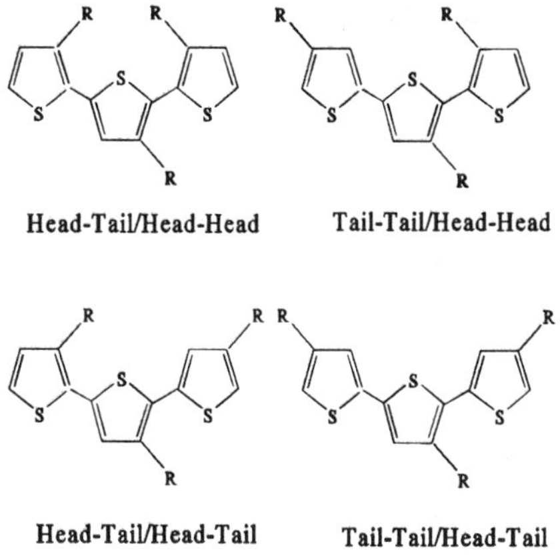

Head-Tail/Head-Head   Tail-Tail/Head-Head

Head-Tail/Head-Tail   Tail-Tail/Head-Tail

Figure 2:   Different regiochemical structure in poly(3-alkylthiophene)s

Highly conjugated and fully substituted poly(3-alkoxy-4-methylthiophene)[*15-17*] have been designed in such a way that the presence of a second substituent was made possible by the introduction of the small oxygen atom in the vicinity of the thiophene backbone[*17*]. Moreover, the asymmetric reactivity of the oxidized monomers[*18-19*] (Figure 3) allowed the preparation of poly(3-alkoxy-4-methylthiophene)s in good yields with a high degree of regioregularity compared to poly(3-alkylthiophene)s which are poorly regioregular when polymerized by oxidative means.

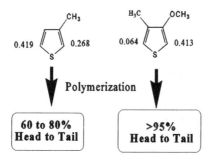

Figure 3:   Effect of the side-chain on the spin density distribution of oxidized thiophene monomers and on the regioregularity of the resulting polymers

The introduction of various substituents on the backbone can not only enhance the processability of some of these polymers but also modulate their electrical, electrochemical and optical properties. Electrochemical redox processes result usually in strong changes in the visible absorption spectra (electochromism), from dark red to light blue in the case of poly(3-alkylthiophene)s[20] from dark blue to transparent light blue in alkoxy-substituted poly(thienylenevinylene)s[21] and poly(3,4-ethylenedioxythiophene)s[22]. On the other hand, the UV-visible absorption characteristics of neutral conjugated polymers can be varied by tuning their conformational structure. It has been shown that the backbone conformation has strong effects on the electronic structure of conjugated molecules and, therefore on their absorption characteristics[23-24]. For example, striking reversible chromic effects[25-29] have been reported in polythiophene derivatives upon heating both in solid state and solution (thermochromism) or when the solvent quality is altered (solvatochromism). The dependence of the electronic structure of conjugated polymers upon their conformation have been fully described[2,24] and can explain the interesting optical effects that have been attributed to a reversible *transition* between a coplanar (highly conjugated) form and a nonplanar (less conjugated) conformational structure of the backbone[30].

Field-induced Chromism in Polythiophene Derivatives

Many experimental results have suggested that the conformational modification of the polythiophene backbone can be induced through order-disorder transitions of the side-chains [30]. It was then postulated that various external stimuli could perturb the side-chain organization and consequently induce some chromic effects. These side-chain transitions can be induced by heating (thermochromism), varying the solvent quality (solvatochromism), ion complexation (ionochromism), photo-induced isomerization (photochromism) and affinity binding (affinity or biochromism) giving rise to a novel class of field responsive materials.

Thermochromism. Neutral poly(3-alkylthiophene)s and poly(3-alkoxy-4-methylthiophene)s exhibit strong chromic effect upon heating both in the solid state and in solution[25-29]. Two types of thermochromic behavior can be observed and are correlated to the substitution pattern of the polymers. As an example of the first type (the two-phase behavior), the temperature dependence of the absorbance of a thin film of poly[3-oligo(oxyethylene)-4-methylthiophene] [31] is shown in Figure 4.

At room temperature this polymer is highly conjugated with an absorption maximum around 550 nm. Upon heating, a new absorption band, centered around 426 nm is increasing while the band at 550 nm is decreasing. This strong blue shift of the maximum of absorption upon heating could be related to a conformational transition from a highly conjugated form (coplanar or nearly planar, deep violet in color) at low temperatures to a less conjugated form (nonplanar and yellow in color) at higher temperatures [30]. At a fixed temperature, there is no evolution as a function of the elapsed time and, since these optical effects are also reversible, they cannot be therefore attributed to a degradation of the polymer.

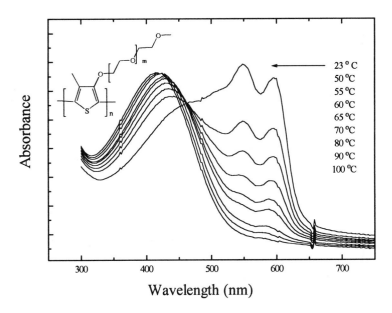

Wavelength (nm)

Figure 4: Temperature-dependent UV-visible absorption spectra of highly regioregular poly(3-(oligo(oxyethylene)-4-methylthiophene) in the solid state

A clear isosbestic point is also observed indicating the coexistence of two phases in the material while it is impossible, however, to determine whether these phases do exist on different parts of the same polymer chain or on different ones.

A similar effect has been observed in solution, both in good and poor solvents [*31*]. In a good solvent (tetrahydrofuran), at room temperature, the maximum absorption of poly[3-oligo(oxyethylene)-4-methylthiophene] takes place at 426 nm, indicating that the polymer chains are already in a twisted conformation. Upon cooling, the color of the solution shifts from yellow at room temperature to violet at -100°C, the transition being fully reversible [*31*]. Poly[3-oligo(oxyethylene)-4-methylthiophene] can be dissolved in poor solvents, such as methanol, but the solution is then violet at room temperature with the maximum of absorption at 550 nm, indicating that the polymer chains are mostly planar. Upon heating, the 426 nm band is increasing while the band at 550 nm is disappearing (Figure 5) in a very similar manner to the situation observed in tetrahydrofuran and in the solid state, although at different temperature ranges [*31*].

Differential scanning calorimetric measurements have revealed well-defined thermal transitions which are well correlated with the observed optical transitions[*29*]. Temperature-dependent X-ray diffraction analyses have revealed a rather amorphous structure for this polymer that is not strongly affected by the heating process[*29*].

118

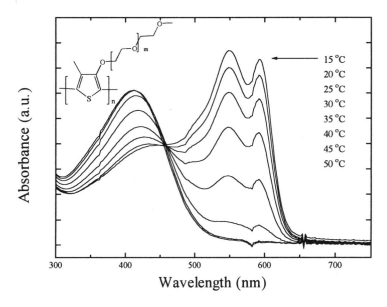

Figure 5: Temperature-dependent UV-visible absorption spectra of highly regioregular poly(3-(oligo(oxyethylene)-4-methylthiophene) in poor solvent (methanol)

This two-phase thermally induced behavior have been also observed in other amorphous poly(3-alkoxy-4-methylthiophene)s and in semicrystalline poly(3-alkylthiophene)s which exhibit a strong chromic transition upon heating, with the occurrence of a clear isosbestic point, indicating the coexistence of two distinct conformational structures [29]. Fairly good correlation have been established between the melting of these polymers and the chromic transitions by DSC, FTIR and X-ray analyses[30], indicating that a similar twisting of the main conjugated chain can be observed in both the amorphous and crystalline phases, although the transition speed and temperature of the crystalline regions are expected to deviate from that of the amorphous zones, explaining the absence of clear isosbestic point in some semicrystalline polythiophene derivatives[30].

This two-phase chromic behavior has been observed in relatively well-defined regioregular polymers, with head-to-tail couplings ranging from 80 to 98%. The head-to-tail structure refers to couplings between the 2-position of a thiophene monomer and the 5-position of a second one, leading to structures shown in Figure 2. Depending on the nature of the monomers used, different regiochemical structures can be obtained that can lead to fairly different properties. By adequately designing the polymerization procedures[32-33] of 3-alkylthiophenes, highly regioregular polymers have been obtained which display a two-phase chromic behavior[29]. Highly regioregular poly(3-alkoxy-4-methylthiophene)s have also been obtained by the simple and straightforward oxidative polymerization of the corresponding monomers. The asymmetric reactivity of these monomers[*19*] leads to a high symmetry along the main polymeric chains, giving rise to a simple way of obtaining highly regioregular functionalized polymers (Figure 3). In contrast to highly regioregular polythiophene derivatives which display a two-phase behavior upon heating, only weak and monotonic shifts of the absorption maximum are observed upon heating of non-regioregular polythiophene derivatives[*30*]. For instance, poly(3-hexylthiophene) containing only 50% head-to-tail couplings, poly(3-dodecyl-2,2'-bithiophene), poly(3-butoxy-3'-decyl-bithiophene) and poly(3',4'-dihexylterthiophene) represent some of the polymers that do exhibit only monotonic shifts of their absorption maxima upon heating without any isosbestic point (Figure 6). Clearly, the lack of regioregularity is responsible for the occurrence of such a monotonic behavior.

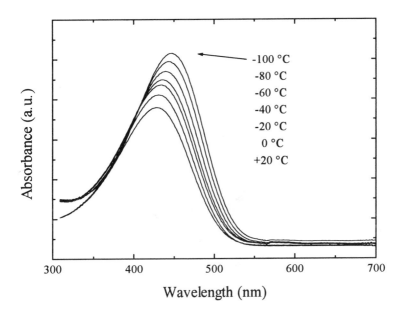

Figure 6:   Temperature-dependent UV-visible absorption spectra of regiorandom poly(3-hexylthiophene) in the solid state

Photochromism. All the results presented above support the assumption that side-chain disordering can be the driving force of the twisting of polythiophene conjugated backbones. It can be postulated therefore that a large range of external stimuli could trigger this kind of chromic effects leading to the utilization of this chromic phenomena as an indirect measurement of various external stimuli. Along these lines, highly regioregular polythiophenes derivatives bearing photo-isomerizable moieties have been designed in order to photo-induce side-chain disordering[*34*].

Azobenzene-substituted polythiophenes has led to the development of novel dual-photochromic polymers, where the isomerization of the photo-active side-chains, modifies not only their own UV-vis absorption but induces also a modification of the optical features associated with the polythiophene conjugated backbone (Figure 7). A trans-cis isomerization reaction of the azobenzene unit is induced as revealed by the decrease of the absorption near 365 nm and the increase of the 325 and 440 nm bands. Moreover, the photo-isomerization of the azobenzene moieties induces also a significant modification of the conjugated backbone optical features. The decrease of the 550 nm band is associated with the growth of a novel absorption band centred around 420-440 nm, which in this case is added to the absorption band related to the n-π* transition of the cis-configuration of the azobenzene moieties.

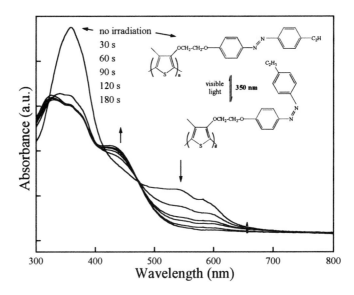

Figure 7:   UV-visible spectra of poly(3-(2-(4-(4'-ethoxyphenyazo)phenoxy) ethoxy)-4-methylthiophene) as a function of irradiation time in chloroform-methanol (1:1) solution at room temperature

This photo-induced chromic effect is reversible and the irradiated polymers slowly recover their initial absorptions features. Moreover, the same effect has been observed in solid state upon irradiation of thin film of azobenzene-substituted polythiophenes, opening the way to the design of novel photochromic polythiophenes that can be useful in a variety of optical devices.

Ionochromism. It is well known that ether and crown-ether moieties can establish some non-covalent interactions with alkali metal cations and it has been postulated that such interactions could modify the side-chain organization, inducing a modification of the backbone conformation and, hence, a modification of its optical characteristics. Interesting ionochromic effects have been observed in some polythiophene derivatives[35-37]; however, by using highly regioregular poly[3-oligo(oxyethylene)-4-methylthiophene][31], it was even possible to determine optically the concentration of some alkali ions in solutions (Figure 8). At a given temperature, some noncovalent interactions between the side chains and alkali metal ions take place and induce a cooperative twisting of the main chain. The addition of increasing aliquots of KSCN to a poly[3-oligo(oxyethylene)-4-methylthiophene] solution in methanol, resulted in a two-phase chromic behavior, with the absorption intensity of the planar form at 550-600 nm decreasing while a new absorption centred around 440 nm is appearing, in a very similar manner to the thermally and photo-induced chromic effects presented above.

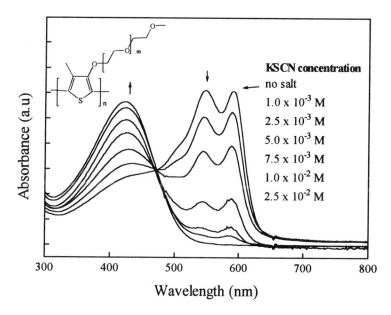

Figure 8:   UV-visible   absorption   spectra   of   poly(3-oligo(oxyethylene)-4-methylthiophene) in methanol with different KSCN concentrations

Other cations (Na^+, NH_4^+) can induce the same effect although with a reduced magnitude (Figure 9). Larger amounts of Na^+ and NH_4^+ are required in order to induce the same effect as K^+, while Li^+ does not produce any measurable effect. The size of the alkali cations seems to be of first importance in the sensitivity of this polymer while the effect of the anion size anion seems to be negligible, KSCN producing almost the same effect as KCl.

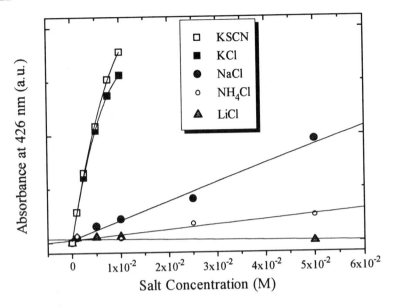

Figure 9:   Variation of the absorbance of the 426 nm band of poly(3-oligo(oxyethylene)-4-methylthiophene) in methanol as a function of various salts at room temperature

Biochromism. From all the knowledge gained in these studies on chromic poly(3-alkylthiophene)s and poly(3-alkoxy-4-metylthiophene)s, it has been thus anticipated that almost any modification of the side chain organization could result in an impressive chromic effect through the alteration of the main chain conformation. If a specific complexation reaction, resulting from recognition or affinity, between functionalized side chains and some targeted chemical or biological moieties can be designed, the side-chain perturbation could thus induce the occurrence of a two-phase chromic behavior in well-defined regioregular polythiophene derivatives and lead thus to a detection mechanism. To test this hypothesis, novel functionalizable polythiophene derivatives have been designed and characterized[38].

The synthetic approach was based on the use of the well known specific complexation between biotin and avidin to trigger the side chain conformational changes in highly regioregular polythiophene derivatives. Novel water-soluble functionalized poly(3-alkoxy-4-methylthiophene)s copolymers have been designed to incorporate some biotin moieties on the side chains. Avidin, a tetrameric protein containing four identical binding sites and forming with biotin an essentially irreversible complex[39] (with a stability comparable to a covalent bond), is then added to an aqueous solution of the copolymer. The copolymer solution, initially violet in color (and thus with the backbone mainly in the planar form), undergoes a dramatic color change to yellow (Figure 10), induced by the alteration of the side chain organization resulting from the complexation of avidin and biotinylated side chains.

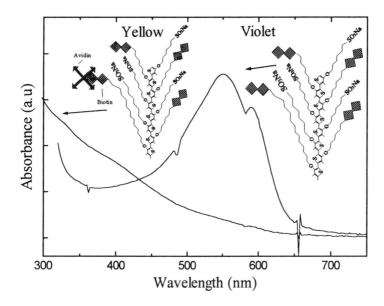

Wavelength (nm)

Figure 10: UV-visible spectra of (a) biotinylated polythiophene derivative without avidin and (b) after the addition of avidin in water at room temperature

The changes of the optical characteristics of the polythiophene upon the complexation of biotin with avidin could be used to visualize and quantify the binding event, opening the way to the development of novel sensors incorporating both the affinity or recognition elements (side chain functions) and the optical probe (conformation changes of the backbone) in the same macromolecular assembly.

Piezochromism. It has been found that the optical absorption characteristics of various poly(3-alkylthiophene)s can be altered as a function of applied pressure[*40*]. A poly(3-dodecylthiophene) thin film[*41*], which is non-planar at high temperatures with the maximum of absorption around 420 nm, undergoes a red-shift with increasing pressure with the maximum of absorption around 520 nm at a pressure of 8 kbar. This red-shift could be explained in terms of increased conjugation length driven by a better packing which compensate the increase of the band gap and the conformational changes in the polymer induced by heating[*30*].

Electrochemical detection. In addition to the optical detection of the interaction of the polymers with some external stimuli, the alteration of the backbone conformation can be detected electrochemically. The electrochemical characteristics of polythiophene derivatives are correlated with the degree of regioregularity and the type of coupling between substituted thiophene rings[*2*]. Polymers with the largest percentage of head-to-tail couplings exhibit higher degree of long range order and conductivity and longer wavelength of maximum of absorpion. These highly regioregular polymers (planar) have smaller band gaps (1.7 eV), 0.3-0.5 eV lower than their regiorandom (non-planar) analogs[42] which should result in a lower oxidation potential for the planar form of the polymer compared to the non-planar conformation.

The cyclic voltammetry of poly(3-(2-methyl-1-butoxy)-4-metylthiophene)[29] (PMBMT), a polymer that can be obtained either in a planar or non-planar form at room temperature, is shown in Figure 11. The planar (violet) form has an oxidation potential of +0.70 V vs SCE while the non-planar (yellow) form has an oxidation potential of +0.88 V vs SCE. At a potential of +0.65 V, there is almost a 200-fold increase in the measured current giving rise to a very simple way of quantifying electrochemically the different field-induced chromic transitions of regioregular polythiophene derivatives[43].

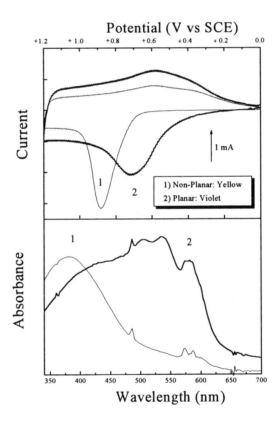

Figure 11: Cyclic Voltammetry of poly(3-(2-methyl-1-butoxy)-4-methylthiophene) PMBMT (a) in planar form and (b) in nonplanar form

OTHER CHROMIC POLYMERS

Polydiacetylenes. Other classes of polymeric materials do show similar optical features. Indeed, thermochromic effect have been first observed in polydiacetylenes[44]. Similarly to polythiophenes, the addition of flexible side chains produced processible and chromic polydiacetylenes[45-49]which undergo color changes when subjected to either heat, pressure or change in solvent quality, with the presence of a clear isosbestic point indicating the coexistence of two different conformational structures, having different optical features (red and yellow). Although the exact mechanism responsible for this phenomena is still subject of debate[50-55] these optical phenomena could be also related to a planar to nonplanar transition of the main conjugated backbone.

For instance, an interesting chromic sensor of influenza A virus has been developed[56-57] using a Langmuir-Blodgett film of a polydiacetylene derivative bearing a recognition site (sialic acid) for the hemagglutin (a surface group on the influenza virus). Binding of hemagglutin to the polydiacetylene thin film induces a strong chromic effect with the color going from blue to red. This approach has been further extended to the detection of other species, by using for example polydiacetylene liposomes incorporating Gm1 ganglioside to colorimetrically detect cholera toxin[58] or to measure optically the concentration of glucose through the conformational changes induced on a lipid polydiacetylene layer bearing an hexokinase enzyme[59].

Polysilanes. Polysilanes are another family of polymers that display strong chromic effects when subjected to certain stimuli[60-66]. Owing to the extensive delocalization of σ electrons along the silicium backbone, polysilanes display an intense absorption band in the near-UV region, which is strongly correlated to the polymer conformation. Thermally-induced chromism in the solid state and in solution have been reported, with good correlations between the side-chain melting and disordering of the polymer backbone from a planar to a non planar conformation. Symmetrically substituted polysilanes exhibit a two-phase chromic behavior, with a clear isosbestic point, upon being subjected to heat, pressure or change in solvent quality[67-70]. On the other hand, a monotonic blue shift of the absorption maximum is observed in unsymmetrically substituted polysilanes when heated[71].

PERSPECTIVES

On the basis of all these results, striking chromic effects can be triggered by a variety of external stimuli, including heat, pressure, electromagnetic radiation, ions, molecules, etc. In some polythiophene derivatives, these field-induced conformational transitions do not only alter the optical features of the materials of interest but their electrochemical characteristics as well, giving rise to the possible development of various devices in different areas, such as in sensing, detection and recognition of numerous chemical and biochemical moieties or solid-state devices.

126

REFERENCES

1. Feast, W.J.; Tsibouklis, J.; Pouwer, K.L.; Groenendaal, L.; Meijer, E.W. *Polymer*, 37, 5017, **1996**
2. Schopf, G.; Koßmehl, G. *Adv. Polym. Sci.*, 129, 1, **1997**
3. Handbook of Conducting Polymers, 2nd Ed.; Skotheim, T.A.; Reynolds, J.R.; Elsenbaumer, R.L.; Eds., *Marcel Dekker, New York*, **1997**
4. Gorman, C.B.; Grubbs, R.H.; in *Conjugated Polymers: The novel science and technology of conducting and nonlinear optically active materials*, Bredas, J.L.; Silbey, R.; Eds, Kluwer Academic Publishers, Dordrecht, The Netherlands, 1992; pp 1
5. Elsenbaumer,R.L.; Jen, K.Y.; Oboodi, O.; *Synth. Met.*, 15, 169, **1986**
6. Sato, M.A.; Tanaka, S.; Kaeriyama, K.; *J. Chem.Soc., Chem. Comm.*, 873, **1986**
7. Sugimoto,R.; Takeda, S.; Gu, H.B.; Yoshino, K. *Chem Express*, 1, 635, **1986**
8. Hotta, S.; Rughooputh, S.D.D.V.; Heeger, A.J.; Wudl, F. *Macromolecules*, 20, 212,**1987**
9. Roncali, J.; Garreau, R.; Yassar, A.; Marque, P.; Ganier,F.; Lemaire, M.; *J. Phys. Chem.*, 91, 6706, **1987**
10. Bryce, M.R.; Chissel, A.; Kathirgamanathan, P.; Parker, D.; Smith, N.M.R.; *J. Chem. Soc., Chem. Com.*, 466, **1987**
11. Leclerc, M.; Diaz, F.M.; Wegner, G.; *Makromol. Chem.*, 190, 3105, **1989**
12 Souto Maior, R.M.; Hinkelmann, K. ; Eckert, H.; Wudl, F.; *Macromolecules*, 23, 1268, **1990**
13. Zagorska, M. and Krische, B.; *Polymer*, 31, 1379, **1990**
14. Sato, M.A.; Morii, H.; *Macromolecules*, 24, 1196, **1991**
15. Feldhues, M.; Kampf, G.; Litterer, H.; Mecklenburg, T.; Wegener, P.; *Synth. Met.* 28, C487, **1988**
16. Leclerc, M.; Daoust, G.; *J. Chem. Soc., Chem. Comm.* 273, **1990**
17. Daoust, G.; Leclerc, M.; *Macromolecules* 24, 455, **1991**
18. Barbarella, G.; Zambianchi, M.; Di Toro, M.; Colonna Jr., M.; Iarossi, D.; Bongini, A.; *J. Org. Chem.*, 61, 8285, **1996**
19. Fréchette, M.; Belletête, M.; Bergeron, J.-Y.; Durocher, G.; Leclerc, M.; *Macromol. Chem. Phys.*, 197, 2077, **1996**
20. Genies, E.; Collomb-Dunand-Sauthier, M.-N.; Langlois,S.; *J. Appl. Electrochem.*, 24, 72, **1994**
21. Eckhardt, H.; Jen, K.Y.; Shacklette, L.W.; Lefrant, S.; *NATO ASI Ser., Ser. E* 182, 305, **1990**
22. G. Heywang, G.; Jonas, F.; *Adv. Mat.*, 4, 116, **1992**
23. Brédas, J.L.; Street, G.B.; Thémans, B.; André, J.M.; *J. Chem. Phys.*, 83, 1323, **1985**
24. Thémans, B.; Salaneck, W.R.; Brédas, J.L.; *Synth. Met.*, 28, C359, **1989**
25. Rughooputh, S.D.D.V.; Hotta, S.; Heeger, a.J.; Wudl, F.; *J. Polym. Sci., Polym. Phys. Ed.*, 25, 1071, **1987**
26. Inganäs, O.; Salaneck, W.R.; Osterholm, J.-E; Laakso, J.; *Synth. Met.*, 22, 395, **1988**

27. Roux, C.; Leclerc, M.; *Macromolecules*, 25, 2141, **1992**
28. Inganäs, O.; *Trends Polym. Sci.*, 2, 189, **1994**
29. Faïd, K.; Fréchette, M.; Ranger,M.; Mazerolle, L.; Lévesque, I.; Leclerc, M.; *Chem. Mater.*, 7, 1390, **1995**
30. Leclerc, M.; Faïd, K.; in *Handbook of Conducting Polymers, 2nd ed.*, Skotheim, T.A.; Reynolds, J.R.; Elsenbaumer, R.L.; Eds., Marcel Dekker, NewYork, p:695, **1997**
31. Lévesque, I.; Leclerc,M.; *Chem Mater.*, 8, 2843, **1996**
32. Chen, T.A.;Rieke, R.D.; *J. Amer. Chem. Soc.*, 114, 10087, **1992**
33. McCullough, R.D.; Lowe, R.D.; *J. Chem. Soc., Chem. Comm.*, 70, **1992**
34. Lévesque, I.; and Leclerc, M.; *Macromolecules*, 30, 4347, **1997**
35. McCullough, R.D.; Williams, S.P.; *J. Amer. Chem. Soc.*, 115, 11608, **1993**
36. Marsella, M.J.; Swager, T.; *J. Amer. Chem. Soc.*, 115, 12214, **1993**
37. Marsella, M.J.; Newland, R.J.; Carroll, P.J.; Swager, T.M.; *J. Amer. Chem. Soc.*, 117, 9842, **1995**
38. Faïd, K.; Leclerc, M.; *J. Chem. Soc., Chem. Comm.*, 2761, **1996**
39. Wilchek, M.; Bayer, E.A.; *Trends Biochem. Sci.*, 14, 408, **1989**
40. Yoshino,K.; Nakajima, S.; Onada, M.; Sugimoto,R.; *Synth. Met.*, 28, C349, **1989**
41. Iwasaki, K.; Fujimoto, H.; Matsuzaki, S.; *Synth. Met.*, 63, 101, **1994**
42. Chen, T.A.; Rieke, R.D.; *Synth. Met.*, 60, 175, **1993**
43. Faïd, K.;Leclerc,M.; *to be published*
44. Exarhos, G. J.; Risen, W.M.; Baughman, R.H.; *J. Amer. Chem. Soc.*, 98, 481, **1976**
45. Patel, G.N.; Chance, R.R.; Witt, J.D.; *J. Chem. Phys.*, 70, 4387, **1987**
46. Plachetta, C.; Rau, N.O.; Hauck, A.; Schulz, S.C.; *Makromol. Chem., Rapid Commun.*, 3, 249, **1982**
47. Rughooputh, S.D.D.V.; Phillips, D.; Bloor, D.; Ando, D.J.; *Polym. Commun.*, 25, 242, **1984**
48. Mino,N.; Tamura, H.; Ogawa, K.; *Langmuir*, 7, 2336, **1991**
49. Wenz, G.; Muller, M.A.; Schmidt, M.; Wegner, G.; *Macromolecules*, 17, 837, **1984**
50. Lim, K.C.; Kapitulnik, A.; Zacher, R.; Heeger, A.J.; *J. Chem. Phys.*, 82, 516, **1985**
51. Taylor, M.A.; Odell, J.A.; Batchelder, D.N.; Campbell, A.J.; *Polymer*, 31, 1116, **1990**
52. Chu, B.; Xu, R.; *Acc. Chem. Res.*, 24, 384, **1991**
53. Nava, A.D.; Thakur, M.; Tonelli, A.E.; *Macromolecules*, 23, 3055, **1990**
54. Variano, B.F.; Sandroff, C.J.; Baker, G.L.; *Macromolecules*, 24, 4376, **1991**
55. Rawiso, M.; Aimé, J.P.; Fave, J.L.; Schott, M.; Muller, M.A.; Schmidt, M.; Baumgartl, H.; Wegner, G.; *J. Phys. Paris*, 49, 861, **1988**
56. Charych, D.H.; Nagy, J.O.; Spevak, W.; Bednarski, M.D.; *Science*, 261, 585, **1993**
57. Spevak, W.; Nagy, J.O.; Charych, D.H.; *Adv. Mater.*, 7, 85, **1995**
58. Pan, J.J.; Charych, D.H.; *Langmuir*, 13, 1365, **1997**
59. Cheng, Q.; Stevens, R.C.; *Adv. Mater.*, 9, 481, **1997**

128

60. Trefonas III, P.; Damewood Jr.,J.R.; West, R.; Miller, R.D.; *Organomettalics*, 4, 1318, **1985**
61. Harrah, L.; Ziegler, J.M.; *J. Polym. Sci.Polym.Lett. Ed.*, 23, 209, **1985**
62. Rabbolt, J.F.; Hofer, D.; Miller, R.D.; Fickes, G.N.; *Macromolecules*, 19, 611, **1986**
63. Lovinger, A.J.; Scilling, F.C.; Bovey, F.A.; Ziegler, J.M.; *Macromolecules*, 19, 2657, **1986**
64. Kuzmany, H.; Rabolt, J.F.; Farmer, B.L.; Miller, R.D.; *J. Chem. Phys.*, 85, 7413, **1986**
65. Schilling, F.C.; Bovey, F.A.; Lovinger, A.J.; Zeigler, J.M.; *Macromolecules*, 19, 2660, **1986**
66. Yuan, C.-H.; West, R.; *Macromolecules*, 26, 2645, **1993**
67. Miller, R.D.; Michl, J.; *Chem. Rev.*, 89, 1359, **1989**
68. Schilling, F.C.; Bovey, F.A.; Lovinger, A.J.; Ziegler, J.M.; *Adv. Chem.*, 224, 341, **1990**
69. Song, K.; Miller, R.D.; Wallraff, G.M.; Rabolt, J.F.; *Macromolecules*, 24, 4084, **1991**
70. Miller, R.D.; Sooriyakumaran, R.; *Macromolecules*, 21, 3120, **1988**
71. Miller, R.D.; Wallraff, G.M.; Baier, M.; Cotts, P.M.; Shukla, P.; Russell, T.C.; De Schryver, F.C.; Declercq, D.; *J. Inorg. Organomet. Chem.*, 1, 505, **1991**

Chapter 9

Dielectric and Electro-Optical Properties of a Ferroelectric Side-Chain Liquid Crystalline Polysiloxane Containing Azobenzene Dyes as Guest Molecules

Rong-Ho Lee[1], Ging-Ho Hsiue[1,3], and Ru-Jong Jeng[2]

[1]Department of Chemical Engineering, National Tsing Hua University, Hsinchu, Taiwan 300, Republic of China
[2]Department of Chemical Engineering, National Chung Hsing University, Taichung, Taiwan 400, Republic of China

Guest-host systems of a ferroelectric side-chain liquid crystalline polysiloxane (PS121A) containing azobenzene dyes have been investigated. Disperse orange 3 (DO3) with a strong dipole moment, and a liquid crystal 4'-(5-hexenyloxy)-4-methoxyazobenzene (HMAB) with a relatively weak dipole moment were used as the guest molecules. The intensity and frequency of the Goldstone mode were increased remarkably for the PS121A/DO3, due to the doping of the DO3 (5 wt. %). The guest molecule with a strong dipole moment results in larger fluctuation of the spontaneous polarization vector in each smectic layer under an applied electric field. As a result, large spontaneous polarization and short response time were obtained for this system. On the other hand, the doping effect of HMAB on dielectric and electro-optical properties was not significant for this system compared to that of the PS121A/DO3. The intensity and frequency of Goldstone mode were slightly increased even with a higher doping level (15 wt. %) of HMAB. PS121A/HMAB exhibited a smaller spontaneous polarization and longer response time than PS121A/DO3 sample. The doping of a suitable amount of the azobenzene dye in the LC phase of the FLCP was helpful for the improvement of the electro-optical properties of such guest-host system.

Ferroelectric liquid crystals (FLC) have attracted attention because of their high speed response and memory effect (*1-3*). The characteristics of fast response and memory effect make them suitable in electro-optical device applications, such as display, light valve and memory devices. Ferroelectric side chain liquid crystalline polymers (FLCPs) exhibit desirable mechanical properties of polymers and electro-optical properties of low molecular weight FLC, which have been investigated extensively

[3]Corresponding author.

(4-5). The excellent film formability and electro-optical properties of FLCP is desirable for large area displays *(6)*.

FLCPs exhibiting a chiral smectic C (S_C^*) phase over a broad temperature range (about 200 °C including room temperature) have been synthesized recently in our laboratory *(7-8)*. FLCPs with such mesomorphic behavior have promising potential for electro-optical applications *(9)*. However, high viscosity reduces the response ability of the mesogenic group in the S_C^* phase toward an applied electric field for this type of FLCPs. This is due to the fact that the long and rigid mesogenic core has a lower mobility, thereby resulting in slow response toward the electric field. Therefore, the improvement of the thermal behaviors and electro-optical properties become a pertinent topic for this type of FLCPs. One of the approaches is the guest-host system consisting of an FLCP and an azobenzene dye *(10)*. In this approach, it is found that the doping of the low molecular weight azobenzene dye into the S_C^* phase would enhance the thickness of the smectic layer, and result in the high mobility of the mesogenic group for the FLCP *(10)*. When a FLCP exhibits high molecular mobility in the fluctuation of the director of the tilt angle (Goldstone mode), the threshold field for switching can be lowered. As a result, the helical structure of the FLCP in the S_C^* phase can be unwound completely. Consequently, high spontaneous polarization (P_s) and short response time (τ) can be obtained.

In this study, we describe here guest-host systems of an FLCP (PS121A) containing azobenzene dyes. Disperse orange 3 (DO3) with a strong dipole moment, and a liquid crystal 4'-(5-hexenyloxy)-4-methoxyazobenzene (HMAB) with a relatively weak dipole moment were used as the guest molecules, respectively. The phase transitions and mesophase of the guest-host FLC polymeric materials were studied using differential scanning calorimeter and optical polarizing microscopy. Dielectric measurements were taken from a low temperature range over the liquid crystal phase. The relaxation behavior of the molecular and collective relaxations will be discussed. Moreover, temperature dependence of the P_s and τ were also taken from the electro-optical measurements. The doping effect of two azobenzene dyes on the dielectric and electro-optical properties of the FLCP was investigated.

Experimental

Chemical structures of PS121A, DO3 and HMAB are shown in Figure 1. The mixtures of the FLCP and azobenzene dyes (FLCP/DO3 or FLCP/HMAB) with different weight ratios (sample DO305: 95/5, DO310: 90/10, DO330:70/30 and HMAB05: 95/5, HMAB10: 90/10, HMAB15: 85/15) were obtained by dissolving the compounds in chloroform. The solution of the mixture was kept in vacuum oven at room temperature to completely dry out chloroform.

The thermal transitions of azobenzene dyes and guest-host FLC polymeric materials were determined by a differential scanning calorimeter (Seiko SSC/5200 DSC). The thermal transitions were read at the maximum of their endothermic or exothermic peaks. Glass transition temperature (T_g) was read at the middle of the change in heat capacity. Heating and cooling rates were 10 °C/min in all of these cases. The transitions were collected from the second heating and cooling scans. A

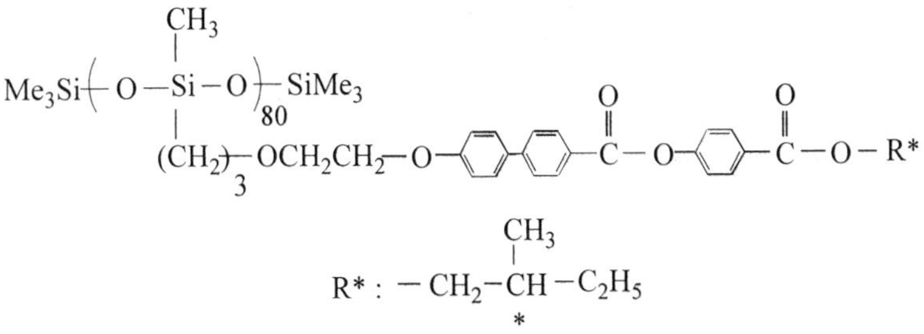

CH$_3$
|
Me$_3$Si$\left(\!$ O$-$Si$-$O$\!\right)$ SiMe$_3$
| $_{80}$
(CH$_2$)$_3$OCH$_2$CH$_2$$-O-$◯◯$-C-O-$◯$-C-O-$R*

R* : $-$CH$_2$$-CH-C_2H_5$
|
CH$_3$
*

(a) PS121A

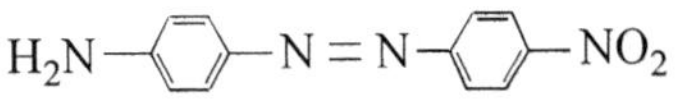

H$_2$N$-$◯$-$N$=$N$-$◯$-$NO$_2$

(b) DO3

CH$_2$=CH$-$(CH$_2$)$_4$$-O-$◯$-N=N-$◯$-OCH_3$

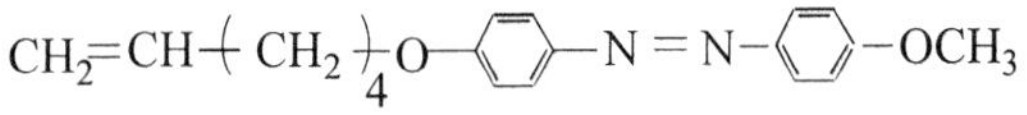

(c) HMAB

Figure 1. Chemical structures of the PS121A, DO3, and HMAB.

Nikon Microphot-FX optical polarized microscope equipped with a Mettler FP82 hot stage and a Mettler FP80 central processor was applied toward observing anisotropic textures. Dielectric spectroscopy was determined on a Novercontrol GmbH. Measurements were performed by a Schlumberger SI 1260 impedance/gain-phase analyzer (frequency: $10^{-1} \sim 10^{6}$ Hz) and a Quator temperature controller. A nitrogen gas heating system ranged from -100 °C to 250 °C was used. The temperature was adjusted within the tolerance of $\pm$ 0.5 °C. The guest-host FLC polymeric materials were sandwiched between two parallel metal electrode plates with a spacer of 50 μm.

The electro-optical measurements were taken using home-made LC cells. The guest-host FLC polymeric materials were sandwiched between ITO coated glass electrodes which were separated by a spacer of 4 μm thickness. The surfaces of ITO glass plates were coated with polyimide (Merck: ZLI-2650) and rubbed in one direction to obtain homogeneous alignment by using rubbing machine (Sigma Koki; RM-50). Homogeneous alignment was achieved by cooling the sample from the isotropic to the S_C* phase with a slow cooling rate under an applied electric field. P_s has been measured by the triangular wave method (11). τ was determined with a photo diode measuring the transmitted light (He-Ne laser light: 632.8 nm) of the sample placed between crossed polarizers. τ is defined as the time required for an intensity change from 10 percent to 90 percent on applying a square wave field. Change of the transmission light intensity through the cell was recorded by a digital storage oscilloscope (Hitachi VC-6025).

Results and Discussion

The thermal transition temperatures and their corresponding enthalpy changes of the PS121A, azobenzene dyes, and guest-host FLC polymeric materials are summarized in Table I. PS121A exhibited a broad temperature range of the S_C* and smectic A (S_A) phases on the heating and cooling scans. No LC phase was observed for DO3, whereas a nematic phase was observed for HMAB on both the heating and cooling scans. Moreover, all of the guest-host FLC polymeric materials exhibited enantiotropic S_A and S_C* phases. No phase separation was observed for these guest-host materials except the sample DO330. For sample DO330, aggregation of DO3 molecules was present, and an additional melting transition at 191.3 °C on the heating scan and a crystallization transition at 167.5 °C on the cooling scan were observed. No phase separation was observed for the guest-host FLC polymeric material HMAB15, even through a large amount of azobenzene dye (15 wt %) was doped. Moreover, the phase transition temperatures decreased with increasing content of the azobenzene dye. The doping of the azobenzene dye into the FLCP seemed to reduce the thermal stability of the LC phases. However, a broad temperature range of the S_C* phase was still present for these guest-host FLC polymeric materials.

The dielectric constant versus temperature and frequency for samples PS121A, DO305, DO330, HMAB05 and HMAB15 are shown in Figure 2. In Figure 2a, the dielectric constant increased with increasing temperature in the S_C* phase. The dielectric constant increased remarkably in the neighborhood of the S_A-S_C* transition temperature because of the presence of the Goldstone and Soft modes. However, the

Table I. Phase transitions and phase transition enthalpies for the samples.

Sample	Phase transitions[a], °C (corresponding enthalpy changes, kJkg^{-1}) $\left(\dfrac{\text{heating}}{\text{cooling}}\right)$
PS121A	G 9.8 S$_C$* 215.2(0.07) S$_A$ 234.6(0.8) I <hr> I 229.2(0.73) S$_A$ 211.7 (0.03) S$_C$* 4.1 G
DO3	K 212.0 I <hr> I 209.0 K
HMAB	K 69.4 (13.8) N 88.8(107.0) I <hr> I 87.1(105.2) N 62.3 (12.4) K
DO305	G 8.1 S$_C$* 160.0(0.6) S$_A$ 230.7(5.9) I <hr> I 229.1(4.1) S$_A$ 164.7 (0.3) S$_C$* 5.8 G
DO310	G 1.9 S$_C$* 134.0(0.4) S$_A$ 225.6(8.1) I <hr> I 224.1(5.5) S$_A$ 129.7 (0.3) S$_C$* 4.6 G
DO330	G 11.5 S$_C$* 141.1 (0.5) S$_A$ 191.3[b] (27.8) I <hr> I 186.3 (5.1) S$_A$ 167.5[b] (17.0) S$_A$ 138.4 (0.4) S$_C$* 5.6 G
HMAB05	G 5.4 S$_C$* 186.2 (0.1) S$_A$ 213.1(5.3) I <hr> I 208.0(4.8) S$_A$ 185.5 (0.1) S$_C$* -1.5 G
HMAB10	G 12.1 S$_C$* 168.2 (0.15) S$_A$ 212.5(6.1) I <hr> I 209.8(5.3) S$_A$ 165.3(0.1) S$_C$* 3.0 G
HMAB15	G 10.2 S$_C$* 152.1 (0.1) S$_A$ 204.5(4.3) I <hr> I 203.1(3.9) S$_A$ 148.7(0.1) S$_C$* 7.0 G

[a] G: glass transition, S$_A$: smectic A phase, S$_C$*: chiral smectic C phase, I: isotropic phase
[b] The melting point of the crystallite in the LC phase

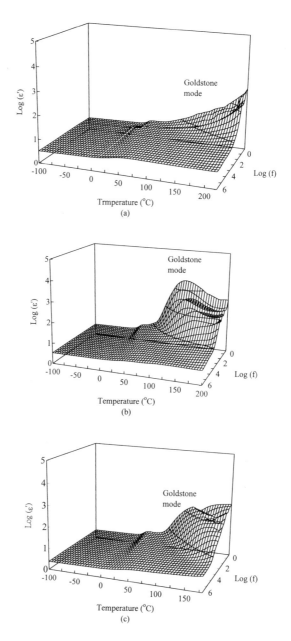

Figure 2. The dielectric constant as a function of temperature and logarithm of frequency for the samples (a) PS121A, (b) DO305, (c) DO330, (d) HMAB05, and (e) HMAB15, respectively.

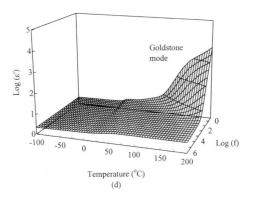

(d)

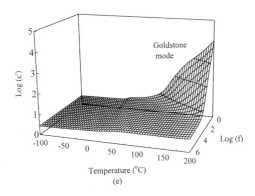

(e)

Figure 2. *Continued.*

relaxation peak (maximum value of the dielectric constant) of the Goldstone mode was not observed due to the fact that relaxation frequency is smaller than 0.1 Hz for PS121A. In Figure 2b, a larger amplitude of the Goldstone mode was observed in the S_C^* phase at a frequency between 0.1 Hz and 100 Hz for sample DO305. The doping of DO3 in the LC phase led to the increase of the thickness of the smectic layer in the S_C^* phase for the FLCP. Subsequently, the loosening of the denser packing of the highly tilted smectic phase resulted in the relaxation frequency of the Goldstone mode shifting toward a higher frequency range. Moreover, the relaxation intensity of the Goldstone mode was also enhanced remarkably due to the existence of the strong dipole-moment molecules of DO3. In Figure 2c, the relaxation frequency also shifted toward the higher frequency range (above 0.1 Hz) for sample DO330. Its amplitude of the Goldstone mode was smaller than that of sample DO305. The formation of the crystalline domain in the LC phase restricted the molecular mobility of the Goldstone mode as the phase separation occurred for sample DO330. On the other hand, the amplitude of the Goldstone mode was increased only slightly for sample HMAB05 and HMAB15 compared to that for the PS121A (Figure 2d and e). The doping effect of HMAB on the Goldstone mode was not significant compared to that of the DO3, even though a larger quantity of HMAB was doped into PS121A. This is possibly due to the fact that the guest molecule with a strong dipole moment could respond toward an applied electric field more rapidly, and subsequently results in faster switching of the mesogenic groups in FLCP. The increase of the relaxation intensity and frequency of the Goldstone mode is important for the electro-optical application of the FLCP. Kalmykov et al. (*12*) have reported that the electro-optical response time is related to the dielectric parameter of the Goldstone mode

$$\tau = \frac{\tau_G \, P_S}{2\varepsilon_0 \, \Delta\varepsilon_G \, E} \tag{1}$$

The response time is dependent on the relaxation time (τ_G) and dielectric relaxation intensity ($\Delta\varepsilon_G$) of the Goldstone mode. This demonstrates that a shorter response time would be obtained for the FLCP which possesses strong relaxation intensity of the Goldstone mode at a high frequency range. This confirms that the doping effect of DO3 on the electro-optical properties was more pronounced than that of HMAB.

The dielectric loss tangent as a function of temperature and frequency for samples PS121A, DO305, DO330, HMAB05 and HMAB15 are given in Figure 3. The molecular relaxations (α-, β-, γ-relaxation) are clearly observed for these samples. The β- and γ-relaxations occurred below T_g, which are corresponding to the reorientation of the mesogenic group and spacer group of the FLCP, respectively. The α-relaxation has been observed at a higher temperature range. This relaxation is the process to be associated with the glass transition. For samples DO305 and DO330, the σ-relaxation was observed above T_g, which is corresponding to the molecular motion of the DO3 (Figure 3b and c) (*10*). For sample DO305, the α-relaxation was covered by the σ-relaxation at the frequency below 10^4 Hz due to the higher intensity of the σ-relaxation. The σ-relaxation temperature shifted toward a

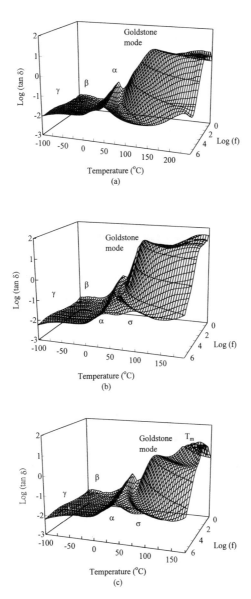

Figure 3. The dielectric loss tangent as a function of temperature and logarithm of frequency for the samples (a) PS121A, (b) DO305, (c) DO330, (d) HMAB05, and (e) HMAB15, respectively.

Continued on next page.

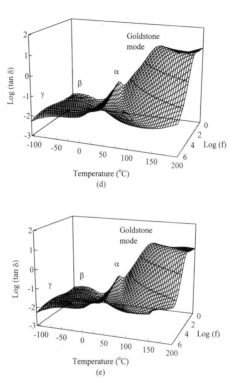

Figure 3. *Continued.*

igher temperature range as the measuring frequency increased. The relaxation ehavior of the α- and σ-relaxations for the sample DO310 was the same as that for ιe sample DO305. For sample DO330, the α-relaxation was not covered by the σ- ·laxation due to the decrease of the intensity of the σ-relaxation. The σ-relaxation ιtensity was reduced because of the occurrence of phase separation. The ;gregation of the DO3 molecules led to the decrease of the molecular mobility. loreover, the α-relaxation coincided with the σ-relaxation at lower frequency range l0³ Hz). Subsequently, these two relaxations separated as the measuring frequency ιcreased. In addition, a melting transition (191.3 ℃) of the crystallites was bserved for the sample DO330. At a higher temperature range, the existence of the ioldstone and Soft modes led to remarkable increase of the dielectric loss tangent in ιe S_C^* phase for PS121A and guest-host FLC polymeric materials. The doping of zobenzene dye resulted in the relaxation frequency of Goldstone mode shifting ιward a higher frequency range (above 10^{-1} Hz). This phenomenon was more gnificant for PS121A/DO3 than for PS121A/HMAB.

Temperature dependence of the relaxation frequency of molecular and collective ·laxations for PS121A and guest-host FLC polymeric materials are shown in Figure 4. straight line similar to the characteristic of Arrhenius plot were obtained except for ιe α- and σ-relaxations. A non-linear character of the α- and σ-relaxations was btained for these samples. The non-linear character indicates that the α-relaxation ɔuld be identified as the glass transition (13). Moreover, FLCP exhibits a larger tilt ηgle at temperatures near T_g in the S_C^* phase. The tilt angle becomes smaller as the :mperature increases. The rotational freedom of azobenzene dye decreases in the ghter pitch (i.e. larger tilt angle) of the S_C^* phase by steric hindrance near T_g (14). ecause of that, the high activation energy of the σ-relaxation was obtained at a lower ·equency range (below 10^4 Hz). The activation energy was decreased with ecreasing steric hindrance at a higher temperature range. This led to non-linear ɦaracter of the σ-relaxation. The activation energies of the dielectric relaxations are ɹmmarized in Table II. The activation energies of the Goldstone mode, σ-, α-, and -relaxations decreased with increasing content of azobenzene dye. This is due to ιe plasticizer effect of the azobenzene dye. The plasticizer effect was not ɾonounced in the activation energies of the γ-relaxation. For sample DO330, the ;gregation of the DO3 and the formation of the crystallites in the LC phase led to the ιcrease of the activation energy for all of the dielectric relaxations. The activation ηergy of the α-relaxation could not be obtained at lower frequency range for samples ʘO305 and D0310 due to the overlapping of the σ-relaxation. However, this was not bserved for the sample DO330 owing to the decrease of the σ-relaxation intensity. ι addition, the doping effect of DO3 on the activation energy of molecular and ɔllective relaxations is more significant than that of HMAB. This is due to the large ipole moment of DO3. The highly polar molecule is favorable to respond toward ιe electric field and subsequently to enhance the mobility of the molecular segments ι its close vicinity.

The electro-optical properties of PS121A and guest-host FLC polymeric ιaterials have been measured. The electro-optical properties of PS121A could not e obtained as a result of a high viscosity and a large tilt angle of molecules in the S_C^*

140

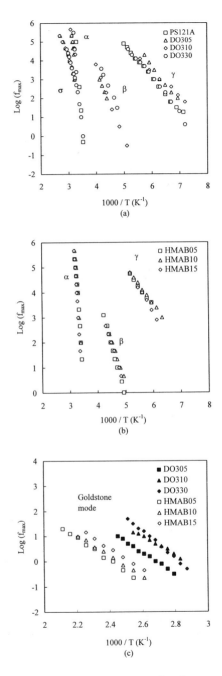

Figure 4. Temperature dependence of the relaxation frequencies for the γ, β, α, and Goldstone mode relaxations of PS121A and guest-host FLC polymeric materials.

Table II. Activation energies (KJmol^{-1}) of the γ, β, α, σ and Goldstone mode relaxations for samples PS121A-HMAB15.

Sample	γ^a	β^b	α^b	σ^b	Goldstone mode^b
			Relaxation		
PS121A	34.0	76.0	418.0^c - 243.2^d	——	——
DO305	31.0	73.2	227.7^c	127.5^c - 65.7^d	102.2
DO310	26.3	61.8	143.8^c	115.7^c - 60.7^d	93.5
DO330	36.1	65.4	216.0^c - 200.0^d	128.7^c - 79.4^d	107.2
HMAB05	33.2	75.8	407.1^c - 243.4^d	——	115.3
HMAB10	30.4	72.0	339^c - 206.0^d	——	106.1
HMAB15	26.5	65.4	225.2	——	104.5

[a] γ-relaxation $\pm$ 4kJmol^{-1}
[b] β, α, σ and Goldstone mode relaxations $\pm$ 15 kJmol^{-1}
[c] The data were calculated at 10^3 Hz
[d] The data were calculated at 10^5 Hz

phase. A long and rigid aromatic mesogenic core with a large tilt angle results in the low rotational freedom of side chain of FLCP. Because of that, the molecules are difficult to be switched in the S_C^* phase by an applied electric field. On the other hand, the doping of azobenzene dye led to the decrease of the rational viscosity of mesogenic group. This would circumvent the switching problem. Because of the doping of azobenzene dyes, the electro-optical properties could be measured from the guest-host FLC polymeric materials. However, this is not the case for DO310, DO330 and HMAB05. For DO330, the phase separation in the S_C^* phase prevent such measurement. For sample DO310, the phase separation was not observed by DSC and POM. However, the aggregation of DO3 was formed in a thin LC cell under the applied electric field as observed by POM. For sample HMAB05, the high viscosity of the denser packing of smectic phase was not effectively reduced by the doping of HMAB (5 wt. %). The temperature dependence of P_s for samples DO305, HMAB10, and HMAB15 is shown in Figure 5. P_s increased with decreasing temperature because of the increase of the tilt angle in the S_C^* phase. After reaching a maximum, P_s then decreased at a lower temperature range. This is due to the fact that the molecules have a higher viscosity at lower temperatures. The molecules could not be switched by the applied electric field, thereby resulting in a smaller P_s value. Moreover, sample DO305 has a larger P_s value compared to samples HMAB10 and HMAB15. This demonstrates that the high dipole moment guest molecules are more helpful for the switching of the mesogenic group in the S_C^* phase. In addition, the temperature dependence of τ for samples DO305, HMAB10, and HMAB15 is shown in Figure 6. τ was increased with decreasing temperature for these samples. Sample DO305 exhibited a shorter τ in the neighborhood of the S_A-S_C^* transition temperature than samples HMAB10 and HMAB15 did. As the temperature decreased, τ of sample DO305 increased rapidly due to the increase of the viscosity. The results indicate that the doping effect of a strong dipole moment guest molecule (DO3) on the electro-optical properties are more pronounced than a relatively weak dipole moment one (HMAB).

Conclusion

The doping of azobenzene dyes in FLCP resulted in a larger fluctuation of the director of the mesogenic groups in each smectic layer under an applied electric field. As a result of that, larger intensity and higher frequency of Goldstone mode relaxation, and better electro-optical properties were obtained. Moreover, the doping effect on the dielectric behavior and electro-optical properties is more significant for the FLCP doped with strong dipole-moment guest molecules compared to that of the FLCP doped with weak dipole moment molecules. The doping of a suitable amount of the azobenzene dyes into the S_C^* phase is helpful for the improvement of the electro-optical properties.

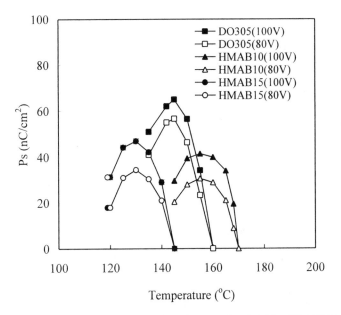

Figure 5. Temperature dependence of P_s for samples DO305, HMAB10, and HMAB15 (4μm, 4 Hz, 80V_{rms}, and 100V_{rms}).

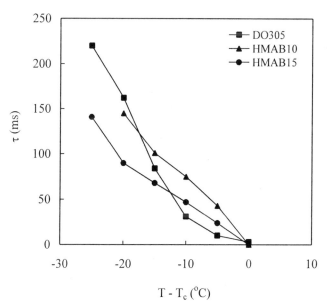

Figure 6. Temperature dependence of τ for samples DO305, HMAB10, and HMAB15 (4μm, 4 Hz, 100V_{rms}, $T_c = T_{SA-SC*}$).

144

Acknowledgments

The authors thank the National Science Council of Taiwan, ROC for financial support (Grant NSC87-2216-E007-032).

Literature Cited

1. Meyer, R. B.; Liebert, L.; Strzelecki, L.; Keller, P. *J. Phys. Paris Lett.* **1975**, *36*, L-69.
2. Clark, N. A.; Lagerwall, S. T. *Appl. Phys. Lett.* **1980**, *36*, 899.
3. Beresnev, L. A.; Blinov, L. M. *Feroelectrics* **1989**, *92*, 335.
4. Zentel, R.; Reckert, G.; Reck, B. *Liq. Cryst.* **1987**, *2*, 83.
5. Parmar, D. S.; Clark, N. A.; Keller, P.; Walba, D. M.; Wand, M. D. *J. Phys. Paris.* **1990**, *51*, 355.
6. Yuasa, K.; Uchida, S.; Sekiya, T.; Hashimoto, K.; Kawasaki, K. *Ferroelectrics* **1991**, *122*, 53.
7. Hsiue, G. H.; Chen, J. H. *Macromolecules* **1995**, *28*, 4366.
8. Chen, J. H.; Hsiue, G. H.; Hwang, C. P. *Chem. Mater.* **1997**, *9*, 51.
9. Dumon, M.; Nguyen, H. T.; Mauzac, M.; Destrade, C.; Gasparoux, H. *Liq. Cryst.* **1991**, *10*, 475.
10. Hsiue, G. H.; Lee, R. H.; Jeng, R. J. *Polymer* **1997**, *38*, 887.
11. Spruce, G.; Pringle, R. D.; *Mol. Cryst. Liq. Cryst.* **1988**, *154*, 307.
12. Kalmykov, Yu P.; Vij, J. K.; *Liq. Cryst.* **1994**, *17*, 741.
13. Malmstrom, E.; Liu, F.; Boyd, R. H.; Hult, A.; Gedde, U. W. *Polym. Bull.* **1994**, *32*, 679.
14. Goodby, J. W. In *Ferroelectric Liquid Crystals Principle, Properties and Application*; Taylor, G. W., Ed.; Gordon and Breach Science Publishers: Philadelphia, **1991**, P207.

Chapter 10

High Dielectric Constant (Microwave Frequencies) Polymer Composites

Shailendra Negi[1], Keith Gordon[1], Saeed M. Khan[2], and Ishrat M. Khan[1,3]

[1]Department of Chemistry, Clark Atlanta University, Atlanta, GA 303124
[2]Department of Electrical Engineering, Kansas State University, Manhattan, KS 66506

Dielectric constants of polymer/salt molecular composites prepared using poly(2-vinylpyridine)/poly(ethyleneoxide) and/or methoxypolyethyleneglycol or tetraethyleneglycol/LiClO$_4$ were determined at microwave frequencies. Change of dielectric constants of composites with varying voltages was studied to determine their possible feasibility as tunable materials. Effect of chain length, functional groups, salt content and T$_g$ on the dielectric constant were determined. Modeling studies at different frequencies were done to evaluate their microwave reflection/attenuation properties.

Polymeric materials are playing an increasingly important role in electronic components, such as antennas, wave guide, electromagnetic interference (EMI) shielding (*1*) and optical materials (*2*). They provide an advantage over conventional materials in terms of lightweight, material/property tuning and combination of various properties in a given matrix. Although applications are found in all region of the electromagnetic spectrum, relatively little work has been done in the microwave region. Polymers with high dielectric constant in the microwave region would be expected to have potential in EMI shields, radar absorbers, hybrid-integrated lines, wave-guides, microwave transmission lines and dielectric resonator antennas (*1,3*). Polymers with inclusions (high dielectric constant inorganic materials) and intrinsically conducting polymers are different types of materials that may be used for the above applications. Certain inorganic materials, like barium titanate, have very high dielectric constants but also have drawbacks, e.g. they are not amenable to property and material tuning required for specific purposes. Inorganic fillers dispersed in a polymer matrix have found use in EMI shields (*4,5*). High dielectric constant materials composed of nickel, silver, copper and nickel coated graphite fibers dispersed in polymers have been also studied (*6*). The metal powder particles are polarizable and increase the permittivity of the composite materials. Properties of polymers with inclusions may be affected by interfacial charge accumulations and are difficult to control. Moreover, large amount of filler material is required to achieve a substantial effect.

[3]Corresponding author.

The intrinsically conducting polymers have desirable features such as lightweight, processability and electrical conductivity and, therefore, they have potential applications as microwave active materials. Furthermore, control of the electronic properties can be done by synthesis, addition of different dopants, and molecular structure modification. To behave as a microwave active material, the system should be able to interact with the field microscopically (microwave radiation). A prerequisite for such interaction is that the system should have polarizable charged groups and/or highly conjugated polarizable functionality. Conjugated polymers, p-doped or n-doped, lead to highly polarized system resulting in high microwave absorption (7).

Polymer films of poly(p-phenylene-benzobis-thiazole) (PBT) (8), doped with iodine show good EMI shielding properties at microwave frequencies, indicated by the small amplitude of field transmitted through the polymer in the frequency range of 2 GHz to 10 GHz. One of the very important polymer systems, polyaniline (PAN), which has high conductivity and dielectric constant, shows high shielding efficiency in the microwave frequency range (9a). On a mass/area basis, the shielding efficiency of conducting PANs is comparable to that of copper. Furthermore, the dielectric constant of PAN can be controlled through chemical processing, molecular weight, doping level, counter ion, solvent etc. Stretching, which leads to higher crystallinity, also changes the dielectric constant. For highly crystalline PAN, increasing the temperature leads to three-dimensional coupled metallic regions, resulting in a dramatic change in the value of the microwave dielectric constant (9b). Hourquebie (10) et. al. have studied a number of conducting polymers (polyanilines, polypyrroles and polyalkylthiophenes) and correlated their dielectric constant values with various parameters, via. nature of dopant, microstructure and monomer structure. Increasing the size of the alkyl group and counter ion results in lowering the values due to decrease in capacitive coupling. A combination of polypyrrole (conducting material) and silicon rubber (lossy material) results in a composite with a high dielectric constant and a high tan δ. These examples show the utility of polymers as unique materials whose microwave properties can be tailored in a number of ways (11).

Microwave active smart materials will find applications in sophisticated technologies in the near future. Such materials will adopt or change their properties in response to an external stimulus in a reversible manner. Kon et. al. (12) have found a conducting system which can be oxidized and reduced reversibly and therefore changes the microwave transmittance from less than 10% to more than 90%. Thus tunable attenuation can be achieved in the microwave region. The microwave characteristics of blends of poly(aniline).HCl-silver-polymer electrolytes can be modulated by changing the oxidation states when applying DC voltage (13).

It is possible to prepare composites of nonconducting polymer/inorganic salt by dispersing inorganic salts in the organic matrix via complexation. The electrical conductivity of such materials has been well established in view of their importance as solid polymer electrolytes in solid state batteries (14,15).

We are interested in systems whose microwave properties may be tunable, i.e. change the value of the dielectric constant on applying small external electrical potential up to 10 volts. Such phenomenon can lead to smart materials (*16*) as mentioned above, and inorganic salt dispersed in an organic polymer matrix might be a suitable system. Earlier studies in microwave tunable material involved intrinsic conducting materials whose microwave properties were modulated by changing the oxidation states of the conducting polymers involved. Our system does not involve any redox reaction, but requires rearrangement of polymer chains and/or ion pairing in the matrix for the said effect. Thus our approach is complimentary to others being currently pursued.

Experimental

Materials: Poly(2-vinylpyridine) (P2VP, MW 200,000, Polysciences) methoxypolyethyleneglycol (MPEG , MW 350), tetraethyleneglycol (TEG, Sigma), and poly(ethylene oxide)(PEO , MW 600,000, Aldrich), $LiClO_4$ (Aldrich) were used as received.

Sample preparations: PEO/P2VP/$LiClO_4$ (*17a,b*) and P2VP/$LiClO_4$ (*17c*) blends were prepared according to literature procedures. Abbreviations of the type PEO(80)/P2VP(20)/$LiClO_4$ indicate blend of $LiClO_4$ containing 80% by weight PEO and 20% by weight P2VP. Blends of PEO/MPEG/P2VP/$LiClO_4$ were prepared by a modified procedure. PEO, MPEG and $LiClO_4$ were dissolved in methanol by heating and stirring vigorously and P2VP in methanol was added and stirred overnight. Solution was evaporated for 2 days at 50 ^{0}C *in vacuo* followed by drying at 80 ^{0}C under vacuum for 2 days. Samples for the measurements were prepared by heat pressing at 50 ^{0}C for 20 minutes. A similar solution blending procedure was applied for the preparation of PEO/TEG/P2VP/$LiClO_4$ blends.

Measurements: Dielectric measurements were carried out using a Hewlett Packard 4192 A Impedance analyzer at interval of 45 minutes at room temperature under nitrogen atmosphere. DSC was performed on a Perkin-Elmer DSC-4 under dry nitrogen using indium as the calibration standard. Thermograms were obtained at a rate of 15 ^{0}C per minute and after the samples were quench cooled at a rate of 320 ^{0}C/minute. After first heating, the next few runs of thermograms indicated essentially the same Tg values and the reported values were obtained from the third run. P2VP/$LiClO_4$ dielectric constant measurements were done in the GHz range in an X-band waveguide (*17c*).

Result and Discussion

Earlier studies in our laboratories have indicated that molecular composites of poly(2-vinylpyridine) and $LiClO_4$ have high dielectric constants (*17c*) in the GHz range as listed in Table I. These composites are amenable to material/property tuning by changing the salt content. Poly(2-vinylpyridine) and lithium salt form a homogenous system at low salt concentrations as indicated by T_g. The nitrogen in the pyridine ring solubilizes the salt via ion-dipole interactions, based on Hard Soft Acid Base (HSAB) principle. One can conceive that applying a voltage across the system could lead ions to rearrange, simultaneously forcing the coordinated polymer chains to rearrange. This can result in a change in the dielectric constant and therefore microwave properties. Thus the system may be tunable by changing applied voltages.

Table I: Relative dielectric constants, tan δ and glass transition temperatures of polymer salt complexes.

P2VP/LiClO$_4^a$	Relative dielectric constant, ε_R^b	Tan δ x 10^4	T_g (^{0}C)
Pure P2VP	7.5	258	110
50	8.5	334	113
20	10.5	87	125
5	16.5	50	133

[a] Mole ratio of 2-vinylpyridine repeating units to lithium perchlorate.

[b] at 12 GHz and $\varepsilon_R = (1+ i \tan \delta)$

Studies showed that the dielectric constants of the system did not change on applying small voltages. The observed glass transition temperatures of the polymer/ salt complexes are high and increase with increasing salt content. Similar trends (i.e. increase in T_g with increase in salt) are observed in most solid polymer electrolyte systems (17,18). Thus, it may be imperative to introduce another component, which makes the system more flexible, i.e. lowers the T_g, and is simultaneously compatible with P2VP/lithium salt system.

Our group has studied PEO/P2VP/LiClO$_4$ blends (17a), prepared by solution blending. Even though PEO and P2VP are not compatible with each other, lithium cation coordinates with both the nitrogen in P2VP and the oxygen in PEO to form a compatible system. Thus lithium perchlorate is effectively dispersed throughout the system in a homogenous fashion. Thermal studies indicate almost a linear change in T_g with change in salt content at low salt concentration and verify the homogeneity of the system. The PEO/P2VP/LiClO$_4$ blends are elastomeric and contain low T_g domains. This indicates that there may be sufficient segmental motions (or flexibility) at room temperature for the chains to respond to external stimulus therefore, making this an appropriate system to study the dielectric constant as a function of voltage. A blend of PEO(80)/P2VP(20)/LiClO$_4$ with an EO/Li$^+$=5 (T_g = -18 ^{0}C) was prepared and dielectric constants were determined at room temperature at 30 minute intervals with varying frequencies and voltages. Preliminary studies show that there was a change in the dielectric constant with voltage at all the frequencies and a maximum change of 3% ($\pm$1%) was observed at 10,000 KHz as shown in Table II. However, the response time was slow.

Table II : Relative dielectric constants of PEO(80)/P2VP(20)/LiClO$_4$; EO/Li$^+$= 5 at various voltages.

Bias (Volts)	Relative dielectric constant, ε_R[a]
0	7.7
2	7.6
5	7.5
10	7.4

[a] at 10 MHz and ε_R = (1+ i tan δ)

To improve the system, methoxypolyethyleneglycol (MPEG, MW 350) which is similar to PEO, was substituted for PEO. MPEG has shorter and more flexible chains and would be expected to have increased flexibility and segmental motions. As a result of the above factors it may be able to accommodate changes in conformations more readily than the corresponding PEO based system. Blends of different compositions of MPEG/P2VP/LiClO$_4$ were investigated. Initially composites with higher content of P2VP and salt were prepared. The T_g's and the dielectric constants of the system are presented in Table III.

Table III. Glass transition temperatures and relative dielectric constants as a function of voltage for blends of MPEG/P2VP/LiClO$_4$.

Blend composition[a]			T_g (^{0}C)	Relative dielectric constant, ε_R[b] at bias voltage shown below					
MPEG[a]	P2VP[a]	EO/Li$^+$		2	5	10	5	2	volts
25	75	8	40						
25	75	5	70	4.3	4.3	4.3	4.3	4.3	
25	75	2	96						
35	65	10	20						
35	65	8	31						
35	65	5	42	5.4	5.4	5.4	5.4	5.4	
35	65	2	91						

[a] weight %.

[b] at 10 MHz and ε_R = (1+ i tan δ)

In view of the dimensional stability, blends of composition MPEG(25)/P2VP(75)/LiClO$_4$ [EO/Li$^+$ = 5] and MPEG(35)/P2VP(65)/LiClO$_4$ [EO/Li$^+$ = 5] were chosen as representative examples for study of change in dielectric constants with voltage variations. Dielectric measurements were done at one-hour intervals. The results show that the dielectric constant remains practically the same on application of an external field and this is most likely related to the fairly high T_g's observed for the system. It became clear that further studies would require composites with lower T_g. Figure 1 shows the DSC thermograms of blends of MPEG(35)/P2VP(65)/LiClO$_4$ and, as expected, an increase in T_g is observed with increasing salt content. The almost linear increase in the T_g values with increasing salt content is consistent with previously observed P2VP/PEO/LiClO$_4$ system (18).

We explored blends PEO/MPEG/P2VP/LiClO$_4$ with much lower content of P2VP. The glass transition temperatures and the dielectric constants are listed in Table IV, and DSC thermograms are shown in Figure 2.

Table IV. Glass transition temperatures and relative dielectric constants as a function of voltage for blends of PEO/MPEG/P2VP/LiClO$_4$.

Blend composition[a]				T_g (°C)	Relative dielectric constant, ε_R[b] at bias voltage shown below.				
PEO[a]	MPEG[a]	P2VP[a]	EO/Li$^+$		2	5	10	5	2 volts
40	40	20	10	-39	10.5	10.3	10.3	10.3	10.2
40	40	20	8	-32					
40	40	20	5	-26	10.8	10.7	10.7	10.6	10.7

[a] weight %

[b] at 10 MHz and ε_R = (1+ i tan δ)

Dielectric values remained almost constant on application of an external field. It may be possible that the coordination of the metal ion is not strong enough with the polymer chains to perturb them sufficiently when the voltage is varied and, therefore, more polar analogue of MPEG, leading to stronger complexation, was chosen to potentially yield better results. Tetraethyleneglycol, which has larger number of hydroxyl group, was selected instead of MPEG. An additional advantage of TEG was that it would enhance the dielectric constant of the systems because of the polar hydroxyl groups (19).

Composites of PEO(40)/TEG(40)/P2VP(20)/LiClO$_4$ were prepared and their T_g's are tabulated in Table V and DSC result are presented in Figure 3.

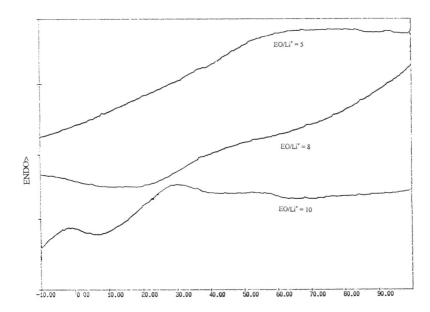

Figure 1. DSC thermograms of MPEG(35)/P2VP(65)/LiClO$_4$ blends at various EO/Li$^+$ ratios.

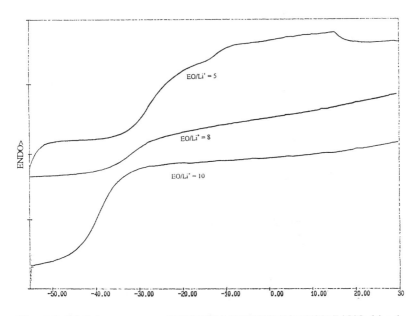

Figure 2. DSC thermograms of PEO(40)/MPEG(40)/P2VP(20)/LiClO$_4$ blends at various EO/Li$^+$ ratios.

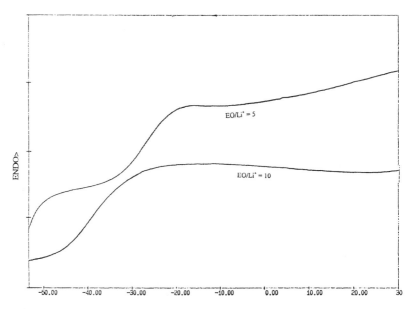

Figure 3. DSC thermograms of PEO(40)/TEG(40)/P2VP(20)/LiClO$_4$ blends at two different EO/Li$^+$ ratios.

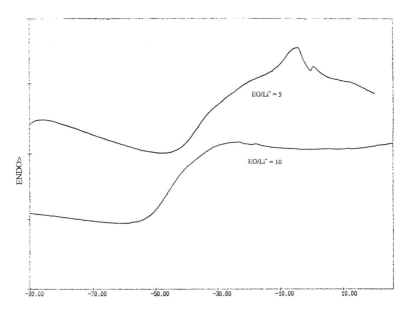

Figure 4. DSC thermograms of PEO(25)/ TEG(55)/P2VP(20)/LiClO$_4$ blends at EO/Li$^+$ ratios of 5 and 10.

Table V. Glass transition temperatures and relative dielectric constants as a function of voltage for blends of PEO/TEG/P2VP/LiClO$_4$.

Blend composition[a]				T$_g$ (^{0}C)	Relative dielectric constant, ε_R[b] at bias voltage shown below				
PEO[a]	TEG[a]	P2VP[a]	EO/Li$^+$		2	5	10	5	2 volts
40	40	20	8	-32					
40	40	20	5	-25	11.6	11.4	11.3	11.3	11.4
25	55	20	10	-47					
25	55	20	5	-35	12.7	12.4	12.2	12.2	12.3

[a]weight %

[b]at 10 MHz and $\varepsilon_R = (1 + i \tan \delta)$

Comparison of the blends PEO(40)/MPEG(40)/P2VP(20)/LiClO$_4$ and PEO(40)/TEG(40)/P2VP(20)/LiClO$_4$ indicates that the blend system have similar T$_g$ values. Thus the effect of smaller size of TEG chain relative to MPEG is probably compensated by greater (or more effective) interaction of blend with the metal ions. As expected, higher salt concentration is reflected in higher T$_g$'s in both series.

Table V also shows that variation of voltage resulted in change of relative dielectric constant and the phenomenon is reversible. Results indicate a change of about 3-5 % and reversible nature of the system was observed for 3 cycles. The blend with smaller EO/Li$^+$ ratio of 10, also shows reversibility although changes are smaller. It should be pointed out that the relative dielectric constants in Tables III, IV and V are determined at interval of 45 minutes after the bias voltage is changed, the systems reached close to initial value on equilibrating overnight at room temperature.

To further increase the amount of TEG, blends of PEO(25)/TEG(55)/P2VP(20)/LiClO$_4$ were prepared. The glass transition temperature of the system is lower than the corresponding system with the 40% TEG (Figure 4). Dielectric measurements of PEO(25)/TEG(55)/P2VP(20)/LiClO$_4$ [EO/Li$^+$ = 5] was investigated as a function of voltage. All these results are shown in Table V. The dielectric constant changes though small (3-5%) were reversible and more pronounced than in the previous case.

Therefore, our observation indicates that higher concentration of hydroxyl groups results in the ability of the system to solubilize higher concentrations of LiClO$_4$ and at the same time the system has a fairly low T$_g$ value. If the molecule is large (polymeric as opposed to oligomeric) with a high hydroxyl group concentration then the material may become crystalline (as in polyvinylalcohol) and the system may not work. Similarly a very high cation concentration could lead to more pseudo cross-linking, resulting in higher T_g and decreased segmental flexibility of the polymer chains. Various parameters thus play important roles and these factors may have opposite effects; one dominating the other at different concentrations. Such systems, therefore, require very fine tuning to get the required desirable properties and our studies are a starting point to prepare polymeric materials where dielectric constants may be modulated with a change of 10% and are reversible for large number of cycles.

Modeling Studies: To study the microwave properties of the blend, PEO(25)/TEG(55)/ P2VP(20)/LiClO$_4$ [EO/Li$^+$ = 5] was chosen and the dielectric constants were measured at various frequencies at intervals of 30 minutes and are listed in Table VI.

Table VI: Relative dielectric constants of PEO(25)/TEG(55)/P2VP(20)/LiClO$_4$; EO/Li$^+$ = 5 at various frequencies.

Frequency (KHz)	Relative dielectric constant, $\varepsilon_R{}^a$	Frequency (KHz)	Relative dielectric constant, $\varepsilon_R{}^a$
.01	-	2000	15.1
.1	5630	3000	13.9
.5	453	5000	12.7
1	195	7000	12.0
10	52.0	9000	11.5
100	29.2	11000	11.2
1000	17.5	13000	11.0

[a] at 0 volt and $\varepsilon_R = (1 + i \tan \delta)$

The reflection properties at normal plane wave incidence (*1, 20*) for the blend reveals that it is a very good reflector under 10 KHz and quite good up to 100 KHz.

Figure 5 shows that at 1 MHz, the reflection coefficient is around 0.7 and at about 10 MHz it drops down to 0.5. The study of its intrinsic attenuation reveals that it will not significantly absorb transmitted fields (Figure 6). This modeling studies do indeed show that the tunable composites are good microwave reflectors.

Conclusion

We were able to prepare low T_g blends with relatively high dielectric constants, which could be modulated reversibly by voltage variation. Dielectric changes in the range of 3-5% were observed as a function of voltage and such effects were

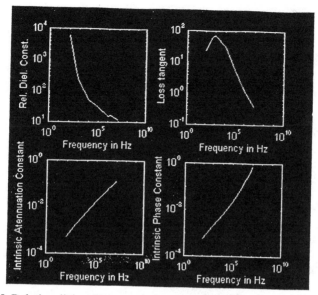

Figure 5. Relative dielectric constant, loss tangent, intrinsic attenuation constant and intrinsic phase constant of PEO(25)/TEG(55)/ P2VP(20)/LiClO$_4$ blend at EO/Li$^+$ ratio of 5.

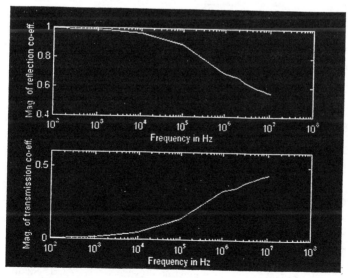

Figure 6. Reflection and transmission properties of PEO(25)/TEG(55)/ P2VP(20)/LiClO$_4$ blend at EO/Li$^+$ = 5.

possible for a few cycles. For a system to be tunable, the important factors are a low T_g, the presence of short flexible molecules, and increased concentration of hydroxyl groups. Elastomeric blends such as $PEO(40)/TEG(40)/P2VP(20)/LiClO_4$ [EO/Li$^+$=5] and $PEO(25)/TEG(55)/P2VP(20)/LiClO_4$ [EO/Li$^+$ = 5] were found to be most promising and may be further modified to obtain reversible tunability up to 10%. The $PEO(25)/TEG(55)/ P2VP(20)/LiClO_4$ [EO/Li$^+$ = 5] blend was a very good reflector at very low frequency. It is expected that more material tuning might result in increase of reflective properties at higher frequencies. If the reflection characteristics are enhanced up to 20 MHz, this material could prove to be a usefulas microwave shields against interference from many common sources like lightning, corona, electric motors etc. Hybridization by mixing this material with metal flakes might further enhance the electrical properties related to electromagnetic shielding (21).

Acknowledgments

Partial support of this work through use of facilities provided and supported by RCMI 5G12RR030602 and NIH/MBRS GM 08247 is gratefully acknowledged.

References

1. a) Lee, R. Q. W.; Simon, R. N. " *Bandwidth Enhancement of Dielectric Resonator Antennas*", 1993. IEEE Antennas and Propagation International Symposium, **1993**, *3*, 1500 (b) Hemming, L. H., "*Architectural electromagnetic shielding handbook*", IEEE Press,1992.
2. Dalton, L. R.; Sapochak, L. S.; Chen, M.; Yu, L. P. " Molecular Electronics and Molecular Electronic Devices," Vol. 2, p. 125-208, ed. K Kienicki, CRC Press, Boca Raton, 1993(b) Marder, S. R.; Cheng, L. T.; Tiemann, G.; Friedlie, A. C.; Blanchard-Desce, M.; Perry, J. W.; Skindhoej, J. *Science*, **1994**, *263*, 511(c) Burland, D. M.; Miller, R. D.; Walsh, C. A. *Chem. Rev.* **1994**, *97*, 31.
3. Wong, T. C. B.; Chambers, B.; Anderson, A. P.;Wright, P.V. " *8 th Int Conf. on Antennas and Propagation*", part 2. IEEE Conf. Publ., **1993**, *379*, 934.
4. Sichel. E., "*Carbon black composites, the physics of electrically conducting composites*", Mercel Dekker Inc., New York, 1982.
5. Emerson and Cummings, "*Microwave Absorbers*," 869 Washington Street, Canton, Massachusetts.
6. (a) Ho, Y.S.; Schoen, P. *J. Mater. Res.*, **1994**, *9*, 246 (b) Chung, K. T.; . Sabo, A.; Pica, A. P. *J. Appl. Phys.*, **1992**, *53(10)*, 6867(c) Bhattacharya, S. K.; Chaklader, A. C. D. *Polym. Plast. Technol. Eng.*,**1982**, *19*, 21(d) Ho, Y. S.; Schoen, P. *J. Mater. Res.*,**1996**, *11*, 469.
7. Theophilou, N. et. al." Highly Conducting Polyacetylene", ACS Meeting, Denver, 1987(b) Naarman, H., "*Proc. International Congress of Synthetic Metals*", ICSM 86, Kyoto, 4A-07, 1986 (c) Bidarian, A.; Kadaba, P. K *J. Mat. Sci. Lett.* **1988**, *7*, 922.

8. Chen, C.; Naishadham, K. Conference *Proceedings-IEEE SOUTHEASTCON*, V-1, IEEE Service Center, **1990**, 38-41

9. Joo, J.; MacDiarmid, A. G. ; Epstein, A. J. *ANTEC*, **1995**, 1672 (b) Epstein, A. J. ; Joo, J.; Kohlman, R. S.; Macdiarmid, A. G. ; Weisinger, J. M.; Min, Y.; Pouget, J. P.; Tsukamoto, J. *Mat. Res. Soc. Proc.*, **1994**, *328*, 145.

10. Hourquebie P.; Olmedo, L. *Synthetic Metals*, **1994**, *65*, 19 (b) Olmedo, L.; Hourquebie, P.; Jousse, F. *Synthetic Metals*, **1995**, *69*, 205.

11. Truong, V. T. ; Codd, A. R.; Forsyth, M. *J. Mater. Sci.*, **1994**, *29*, 4331.

12. Rupich, M. W. ; Liu, Y. P.; Kon, A. B. *Mat. Res. Soc. Symp. Proc.*, **1993**, *293*, 163.

13. Barnes, A.; Despotakis, A.; Wright, P. V.; Wong, T. C. P.; Chambers, B.; Anderson, A. P. *Electronics Letter*, **1996**, *32*, 358.

14. a) Fenton, D. E.; Parker, J. M.; Wright, P. V. *Polymer*, **1973**, *3*, 589 (b) Wright, P. V. *Brit. Polym. J.*, **1975**, *7*, 319 (c) Armand, M. B. *Annu. Rev. Mater. Sci.*, **1986**, *16*, 245.

15. a) Armand, M. B. *Solid Satte Ionics*, **1983**, *9 & 10*, 745 (b) Duval, M.; Gauthier, M.; Belanger, A.; Harvey, P. E.; Kapfer, B.; Vassort, G. *Makromol. Chem., Macromol. Symp.*, **24**, 151 (1989).

16. Khan, S. M.; Negi, S.; Khan, I. M. *Polymer News*, **1997**, *22*,414-418 and references therein.

17. a) Li, J.; Pratt, L. M.; Khan, I. M. *J. Polym. Sci. Chem. Ed.*, **1995**, *33*, 1657 (b) Li, J.; Khan, I. M. *Macromolecules*, **1993**, *26*, 4544 (c) Li, J.; Khan, S. M.; Khan, I. M. *Polym. Prep.*, **1995**, *36*, 388.

18. Khan, I. M.; Li, J.; Arnold, S.; Pratt, L. M. *Electrical and Optical Polymer Systems: Fundamental, method and Application,* Wise, D.L.; Trantolo, D. J.; Wnek, G. E.; Cooper, T. M.; Grever , J. D.(Editor), Marcell Dekker, NY, 1998, pp 331-358.

19. Smith, V. K.; Pollard, J. F.; Ward, T. C.; Graybeal, J. D.; Ma, J. J.; Quirk, R. P. *Polym. Prep.*, **1987**, *28*, 326.

20. Balanis, C. A. *"Advanced Engineering Electromagnetic "*, Wiley,1989.

21. Pohl, H. A. *"Dielectrophoresis"*, Cambridge University Press, New York , 1978.

PHOTORESPONSIVE POLYMERS: NONLINEAR OPTICAL AND PHOTOREFRACTIVE

Chapter 11

Identification of Critical Structure–Function Relationships in the Preparation of Polymer Thin Films for Electro-Optic Modulator Applications

A. W. Harper[1], F. Wang, J. Chen, M. Lee, and L. R. Dalton

Loker Hydrocarbon Research Institute, University of Southern California, Los Angeles, CA 90089–1661

The past decade has witnessed the design and preparation of second order nonlinear optical chromophores characterized by ever increasing molecular hyperpolarizability; indeed, values of $\mu\beta$ (where β is the hyperpolarizability and μ is the dipole moment) greater than 1×10^{-44} esu (at 1.9 μm wavelength) have been realized. However, chromophores which exhibit large hyperpolarizabilities often exhibit large polarizability and dipole moments leading to strong intermolecular electrostatic (London forces) interactions. Such intermolecular interactions oppose the translation of large observed molecular optical nonlinearities into large macroscopic optical nonlinearities by opposing electric field poling induction of the acentric chromophore order required for such macroscopic optical nonlinearity. In this communication, it is demonstrated that the attenuation of poling-induced order is strongly influenced by chromophore shape and that simple derivatization of chromophores with "steric" buffer substituents leads to dramatic improvement in poling-induced acentric order and associated macroscopic optical nonlinearity. Materials appropriate for device fabrication and characterized by electro-optic coefficients in the range 25-55 pm/V (at 1.3 μm operational wavelength) have been developed. Since electrostatic interactions can lead to chromophore aggregation resulting in light scattering and inhibition of thermosetting lattice hardening reactions, it is not surprising that modified chromophore materials also exhibit reduced optical loss and improved thermal stability of optical nonlinearity induced by thermosetting reactions.

A crucial objective of research involving second order organic nonlinear optical materials is the realization of polymeric electro-optic waveguides characterized by electro-optic coefficients (at 1.3 μm communication wavelength) of greater than 30 pm/V and which are stable for long periods of time (e.g., 1000 hours) at temperatures

[1]Current address: Department of Chemistry, Texas A&M University, College Station, TX 77843–3255.

on the order of 100°C. Such values of electro-optic coefficient are necessary for utilization of digital level drive voltages (without amplification) with standard electro-optic modulator device (Mach-Zehnder interferometer, birefringent modulator, and directional coupler) configurations. Materials must also exhibit modest (1 dB/cm) to low optical loss at the operational wavelength and must be amenable to low loss integration with semiconductor VLSI electrical circuitry and silica fiber optic transmission lines.

To satisfy device-related materials requirements, chromophores must exhibit large molecular hyperpolarizabilities and must be incorporated in high number density into thermally stable acentric chromophore lattices defined by a high degree of chromophore order. Realization of approximately 10% of the potential macroscopic optical nonlinearity of a chromophore requires chromophore loading on the order of 30% (by weight) and an order parameter, $<\cos^3\theta>$, on the order of 0.3. Clearly, there is not much opportunity to compromise on chromophore optical nonlinearity, chromophore loading, and/or poling efficiency and still achieve acceptable electro-optic coefficients. Since low optical loss and robustness are also critical material requirements, the chromophore must exhibit good thermal, electrochemical, and photochemical stability and must exhibit good solubility in acceptable spin casting solvents. The chromophore must be capable of being poled with good efficiency and coupled to a high glass transition lattice subsequent to poling. Not only must the chromophore have significant mobility under poling but also the chromophore/polymer matrix must be resistant to corona damage and to phase separation of components during poling and hardening. The resulting nonlinear optical material must be compatible with fabrication of buried channel waveguide structures by reactive ion etching and the deposition of compatible cladding layers.

An introduction to high μβ chromophores.

A number of researchers, including Marder and Perry (Cal Tech/JPL), Marks and Ratner (Northwestern), and Alex Jen (formerly at EniChem and now at Northeastern Univ.), have advanced the synthesis of improved chromophores (which we shall collectively refer to as "high μβ chromophores") exploiting insights gained from quantum mechanical calculations. As we (1) and others (2) have reviewed this work elsewhere, we will not review this work here other than to present selected data shown in Table I. The data shown in this table illustrate the systematic increase in optical nonlinearity which has been achieved relative to standard stilbene and azobenzene (disperse red, DR) chromophores (upper left, Table I). If high μβ chromophores could be incorporated in high number density into acentric macroscopic lattices with a high degree of acentric order, it is clear that very large macroscopic optical nonlinearities (electro-optic and second harmonic generation coefficients) would result. Of course, such chromophores must exhibit appropriate chemical and thermal stability and in some cases the introduction of acceptor groups to enhance optical nonlinearity has resulted in an unwanted sensitivity to nucleophilic attack.

The research that we report here focuses upon the translation of promising high μβ chromophores into electro-optically active materials suitable for fabrication of prototype devices. Our initial focus is a review of problems associated with intermolecular chromophore electrostatic interactions and how these can be minimized by simple structural modification of chromophores. We summarize our experiences with defining maximum achievable optical nonlinearity, minimum optical loss (including processing associated loss), and maximum material stability. This

Table I. Representative High μβ Chromophores

Chromophore	μβ (10⁻⁴⁸ esu)	μβ/MW	Chromophore	μβ (10⁻⁴⁸ esu)	μβ/MW
	510	1.3		2464	5.5
	570	1.7		2600	7.4
	482	1.8		2910	8.6
DR	586	2.2		4130	8.8
	904	3.0	TCI	6144	10.2
ISX	1960	4.0		7100	14.1
	2480	4.1		8640	14.2
	1300	4.3		6200	16.9
	2400	5.1		15000	27.7

research has already permitted a new generation of electro-optic modulators to be fabricated from polymeric materials. These modulators are capable of broadband operation (to above 100 GHz), driven by digital level drive voltages, and characterized by optical loss on the order of 1 dB/cm. The thermal stability achieved with high μβ chromophores has been found to be comparable to that previously obtained for azobenzene chromophores.

Translating microscopic optical nonlinearity to macroscopic optical nonlinearity.

The effective electro-optic coefficient, r, of a material is related to chromophore number density, N, chromophore molecular first hyperpolarizability, β, and acentric order parameter, $<\cos^3\theta>$, by

$$r = 2fN\beta<\cos^3\theta>/n^4 \tag{1}$$

where f is a frequency-dependent local field factor which takes into account the dielectric matrix that contains the chromophore and n is the index of refraction. The expression for electro-optic coefficient, r, can also be expressed in terms of chromophore weight fraction, w, by noting that $N = wN'\rho/MW$ where N' is Avogadro's number, ρ is material density, and MW is chromophore molecular weight. In the absence of chromophore-chromophore interactions (the independent particle or "gas" model), the chromophore acentric order parameter arising from electric field poling can be expressed as

$$<\cos^3\theta> = \mu f' E_p/5kT \tag{2}$$

where μ is the chromophore dipole moment, E_p is the applied electric poling field, f' is a local field factor ($F = f'E_p$ is the actual poling field felt by the chromophore), k is the Boltzmann constant and T is the Kelvin temperature. From the above, it is clear that within the approximation of treating chromophores as independent particles that predicted macroscopic optical nonlinearity should scale as $\mu\beta/MW$ (see Table I), $<\cos^3\theta>$ should be independent of N, and r should increase linearly with N. That this is not so is now obvious from a number of experimental observations but the point is conveniently illustrated in Figure 1.

As is demonstrated in Figure 1, electro-optic coefficients increase linearly with chromophore number density (with a slope proportional to $\mu\beta$) only at low chromophore loading. At higher concentrations, deviation from linearity is observed and the problem becomes more severe as chromophores with higher $\mu\beta$ are studied. In particular, maximum in plots of r versus N are observed and these shift to lower N for the higher $\mu\beta$ chromophores. The problem can be traced to increasing chromophore electrostatic interactions as is illustrated in Table II. In this Table, we illustrate the variation of dipole moment, polarizability, and ionization potential (the parameters which define chromophore-chromophore electrostatic interactions--London forces) with molecular hyperpolarizability.

While these simple correlations are suggestive, it is necessary to analyze the problem in a more quantitative manner. We have done this by re-examining and extending Debye-London theory (3). In particular, we have analyzed the problem of chromophores interacting with an externally applied poling field, interacting with each other, and experiencing thermal collisions which act to randomize their order with respect to the poling field axis system. Two levels of analysis have been carried out. In our most approximate treatment, we follow the assumptions of London (4) and treat chromophores as hard spherical objects and average out the anisotropic dipole-dipole, induced-dipole, and dispersion interactions between chromophores. We also make a high temperature approximation and carry out a power series expansion in the ratio of chromophore/poling field interaction energy to thermal

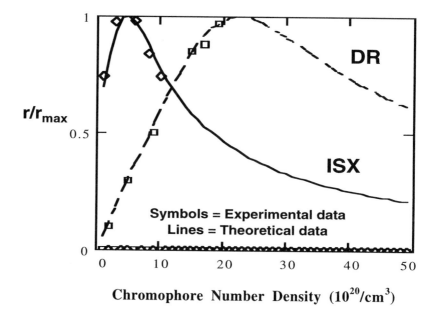

Chromophore Number Density $(10^{20}/cm^3)$

Figure 1. Normalized electro-optic coefficients versus chromophore number density for the disperse red (DR) and isoxazolone acceptor (ISX) chromophores of Table I.

Table II. Dipole moments, polarizabilities, ionization potential, and hyperpolarizabilities for some representative chromophores.

Compound	Structure	μ (Debye)	α $(10^{-23} cm^3)$	I $(10^{-19} J)$	β_0 $(10^{-30} esu)$
DMNA	Me₂N—⟨⟩—NO₂	6.4	2.2	9.04	9.7
DANS	Me₂N—⟨⟩—⟨⟩—NO₂	6.6	3.4	8.70	55
DR	Me₂N—⟨⟩—N=N—⟨⟩—NO₂	7.0	3.8	8.26	55
ISX		8.0	7.6	7.90	153
FDCV		9.0	10.3	7.77	350

α = molecular polarizability, I = ionization potential, β = molecular first hyper-polarizability, and μ = molecular dipole moment.

energy and we retain only the first term. This approach leads to the following analytical result:

$$<\cos^3\theta> = (\mu f'E_p/5kT)[1 - L^2(W/kT)] \tag{3}$$

where W is the London electrostatic interaction energy expression given by

$$W = (1/R^6)\{(2\mu^4/3kT) + 2\mu^2\alpha + 3I\alpha^2/4\} \tag{4}$$

where R is the average distance of separation between chromophores. W will exhibit a quadratic dependence on chromophore number density N.

It is clear even from this highly approximate analysis that an attenuation of poling-induced acentric order (and hence electro-optic coefficient, r) can be expected at high chromophore concentrations for chromophores with significant values of μ, α, and I. A more informative illustration of this point is given in Figure 2 which shows calculated variation of acentric order parameters with average chromophore separation (related to chromophore concentration, i.e., N is proportional to R^{-3}) for the chromophores and electrostatic interactions of Table II.

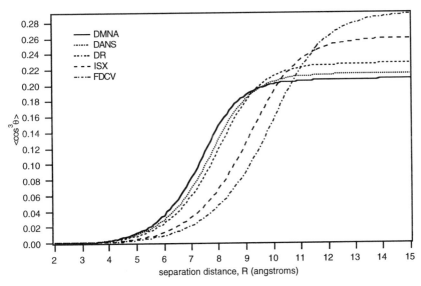

Figure 2. Variation of acentric order parameter with average separation distance between chromophores is shown for the chromophores (and electrostatic interactions) of Table II.

As electrostatic interaction, W, increases an attenuation of acentric order onsets at ever increasing R (or decreasing concentration). Moreover, this Figure illustrates that acentric order is rapidly attenuated as a function of R.

A more sophisticated analysis is required for quantitative simulation of experimental data. In particular, we explicitly consider both chromophore shape (within a hard object approximation) and do not average out anisotropic intermolecular interactions. As discussed elsewhere (3), this approach leads to reasonably good simulation of data particularly when cases of non-transient aggregation are avoided. Examples of the quality of agreement between theory and experiment are shown in Figure 1 for the DR and ISX chromophores. This level of treatment suggests that the problem of attenuation of poling-induced electro-optic activity at high chromophore concentrations is particularly problematic for prolate ellipsoidal chromophores. The close approach along minor axes of the prolate ellipsoids particularly favors centric aggregation. Electrostatic interaction energies for such centric aggregates can reach many times kT which can lead to non-transient aggregation. If aggregate size approaches the wavelength of light then light scattering can be observed. As is illustrated in Figure 3, this is indeed observed in a number of samples and the onset of

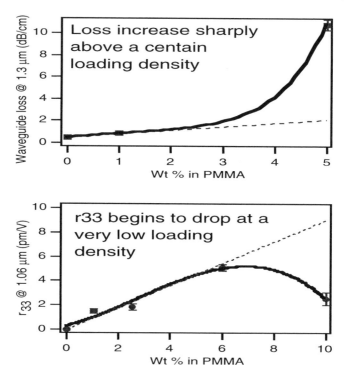

Figure 3. Correlated variation of waveguide loss (upper) and electro-optic coefficient (lower) versus chromophore loading (wt. % in PMMA) for the TCI chromophore of Table I is shown. The behavior reflects the formation of non-transient aggregates.

such scattering corresponds to the onset in the attenuation of electro-optic activity. To simulate such effects, molecular dynamics must be added to the calculations.

Our point in this communication is not to discuss the quantitative simulation of a large number of effects that can be attributed to chromophore-chromophore electrostatic interactions but rather to explore a simple prediction from the more rigorous theoretical analysis, namely, that attaching steric substituents to chromophores to increase the dimensions of the minor axes of prolate ellipsoidal chromophores should decrease chromophore-chromophore electrostatic interactions. Here we report are first crude efforts to exploit this observation.

Reduction in chromophore chromophore electrostatic interactions by derivatization with bulky substituents.

A comparison of theory and experiment establishes that it does not make sense to derivatize chromophores such as the disperse (DR) and DANS other than to improve their solubility. The attenuation of electro-optic activity is not a problem for these systems. We can gain only for those systems where optical nonlinearity is seriously attenuated and where we are in the steep region of the acentric order parameter attenuation as a function of variation of chromophore minor axes dimensions. For such cases, a change of minor axes dimensions by 1 or 2 angstroms can lead to factors of 2 and 3 increase in maximum realizable electro-optic coefficient.

To date, we have attempted only minor chromophore modifications which can be effected in a cost effective manner necessary to produce the quantities of materials (grams to kilograms) necessary for prototype device studies (Figure 4).

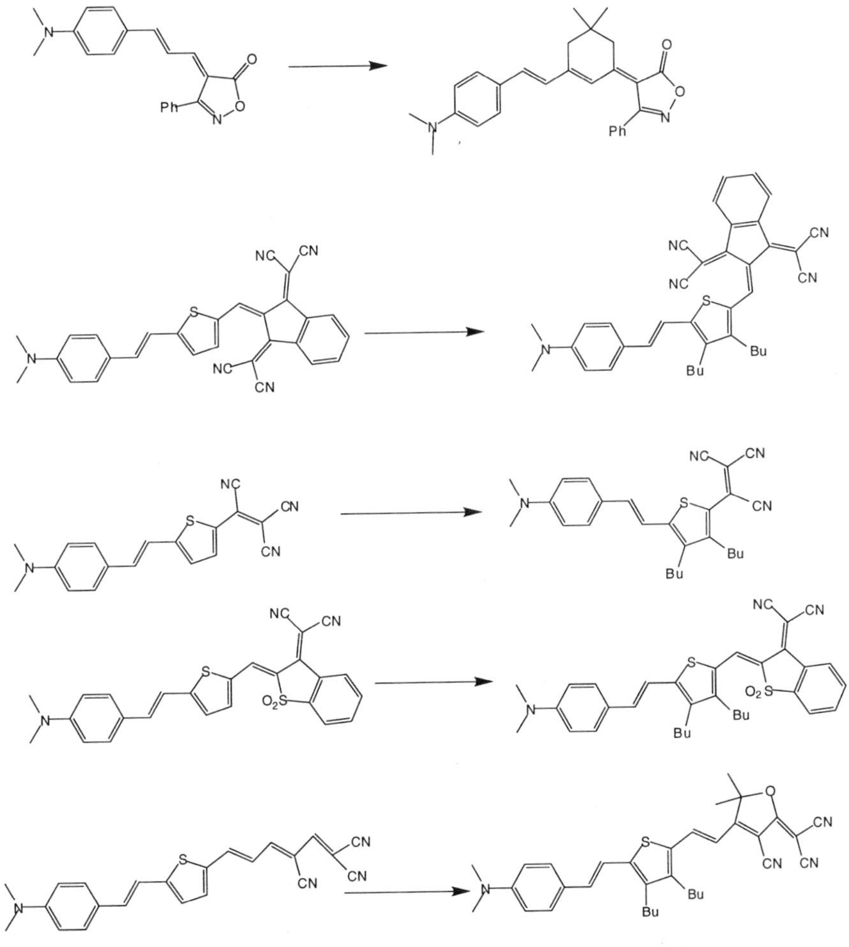

Figure 4. Illustration of derivatization of chromophores with bulky substituents.

Minor modification of the chromophore structure also is likely to result in the least perturbation of chromophore electronic structure so that we do not have to worry about changing values of μ and β by the modification. In experimental practice, we have used the criteria of insignificant differences between the wavelengths of the absorption maxima in derivatized and underivatized chromophores as an indication of minimum influence of derivatization on chromophore π electron structure.

In all of the cases shown in Figure 4, derivatization has resulted in at least a factor of two increase in maximum obtainable electro-optic coefficient. Best results to date have been obtained for the last entry in Figure 4 where we observed an electro-optic coefficient of 55 pm/V (at 20 and 24 weight percent chromophore loading) and an optical loss of 0.75 pm/V (at 1.3 μm). Data for the first entry of Figure 4 (using an isophorone group to sterically hinder close approach of chromophores) are shown in Figure 5 and illustrate that the isophorone group

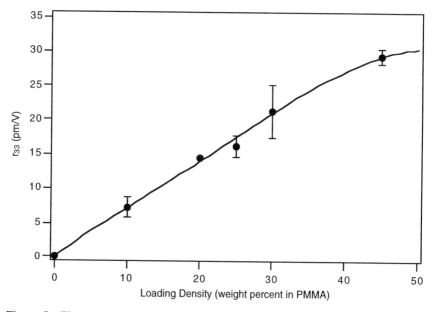

Figure 5. Electro-optic coefficient versus chromophore loading data obtained for the first derivatized chromophore of Figure 4. Linearity of the graph of r versus w (N) is increased to approximately 40 wt % by derivatization with an isophorone group to sterically hinder chromophore-chromophore close approach.

effectively eliminates the problem of attenuation of electro-optic activity by intermolecular chromophore electrostatic interactions. The linearity of r versus N has been extended to 40% (wt) chromophore loading. Essentially, most of the realizable (by electric field poling) optical nonlinearity of the chromophore has been achieved. The optical loss of this chromophore at high loading was found to be 0.9 dB/cm (1.3 μm). An analog of this chromophore where the isophorone group has been replaced by a verbenone group has also been made with comparable results. Both composite materials where the chromophores of Figure 4 are incorporated into polymer matrices such as polymethylmethacrylate (PMMA) and hardened (thermosetting DEC chromophore) polymer materials with the chromophore covalently to attached to a three dimensional lattice have been prepared and examined. Comparable results in terms of optical nonlinearity are observed for the two approaches. With thermosetting processing of hardened polymer materials, care must be exercised to avoid optical losses due to processing, e.g., phase separation of components. However, with care materials have been fabricated into devices characterized by digital level drive voltages ($V_\pi = 7$ volts or less), optical loss on the order of 1.0 dB/cm, and thermal stability to 100°C. For example, the isopherone and verbenone analogs of the first chromophore of Figure 4 yield simple birefringent modulators with $V_\pi = 7$ volts. The DEC (double end crosslinkable) chromophore versions of these chromophores yield thermal stabilities comparable to those obtained for DEC disperse red chromophores (1).

The second chromophore (at 56% by weight loading into PMMA) of Figure 4 exhibits a increase in optical nonlinearity form 21 pm/V to 42 pm/V as a function of derivatization of the thiophene group with butyl groups.

Processing issues: The integration of polymeric modulators with silica fibers and VLSI electronic circuitry.

We have carried out extensive investigations of the sources of poling induced optical loss and of losses associated with the fabrication of buried channel waveguide structures and the intergration of polymeric modulator waveguides with silica optical fibers.

Poling induced losses can typically be associated with surface damage from the corona field or with phase separation (or aggregation) occuring during poling and hardening. Corona damage can be mimimized by maintaining appropriate material hardeness during poling. The use of stepped electric field/poling temperature profiles during poling and thermosetting processing is very effective in reducing poling induced surface damage and phase separation.

To avoid losses associated with phase separation during thermosetting lattice hardening (to lock-in poling-induced order) reactions, it is important to maintain a stoichiometric ratio of reacting components (DEC chromophores and crosslinking reagents). Maintenance of stoichiometric ratios, of course, means avoiding side reactions such as with atmospheric water. When appropriate attention is given to these issures, optical loss on the order of 1 dB/cm (or even less can be realized). However, reactions with atmospheric water, corona surface damage, etc. In carelessly handled samples can easily lead to optical losses above 5 dB/cm.

With appropriate care optical losses due to wall roughness from pitting during reactive ion etching can be reduced to insignificant values (e.g., 0.01 dB/cm).

Recently, optical loss due to optical mode size mismatch between polymer EO waveguide modes and silica fiber modes have been dramatically reduced exploiting reactive ion etched tapered transitions (5). This achievement together with the demonstration of integration of polymeric electro-optic circuitry with VLSI semiconductor electronics (6) and very large polymeric modulator bandwidths (7) represent dramatic advances in the utilization of polymeric electro-optic materials.

References

1. Dalton, L. R.; Harper, A. W.; Ghosn R.; Steier, W. H.; Ziari, M.; Fetterman, H.; Shi, Y.; Mustacich, R.; Jen, A. K. Y.; Shea, K. J. *Chem. Mater.* **1995**, *7*, 1060.
2. Kanis, D. R.; Ratner, M. A.; Marks, T. J. *Chem. Rev.* **1994**, *94*, 195.
3. Dalton, L. R.; Harper, A. W.; Robinson, B. H. *Proc. Natl. Acad. Sci. USA* **1997**, *94*, 4842.

4. London, F. *Trans. Faraday Soc.* **1937**, *33*, 8.
5. Chen, A.; Chuyanov. V.; Marti-Carrera, F. I.; Garner, S.; Steier, W. H.; Chen, J.; Sun, S.; Dalton, L. R. *Proc. SPIE* **1997**, *3005*, 65.
6. Kalluri, S.; Ziari, M.; Chen, A.; Chuyanov, V.; Steier, W. H.; Chen, D.; Jalali, B.; Fetterman, H. R.; Dalton, L. R. *IEEE Photonics Tech. Lett.* **1996**, *8*, 644.
7. Chen, D.; Fetterman, H. R.; Chen, A.; Steier, W. H.; Dalton, L. R.; Wang, W.; Shi, Y. *Proc. SPIE* **1997**, *3007*, 314.

Chapter 12

Novel Side Chain Liquid Crystalline Polymers for Quadratic Nonlinear Optics

F. Kajzar[1], Gangadhara[2], S. Ponrathnam[2], C. Noël[2], and D. Reyx[3]

[1]LETI (CEA–Technologies Avancées, DEIN/SPE), CE Saclay 91191 Gif sur Yvette Cedex, France
[2]ESPCI, 10, rue Vauquelin, 75231 Paris Cedex 05, France
[3]Laboratoire de Chimie et de Physique des Matériaux Polymères UMR 6115, Université du Maine, 72017 Le Mans Cedex, France

The synthesis of a new class of polymaleimide - based side chain liquid crystal polymers for quadratic nonlinear optical (NLO) applications is reported. The active chromophore is a cyanobiphenyl. The thermal behavior was established by differential scanning calorimetry, optical microscopy and X-Ray diffraction. Depending on the length of the spacer and the chromophore concentration they exhibit smectic and/or nematic or liquide crystalline phases. The active chromophores were oriented by the standard corona poling technique and the degree of axial ordering was determined as a function of poling conditions by linear optical absorption coefficients The second order NLO susceptibility tensor component coefficients d_{31} and d_{33} were measured by the second harmonic generation technique A gain in net polar ordering was observed when starting with a liquid crystalline system. The ratio d_{33}/d_{31} was found to be much larger than 3 in agreement with the molecular statistical models for static electric field poling of liquid crystals. These new polymers are of particular interest: both the liquid crystalline character and the high glass transition temperatures (up to 160°C) contribute to the enhanced orientational stability of NLO active chromophores in these systems.

Side chain polymers (SCP's) with the required noncentrosymmetric structure induced by static electric field poling[1] or all optical poling[2] have emerged as a new class of interesting materials for applications in second - order nonlinear optics (NLO). Frequency conversion through the use of frequency doubling or parametric oscillators (OPO's), optical storage and optical image processing through the photorefractive effect, sensors and ultrashort electric pulse generation through the optical rectification effect are often invoked. SCP's combine the good optical quality of the usually σ-bond polymer chain with the high value of first hyperpolarizability β of charge transfer (CT) chromophore side groups. Even if the figures of merit of poled SCP's are now nearly on a par with or better than that of the reference single-crystal LiNbO$_3$ for applications such as electro-optic modulation, thus making possible a high modulation rate (in excess of 100 GHz[3,4]), there exist still possibilities to improve their NLO response as well as their thermal and temporal stability. It is worth noting that unlike inorganic ferroelectric crystals, where the electro-optic response is dominated by phonon contributions, the electro-optic effect in SCP's arises in the electronic structure of the individual chromophores, yielding large EO coefficients with little dispersion from DC to optical frequencies, low dielectric constants and fast response times. Poled polymers have been demonstrated to exhibit EO coefficients larger than that of LiNbO$_3$, coupled with a dielectric constant nearly an order of magnitude smaller. The low dielectric constant could lead to an improvement of more than a factor of 10 in the bandwidth - length product over current LiNbO$_3$ devices.

The quantity of interest for device applications is the macroscopic second order NLO susceptibility $\chi^{(2)}(-\omega_3;\omega_1,\omega_2)$; where ω_1,ω_2 are the angular frequencies of the two input beams and $\omega_3 = \omega_1 + \omega_2$ the frequency of the output beam. For a single crystal the relationship between the macroscopic $\chi^{(2)}(-\omega_3;\omega_1,\omega_2)$ susceptibilty and the microscopic (molecular) $\beta(-\omega_3;\omega_1,\omega_2)$ first hyperpolarizability is given by

$$\chi_{IJK}^{(2)}(-\omega_3;\omega_1,\omega_2) = \sum_n N^{(n)} \sum_{ijk} f_i^{(n)\omega_1} f_j^{(n)\omega_2} f_k^{(n)\omega_3} a_{iI}^{(n)} a_{jJ}^{(n)} a_{kK}^{(n)} \beta_{ijk}^{(n)}(-\omega_3;\omega_1,\omega_2) \quad (1)$$

where $N^{(n)}$ is is the number density of molecular species (n) and the f's are appropriate local field factors. For small molecules, $f^{(n)\omega}$ can be approximated using the well known Lorentz-Lorenz formula

$$f^\omega = \frac{\varepsilon^\omega + 2}{3} \quad (2)$$

where $\varepsilon^\omega = (n^\omega)^2$ is the dielectric constant and n^ω is the refractive index at optical frequency ω. The a's in Eq. (1) are the Wigner rotation matrices.

In polymeric materials, where the location and orientation of each molecule are not known, we must consider an average over the orientational distribution and the local field factors. The summation over individual molecules for the macroscopic second order NLO susceptibility, as required in single crystals, is replaced by a thermodynamic average

$$\chi_{IJK}^{(2)}(-\omega_3;\omega_1,\omega_2) = NF < \beta_{ijk}(-\omega_3;\omega_1,\omega_2) >_{IJK} \quad (3)$$

where $<\beta_{ijk}>_{IJK}$ is the IJKth component of the orientational average of the molecular first hyperpolarizability. For poled polymers functionalized with 1-D CT molecules and with point symmetry ∞mm, the number of independent non zero $\chi^{(2)}$ tensor elements, provided that Kleinman's conditions are satisfied, reduces to two. These can be simply written as:

$$\chi_{333}^{(2)} = \chi_{ZZZ}^{(2)} = NF\beta_{zzz} < \cos^3 \theta > \tag{4}$$

and

$$\chi_{311}^{(2)} = \chi_{ZXX}^{(2)} = NF\beta_{zzz} < \frac{1}{2}\cos \theta \sin^2 \theta > \tag{5}$$

where the xyz coordinates refer to the molecular coordinate system with the principal first hyperpolarizability tensor component, β_{zzz}, lying along the z - axis and the macroscopic axes are designated as 1,2,3 = X,Y,Z, the poling field being applied along the 3 (Z) direction assumed to be perpendicular to the thin film surface. θ is the angle between the molecular long z - axis and the poling field and the expressions in brackets are the polar ordering parameters. One can see from Eqs. (4)-(5) that in order to

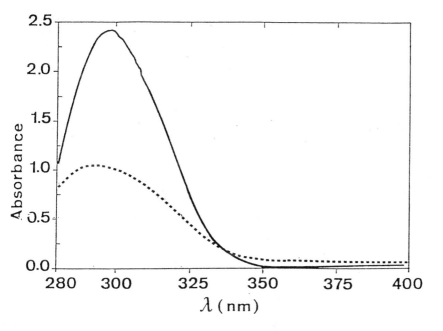

Fig. 1. Optical absorption spectra of a bicyanophenyl - based SCLCP thin film before (solid line) after (dashed line) poling

174

achieve maximum values for the $\chi^{(2)}$ components, the degree of polar order obtained during the poling process has to be optimized.

In this context liquid crystals (LC's) are of particular interest because of their properties of self organization. These give rise to systems with high axial order, which can be used to enhance the polar order as predicted by a number of molecular statistical models[5-8]. Of these materials side chain LC polymers seem most attractive because of their inherent tailorability - that is, the ability to alter material properties to fit specific needs such as: optical clarity, ease of processability, reasonable mechanical properties and easy coupling to an external electric field. Quite recently, Gonin et al[9,10] demonstrated experimentally that side chain LC polymers can exhibit a gain in polar ordering by a factor of 2.3-3.3 over the isotropic counterparts.

In subsequent studies[11-17] we reported a systematic investigation of new side chain LC polymers aimed at providing structure property relationships and reaching a better understanding of the effects of their LC behavior on their polarization, NLO properties and temporal stability. The polymers studied each have the same cyanobiphenyl-based active chromophore. It should be noted that the 4-cyanobiphenyl group is an excellent candidate for blue conversion. Its main advantages lie in a wide transparency range (cf. Fig. 1) with an optical absorption cut-off at ca. 350 nm and a noticeable first hyperpolarizability $\beta = (16\text{-}20)\text{x}10^{-30}$ esu[18]. This chromophore was attached directly or through spacers to different polymer backbones such as polyepichlorohydrin (PECH)[9], polybischloromethyloxetane (PBCMO)[12,13], polyacrylate (PA)[10,11], polymethacrylate (PMMA)[10,11] and poly-p-chloromethylstyrene (PCMS)[16]. Depending on the stiffness of the backbone, the length and flexibility of the spacer and the chromophore content, the resulting polymers were isotropic or nematic. Thin films of selected polymers were oriented by a corona discharge poling process and the second order NLO properties were studied by second harmonic generation (SHG). First, we showed that both $\chi^{(2)}_{ZZZ}$ and $\chi^{(2)}_{ZZZ} / \chi^{(2)}_{ZXX}$ track with the axial order parameter, $<P_2>$, as predicted by the molecular statistical models[5-8]. We were able to obtain $\chi^{(2)}_{ZZZ}$ values up to 35.6 pm/V at $\lambda = 1.064$ μm, thus exceeding many previously reported for off-resonant SHG measurements on isotropic side chain functionalized polymers[19]. We also found an enhanced temporal stability. The second - order nonlinearities showed no significant decay at ambient conditions for the polymers exhibiting glass transition temperatures (T_g) higher than 100°C. This result was attributed to peculiar features of the molecular dynamics of side chain LC polymers, namely the orientational motions of the chromophores about the polymer backbone (δ - mode)[15,20]. These motions are highly cooperative and require more free volume than that characteristic of the main chain segmental motions (α - mode). The temperature at which they cease to exist has been estimated to be $T_g - T_o(\delta) \approx 56$ K[21]. However, the glass transition temperatures of the side chain LC polymers we have synthesized so far are lower than ca. 140°C, thus limiting their thermal stability. This is a severe limitation because considerations of potential use must acknowledge certain thermal requirements.

In this paper we describe our progress towards the development of NLO side chain polymers exhibiting LC behavior and high T_g. Polymers with maleimide units along the backbone have been chosen because it is well known that high dipole moment of the maleimide unit at right angle to the backbone and its cyclic structure impart great stiffness to polymer chain, thus resulting in high T_g. The cyanobiphenyl - based chromophore was attached to the polymer backbone through oligomethylene spacers in order to decouple the chromophores from the main chain and thus to enhance the possibility of self organization into an anisotropic mesophase.

Experimental

Materials: p-(4'-Cyanobiphenyl-4-yloxy)-methylstyrene (**CBMS**)[16] , poly[p-(4'-cyanobiphenyl-4-yloxy)methylstyrene] (**PCBMS**)[16], exo-7-oxabicyclo[2,2,1] hept-5-ene-2,3-dicarboximide[22], n-(4'cyanobiphenyl-4-yloxy)alkan-1-ols[23] (**CBA$_n$**) (n=2-8) and poly(styrene - alt - maleimide)[24] were prepared as described in the literature.

Synthesis of monomers : N-[n-(4'-cyanobiphenyl-4-yloxy)alkyl]maleimides (**M$_n$**) were prepared via the Mitsunobu reaction previously described for the condensation of alcohols with maleimide[25] (yield : 90-95%) and subsequent deprotection of the maleimide double bond by a retro-Diels-Alder reaction (yield : 80-95%) as outlined in Scheme 1.

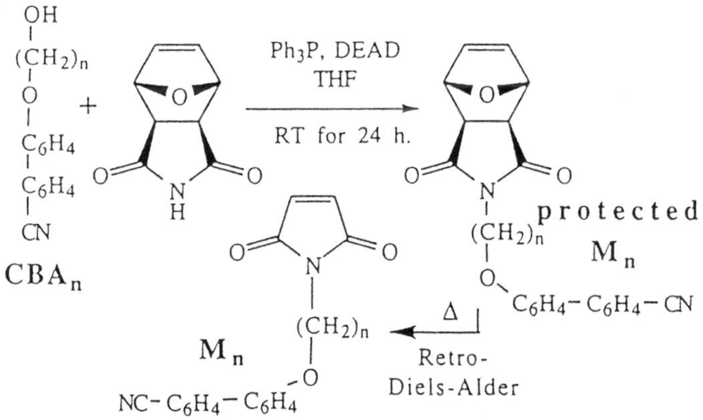

Scheme 1

Synthesis of polymers: Poly {N-[n-(4'-cyanobiphenyl-4-yloxy) alkyl] maleimide }s (**PM$_n$**) and poly[p-(4'-cyanobiphenyl-4-yloxy) methylstyrene-co-{N-[3-(4'-cyanobiphenyl-4-yloxy)propyl] maleimide}] [**P(CBMS-co-M$_3$)**] were obtained by free radical polymerization (AIBN, 2 mol%) in DMSO ([monomer]=0.6 mol/L) under vacuum in sealed tubes at 70°C for 48 h.

Poly [styrene - alt - {N - [3-(4'-cyanobiphenyl-4-yloxy) propyl] maleimide }] [**P(S-alt-M₃)**)] was obtained using 50% excess of both the alcohol and the Mitsunobu reagent. Diethyldicarboxylate (DEAD) in dry DMF (0.3 mol/L) was added dropwise to a DMF solution of poly(styrene - alt - maleimide) ([maleimide unit] = 0.1 mol/l), **CBA₃** and triphenylphosphine (**Ph₃P**) maintained at 0°C. The reaction mixture was stirred at room temperature for 48h and then concentrated The polymer was precipitated into methanol and purified by further reprecipitation into methanol

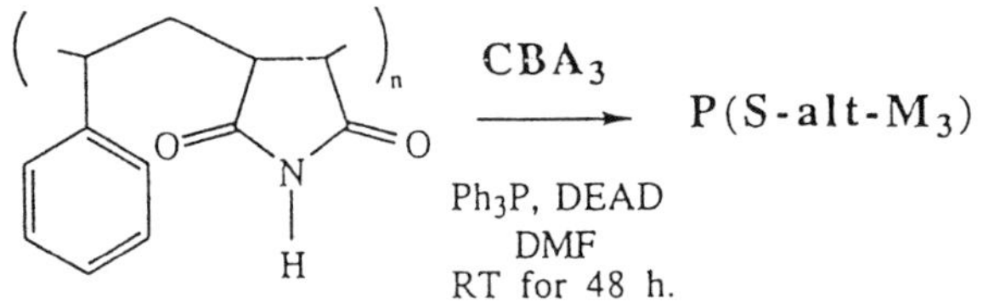

The polymers had all the expected IR and NMR characterizations.

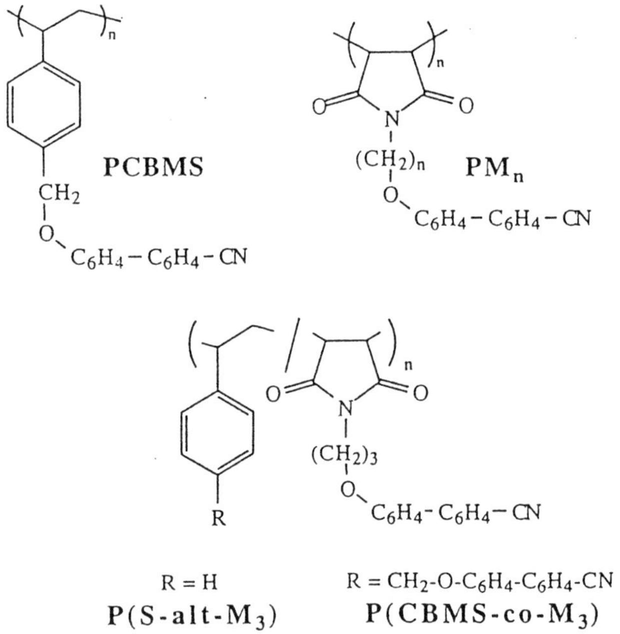

Scheme 2

Techniques : The transition temperatures were measured using a differential thermal analyzer (Du Pont 1090) operating at 20°C/min. Optical observations were made using a polarizing microscope (Olympus BHA-P) equipped with a Mettler FP52 hot-stage and a FP5 control unit. UV spectra were recorded using an Uvikon 810 spectrometer. Polymeric films were prepared by spin-coating solutions (ca. 60g/L) of the polymers in either 1,1,2,2 - tetrachloroethane or NMP/1,1,2,2 - tetrachloroethane mixture onto cleaned glass plates followed by the solvent evaporation *in vacuo*. The standard corona poling technique (cf. Fig. 2) was used to orient the chromophores. The second order nonliner optical properties of poled films were studied by the transverse second harmonic generation (SHG) technique.

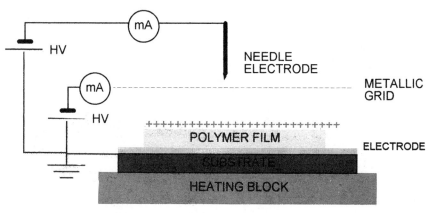

Fig. 2. Schematic representation of the corona poling apparatus.

RESULTS AND DISCUSSION

Thermal behavior

There is no real evidence for LC properties in **PCBMS** (T_g = 132°C) and **P(S-alt-M₃)** (T_g = 130°C). Microscopic observations show that these polymers give a clear isotropic phase above T_g. The present results suggest that **PCBMS** should be considered to be polyethylene with the benzyl ether of 4 - hydroxy-4'-cyanobiphenyl as the chromophore directly attached to the backbone instead of polystyrene with the 4 - cyanobiphenyl unit linked to the backbone through a - O - CH₂ spacer. Without the presence of a flexible spacer a LC phase cannot form at all because the backbone restricts the orientation of the mesogenic side - chain units. It should be noted , however, that by pressing the cover glass of the microscopic preparation, transient birefringence appears in the sample (cf. Fig. 3). This effect is observable over 20°C, although it becomes weaker and weaker as the temperature is raised. It may be due to polymer alignment under press.

In **P(S-alt-M₃)** the molar ratio of mesogenic to non-mesogenic units is clearly too low for LC phase formation. Indeed, it is well known that polymers lose their mesogenic behavior below a minimum mesogen concentration, which depends on the nature of both monomers. We showed that PECH[11] and PBCMO[12,13] require a minimum 4-bicyanophenyl - based mesogen concentration of about 65 mol % to form a nematic phase.

P(CBMS - co - M₃) has a glass transition temperature T_g of 142°C and forms a nematic phase over a broad temperature range ($T_{NI} \approx$ 220°C)(cf. Fig. 4). Except the functionalized polymaleimide **PM₂**, with spacer of only two methylene units, which is

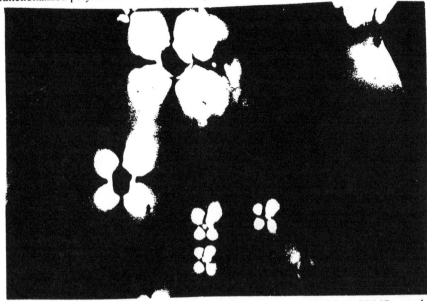

Fig. 3. Transient birefringence obtained under pressure at 145°C in PCBMS sample. Crossed polarizers.

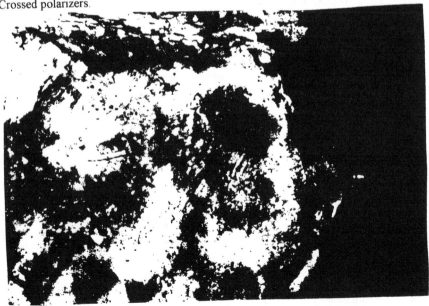

Fig. 4. Nematic phase of P(CBMS - co - M₃) at 200°C. Crossed polarizers.

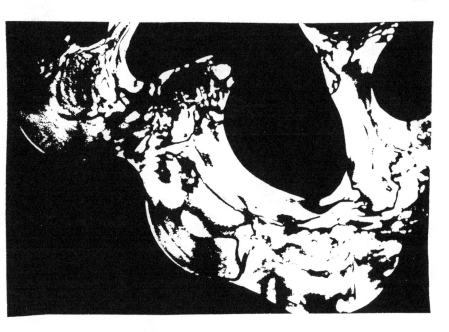

Fig. 5. Nematic phase of PM3 at 260°C. Crossed polarizers.

crystalline, all the synthesized polymaleimides PM_n (n = 3 - 8) exhibit a smectic and a nematic phase (cf. Fig. 5). Up to now the smectic phase remains unidentified. It should be noted that the S/N and N/I transition temperatures show even-odd effects, the transition temperatures being higher for odd spacers (cf. Table 1). For short spacers, the glass transitions were observed in the temperature range 110 - 150°C.

Table 1

Dependence of transition temperature on the spacer length for PM_n.

Spacer length (n)	Transition temperature (°C)[a]
2	T_g = 150, T_m = 240, T_{NI}>300[b]
3	G 130 S 195 N 280 I
4	G 115 S 155 N 225 I
5	G 75 S 180 N 265 I
6	G 76 S 115 N 136 I
7	G 61 S 172 N 220 I
8	G 43 S 120 N 185 I

a - from DSC analysis
b - G: glassy state, S: smectic phase, N: nematic phase, I: isotropic liquid

Electric field - induced alignment

Theoretical background. Placed in the poling field, E_p, the chromophores will tend to lower their energy by rotating so that the dipole axis lies parallel to the field. This will lead to both axial and polar ordering. The level of orientation can be described by the order parameters, $<P_l>$, defined as

$$< P_l >= 2\pi \int_0^\pi G(\theta) P_l(\cos\theta) \sin\theta d\theta \tag{6}$$

where θ gives the deviation of the molecular long axis from the field direction, the $P_l(\cos\theta)$ are Legendre polynomials and $G(\theta)$ is the orientational distribution function. The odd order parameters $<P_{2n+1}>$ characterize the polar order while the even order parameters give the degree of axial ordering. Usually, $<P_2>$ and $<P_4>$ are sufficient to describe the state of axial order in uniaxial systems. They are given by

$$< P_2 >= \frac{1}{2} < 3\cos^2\theta - 1 > \tag{7}$$

and

$$< P_4 >= \frac{1}{8} < 35\cos^4\theta - 30\cos^2\theta + 3 > \tag{8}$$

In Boltzmann statistics, the orientational distribution function is given by

$$G(\theta, E_p) = \frac{e^{\frac{U(\theta)}{kT_p}}}{2\pi \int_0^\pi e^{\frac{U(\theta)}{kT_p}} \sin\theta d\theta} \tag{9}$$

where $U(\theta)$ is the molecular energy in the electric field, T_p is the poling temperature and k the Boltzmann constant. The equilibrium state reached after a certain poling time, t_p, willd depend on the value of the molecular energy relative to that of the thermal disordering energy, kT_p.

In conjugated rod type molecules, such as 1-D CT molecules, there is greater opportunity for electron transfer along the molecule than normal to it, so that the molecules are characterized by a strong anisotropy in the polarizability, $\Delta\alpha$, and the orientation process has two components. The dominant contribution is from the interaction of the applied field with the permanent dipole moment, μ_0, which leads to polarization through molecular rotation. The corresponding energy term

$$U_{E_p}(\theta) = -\mu_0 E_p \cos\theta \tag{10}$$

will be lowered when the dipole axis (i.e., the rod axis for 1-D CT molecules) lies parallel to the field with the positive end facing the negative electrode and *vice-versa*. The second contribution results from the formation of an induced dipole which will give rise to molecular orientation by virtue of the anisotropy in its polarizability. The θ - dependent part of the energy term connected with this induced dipole moment

$$U_i(\Theta) = -\Delta\alpha E_p^2 P_2(\cos\theta) / 3 \tag{11}$$

in which $P_2(\cos\theta)$ is the second-order Legendre polynomial $(P_2(\cos\theta) = \frac{1}{2}(3\cos^2\theta - 1)$ will be lowered whichever way up the molecule is ($\theta = 0$ or π). It is worth noting that the first process will lead to polar orientation of the chromophores while the second will lead to axial ordering.

So far we have assumed the medium to be isotropic in character before poling. It should be noted, however, that the factors which enhance the second-order NLO activity in efficient chromophores (high axial ratio, planarity, charged end groups, ...) are essentially those factors which encourage liquid crystallinity in small - molecule materials. In other words, in concentrated systems, CT conjugated rod molecules should experience a certain tendency towards mutual axial alignment or even possess mesogenic properties themselves as observed for the 4 - cyanobiphenyl derivatives.. Such a CT rod - shaped molecule, placed in an environment of neighbors already axially ordered (director along the Z-axis), will have a tendency to align with the director. Its energy will be a function of its orientation with respect to the director and of the degree of preferred orientation of all the molecules expressed as the order parameter, $<P_2>$. According to Maier and Saupe[26,27] the orientation dependent part of its potential energy can be written as

$$U_O(\Theta) = -\zeta < P_2 > P_2(\cos\theta) \tag{12}$$

where ζ is a constant which scales the magnitude of the interaction. It will be lowered when the chromophore is parallel to the director. Bearing in mind that the LC director is a nonpolar axis resulting from gross alignment due to highly directional (non dipolar) Van der Waals interactions at the molecular level, this will lead to axial ordering.

The molecular statistical models[5-8] developed so far for describing electric field poling of NLO molecules have the same expression for the energy term connected with the permanent dipole moment in the electric field, $U_{E_p}(\theta)$ (cf. Eq. (10)). Only the model by Van der Vorst and Picken[7,8] takes into account the θ - dependent part of the energy term for the linearly induced dipole moment, $U_i(\theta)$ (cf. Eq. (11)). The models differ in their choice of $U_0(\theta)$. The isotropic model assumes no initial axial order ($<P_2>$ = $<P_4>$ = 0) while the Ising model assumes perfect axial order ($<P_2>$ = $<P_4>$ = 1) before the application of the field. Singer, Kuzyk and Sohn[6] do not introduce an additional energy term. They consider that the intrinsic axial order of the LC « host » is transferred to the NLO « guest » molecules. The axial ordering existing independently of the poling field is specified by the order parameters $<P_2>$ and $<P_4>$

which must be determined experimentally at zero field strength. By contrast, Van der Vorst and Picken[7,8] take explicitly into account the intrinsic mesogenic properties of the NLO molecules themselves through the effective single particle energy $U_0(\theta)$ originally used by Maier and Saupe[26,27] in their theory for nematic LC's. The expressions for $<P_2>$, $<\cos^3\theta>$ and $<\cos\theta\sin^2\theta>/2$, as obtained in the four models are given in Table 2.

Table 2.

Orientational averages $<\cos^3\theta>$, $<\cos\theta\sin^2\theta>/2$ and $<P_2>$ in the four statistical models

Model	$<\cos^3\theta>$	$<\sin^2\theta\cos\theta>/2$	$<P_2>$
Ising	u^a	0	1
Isotropic	$u/5^a$	$u/15$	$\approx u^2/15$
SKS	$u(\dfrac{1}{5}+\dfrac{4}{7}<P_2>+\dfrac{8}{35}<P_4>)$	$u(\dfrac{1}{15}+\dfrac{1}{21}<P_2>-\dfrac{8}{70}<P_4>)$ No analytical formula[cc]	b)
MSVP	No analytical formula[cc]		c)

a) $u = \mu_0 E_p/kT$.
b) The order parameters $<P_2>$ and $<P_4>$ are assumed to be independent on the electric field strength and are to be determined at $E_p = 0$.
c) The order parameter $<P_2>$ has to be calculated selfconsistently by numerical methods.

Development of axial order
Changes in the optical absorption spectra were used to probe the development of axial order as a function of poling time. If the primary electronic transition moment is directed along the molecular axis, then the absorbance will decrease as the molecules align parallel to the poling field, i.e. perpendicular to the film surface (cf. Fig. 1).) The absorbance will decrease because the average number of chromophores with transition moment perpendicular to the plane of the electric vector has increased. In the transmission mode, $<P_2>$ is given by

$$< P_2 >= 1 - \frac{A_\perp}{A_0} \tag{13}$$

where A_0 and $A_\perp$ are the absorbances of the film before and after poling.
    For the isotropic polymers, **PCBMS**[16] and **P(S - alt - M$_3$)**, the axial order develops rapidly as soon as the kinetic limitations on the polymer mobility are relieved, i.e. as soon as T_g is reached. As shown in Fig. 6, the order develops rapidly at first

with time and then levels off toward a maximum value $<P_2>_{max}$. The saturation limits $<P_2>_{max}$. lie in the range 0.3 - 0.35 at temperatures close to T_g. These data compare

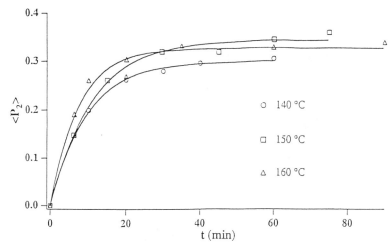

Fig. 6. Development of orientation with time in P(S-alt-M₃). Applied voltage 6kV.

nicely with values reported for other isotropic NLO functionalized polymers[28]; It should be noted, however, that above a critical poling temperature (ca. T_g + 20°C) the plateau value falls. Such a behavior is expected from the consideration of equilibrium between the competing rates of alignment due to the directing electric field and thermal randomization.

As already noted for other side chain LC polymers[11], the as - prepared films obtained from the LC polymers P(CBMS-co-M₃) and PM₃ are not birefringent. Upon spin-coating from isotropic solutions, the polymer chains are frozen in a disordered, optically isotropic state because of the rapid removal of the solvent. In this metastable glassy state, mesogenic chromophores are restricted from ordering to the thermodynamically favored LC phase by the kinetic limitations imposed on polymer chain mobility at room temperature. Once the temperature is above T_g, the polymer is sufficiently mobile to rearrange to a more ordered state, and as poling time and/or poling temperature are increased the sample morphology evolves towards the stable LC phase. Therefore, it is necessary to pole these systems at temperatures above T_g if we want to derive benefit from their LC character.

As shown in Fig. 7, for the nematic polymer P(CBMS-co-M₃) (order parameters up to 0.55 could be obtained. This value is larger than those reported in the present work for PCBMS) and P(S-alt-M₃) and in the literature for other isotropic side-chain functionalized NLO polymers. Clearly, heating as-spin coated films above T_g relieves the kinetic constraints on the cooperative motions of the polymer chains, thus allowing nematic order to develop. In the presence of the poling field, the nematic phase is formed aligned, i.e. free from domain - induced light scattering. The development of axial order could be described by a single exponential

184

$$< P_2 >=< P_2 >_{max} \left[1 - e^{-\frac{t}{\tau}} \right] \qquad (14)$$

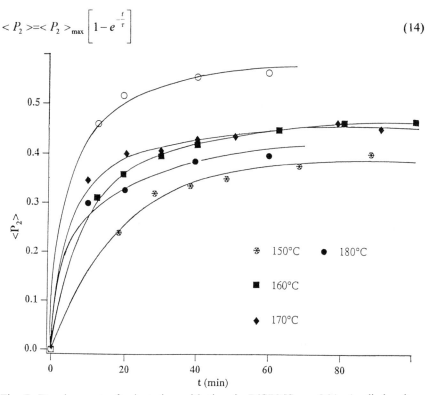

Fig. 7. Development of orientation with time in P(CBMS-aco-M₃). Applied voltage 6kV (open circles correspond to experimentl data obtained at 160°C with U_p = 7 kV).

where τ is a constant characterizing the rate of orientation for a given set of poling conditions. It decreases rapidly above T_g. The value of $<P_2>_{max}$, the plateau level of orientation reached after long times, decreases as the poling temperature increases. This behavior was predicted by the mean field theories of nematics[26,27,29]. These indicate that $<P_2>$ in the nematic phase decreases with increasing values of the reduced temperature $T^* = T/T_{NI}$ until it reaches a critical value $<P_2>_{NI}$ when the phase transforms to the isotropic one. At the transition an average value of $<P_2>_{NI}$ appears to be approximately 0.3 for side chain polymers[30]. The saturation limit $<P_2>_{max}$ is also sensitive to the strength of the poling field used. As shown both theoretically and experimentally, when a LC is subjected to a strong electric field the nematic/isotropic transition is shifted to higher temperature. The stronger the electric field, the higher will be T_{NI}. As indicated before, this will correspond to lower values of $T^* = T_p/T_{NI}$. and, as consequence, to larger values of $<P_2>_{max}$. Interestingly, it was found that specimens which have been held in the field for short periods subsequently relaxed to a state with a lower value of $<P_2>$ when the field was removed while those held for longer times retained the perfection of axial order in the glassy state.

Similar results were obtained for the LC polymer, PM3, when poled at and above T_{SN} (i.e. in the temperature range corresponding to the thermal stability range of the nematic phase) and slowly cooled to room temperature in the presence of the poling field (cf. Fig. 8). If, however, the specimens were poled at temperature between T_g and T_{SN}, then the transformation from the metastable isotropic state to the thermodynamically favored smectic state led to the formation of small birefringent smectic domains of different orientations as could be easily observed by polarizing microscopy. In this case, the measured order parameter $<P_2>$, i.e. the order parameter of the molecules with respect to the external « global » director (the direction of the poling field) is given by the simple convolution of the order parameter, $<P_2>_m$,

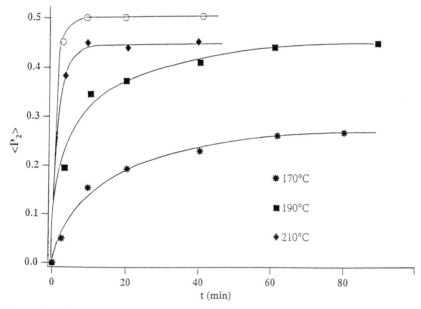

Fig. 8. Development of orientation with time in PM3. Applied voltage 6kV (open circles correspond to experimentl data obtained at 210°C with $U_p = 7$ kV).

characterizing the quality of the molecular orientation within each domain with respect to the director for that domain and the order parameter, $<P_2>_d$, characterizing the orientation distribution of the domain directors about the global director

$$< P_2 > = < P_2 >_m \bullet < P_2 >_d \tag{15}$$

where $< P_2 >_m$ is typical of the liquid crystalline phase, and, in the case of many polymeric smectic systems, higher than 0.7-0.8. Thus, the low degree of preferred orientation determined on a global scale ($< P_2 >_{max} \approx 0.25$-0.3) results from the orientation distribution of the domain directors. From the present data, $< P_2 >_d$ should lie between 0.3 and 0.4. Compared to nematics, the much more rigid smectics require considerably higher voltages for their deformation.

Second order NLO properties.

The evaluation of the NLO properties of the synthesized polymers is in its early stages, but some results can be reported. The NLO properties, imparted by corona poling, were determined by SHG in transmission at the fundamental wavelength of 1.064 μm. Under these experimental conditions, the absorption of the polymers at the harmonic wavelength (0.532 μm) is negligible and the values of the SHG coefficients d_{33} (= $\chi^{(2)}_{333}/2$) and d_{31}(= $\chi^{(2)}_{311}/2$) given in Table 3 are essentially not resonance enhanced. For the isotropic polymer PCBMS[16], the values obtained at the plateau level of orientation (<P_2> = 0.3) for d_{33} and d_{31} were 12.8 and 2.13 pm/V, respectively. These data give a ratio d_{33}/d_{31} of 6 , which exceeds the theoretical value of 3 predicted for isotropic systems (cf. Table 2).

Table 3

Poling conditions and NLO coefficients for a number of samples (λ = 1.064 μm).

Polymer	Film thickness (A°)	U_p(kV)	T_p(°C)	t_p (min)	<P_2>	d_{33} pm/V	d_{31} pm/V	a=d_{33}/d_{31}
P(S-alt-M$_3$)	2040	6	140	60	0.3	13.06	2	6.5
	2600	6	140	90	0.33	24.15	1.97	12.2
	1980	6	150	75	0.36	19.6	22.11	9.3
P(CBMS-co-M$_3$)	2500	6	150	110	0.4	17.9	1.53	11.7
	1500	6	160	110	0.52	30.5	2.23	13.6
	1400	6	170	80	0.48	25	1.89	13.2
PM$_3$	1500	6	210	40	0.25	15.6	1.67	11.7

One of the assumptions made in the isotropic model, that is the chromophores can rotate freely, may be responsible for this discrepancy. When the NLO chromophores are linked to the polymer backbone either directly or through short spacers, the motions of both the main chain and the side groups are coupled. This results in conformational constraints on chromophore ordering. Herminghaus et al.[31] and Robin et al[32]. developed simple models which take into account the effects of such conformational constraints and yield d_{33}/d_{31} ranging from 3 to 6. Similarly, d_{33} and d_{31} were found to be 13.06 and 2 pm/V, respectively, for the **P(S - alt - M$_3$)** films characterized by an order parameter <P_2> of about 0.3. However, as <P_2> increases, so do d_{33}/d_{31}. Clearly for the range 0.3<<P_2><0.4 one obtains systems on the borderline between isotropic and LC behavior. According to the Maier - Saupe theory[26,27] for low molar mass nematics, <P_2> changes discontinuously from zero at the isotropic/nematic transition to 0.429. However, Luckhurst[29] has shown that the Maier - Saupe theory results in an overestimate of about 25% in <P_2>$_{NI}$. In addition, comparison of the temperature dependence of <P_2> for LC monomers and corresponding side chain polymers has demonstrated that the absolute values of <P_2> related to the reduced temperature T/T_{NI} are always lower for the polymers. As already mentioned, a good estimate of $<P_2>_{NI}$ appears to be approximately 0.3 for side chain polymers[30].

In agreement with the theoretical predictions[5-8], the results listed in Table 3 show that liquid crystallinity makes higher anisotropy for the second order susceptibility tensor $\chi^{(2)}$ since d_{33} is much larger than $3d_{31}$ for all the **P(CBMS-co-M₃)** samples investigated. As already observed for other 4 - cyanobiphenyl - based nematic side chain polymers[10,11,13,14], both d_{33} and d_{33}/d_{31} increase as $<P_2>$ increases. d_{33} values range from 18 to 30 pm/V, the smaller value determined at $<P_2> = 0.4$ and the larger at $<P_2> = 0.58$. Taking into account the results obtained for the wavelength dispersion of the SHG coefficients of 4-cyanobiphenyl - based nematic copolyethers[9], it should be possible to reach values of d_{33} in the range 30-50 pm/V at 0.777 μm.

Until now we did not succeed in obtaining order parameters higher than 0.25 while maintaining good quality PM3 samples. This may be due to conductivity problem at the high temperature required to pole the system in the thermal stability range of the nematic phase. Further purification of the polymer is needed to minimize the high temperature film conductivity. Nevertheless, a d_{33} value of 15.3 pm/V was measured.

Preliminary results indicate that attachment of the 4-cyanobiphenyl based chromophore to the polymaleimide backbone gives systems that retain the majority of imparted polarization if they have been held in the poling field for long periods. Typically, at ambient conditions, changes of the SH coefficients were less than 10% over 50 days.

Acknowledgments

This research was supported by the Indo-French Center for the Promotion of Advanced Research, through Research project n° 1008 - 1.

References

1. Mortazavi M. A., Knoesen A., Kowel S. T., Higgins B. G. and Dienes A., *J. Opt. Soc. Am.* B, **6**, 733(1989).
2. Charra F., Kajzar F., Nunzi J. M., Raimond P. and Idiart E., *Opt. Lett.*, **18**, 941(1993).
3. Dalton L. R., Harper A. W., Wu B., Ghosn R., Lanquindanum J., Liang Z. and Xu C., *Adv. Mater.*, **7**, 519(1995).
4. Dalton L. R. and Harper A. W., *Photoactive Organic Materials for Electro-Optic Modulator and High Density Optical Memory Applications*, in « *Photoactive Organic Materials. Science and Application* », F. Kajzar, V. M. Agranovich and C. Y.-C. Lee Eds, NATO ASI Series, vol. **9**, Kluwer Academic Publishers, Dordrecht 1996, pp.183-198.
5. Meredith G. R., Van Dusen J. G. and Williams D. J., *Macromolecules*, **15**, 1385(1982).
6. Singer K. D., Kuzyk M. G. and Sohn J. E., *J. Opt. Soc. Am. B*, **4**, 968(1987).
7. Van der Vorst C. P. J. M. and Picken S. J., *Proc. SPIE*, **866**, 99(1987).
8. Van der Vorst C. P. J. M. and Picken S. J., . *J. Opt. Soc. Am.* B, **7**, 320(1990).

188

9. Gonin D., Noël C., Le Borgne A., Gadret G. and Kajzar F., *Makromol. Chem., Rapid Commun.*, **13**, 537(1992).

10. Gonin D., Guichard B., Noël C. and Kajzar F., *Macromol. Symp.*, **96**, 185(1995).

11. Gonin D., Guichard B., Noël C. and Kajzar F., « *Highly Efficient Liquid Crystal Polymers for Quadratic Nonlinear Optics* », in « *Polymers and Other Advanced Materials : Emerging Technologies and Business Opportunities* », P. N. Prasad et al Ed.s, Plenum Press, New York 1995, pp. 465-483.

12. Guichard B., Noël C., Reyx D. and Kajzar F., *Macromol. Chem. Phys.*, **197**, 2185(1996).

13. Guichard B., Poirier V., Noël C., Reyx D., Le Borgne A., Leblanc M., Large M. and Kajzar F., *Macromol. Chem. Phys.*, 197, 3631(1996).

14. Gonin D., Guichard B., Large M., Dantas de Morais T., Noël C. and Kajzar F., *J. Nonl. Opt. Phys. and Mat.*, **5**, 735(1996).

15. Dantas de Morais T., Noël C. and Kajzar F., *Nonlinear Optics*, **15**, 315(1996).

16. Gangadhara, Noël C., Ching K. C., Large M., Reyx D. and Kajzar F., *Macromol. Chem. Phys.*,**198**, 1665(1997).

17. Noël C. and Kajzar F., Proceedings of the Fourth International Conference on Frontiers of Polymers and Advanced Materials , Cairo, January 4-9, 1997, P. N. Prasad, J. E. Mark, S. H. Kandil and Z. Kafafi Eds, Plenum Press, New York, in press..

18. Gonin D., Noël C. and Kajzar F., *Nonlinear Optics*, **8**, 37(1994).

19. Firestone M. A., Park J., Minami N., Ratner M. A., Marks T. J., Lin W. and Wong G. K., *Macromolecules*, **28**, 2247(1995).

20. Gonin D., Noël C., and Kajzar F., in « *Organic Thin Films for Waveguiding Nonlinear Optics: Science and Technology* », F. Kajzar and J. Swalen Eds.; Gordon & Breach Sc. Publ., Amsterdam 1996, p. 221 - 288.

21. Moscicki J. K., in « *Liquid Crystal Polymers: from Structures to Applications* » A. A. Collyer Ed., Elsevier Applied Science, London 1992, Chapter 4, pp. 143-236.

22. Kwart H. and Burchuk I., *J. Am. Chem. Soc.*, **74**, 3094(1952).

23. Percec V. and Lee M., *Macromolecules*, **24**, 1017(1991).

24. Abayasekara D. R. and Ottenbrite R. M., *Polymer Prep.*, **26**, 285(1985).

25. Walker M. A., *Tetrahedron Letters*, **35**, 665(1994).

26. Maier W. and Saupe A., *Z. Naturforsch.*, **14a**, 882(1959).

27. Maier W. and Saupe A.., *Z. Naturforsch.*, **15a**, 287(1960).

28. Xie H. - Q, Huang X. - D. and Guo J. - S., *Polymer*, 37, 771(1996).

29. Luckhurst G. R., « *Molecular Field Theories of Nematics* », in « *The Molecular Physics of Liquid Crystals* », G. R. Luckhurst and G. W. Gray Eds., Academic Press, London 1979, pp. 85-119

30. Finkelmann H. and Rehage G., *Adv. Polym. Sci.*, **60-61**, 99(1984)

31. Herminghaus S., Smith B. A. and Swalen J. D., *J. Opt. Soc. Am.* B, **8**, 2311(1991).

32. Robin P., Le Barny P., Broussoux D., Pocholle J. P. and Lemoine V., in « *Organic Molecules for Nonlinear Optics and Photonics* », J. Messier, F. Kajzar and P. N. Prasad Eds., Kluwer Academic Publ., Dordrecht 1991, p. 481.

Chapter 13

Dielectric Relaxation and Second-Order Nonlinearity of Copolymethacrylates Containing Tolane-Based Mesogenic Groups

Ging-Ho Hsiue[1], Ru-Jong Jeng[2], and Rong-Ho Lee[1]

[1]Department of Chemical Engineering, National Tsing Hua University, Hsinchu 300, Taiwan, Republic of China
[2]Department of Chemical Engineering, National Chung Hsing University, Taichung 400, Taiwan, Republic of China

A series of copolymethacrylates with different contents of tolane-based mesogenic group have been synthesized. The mesogenic group content was characterized with ^{1}H-NMR. The phase behaviors were determined by the differential scanning calorimeter and optical polarizing microscopy. A smectic A phase was obtained when the mesogenic group content was increased up to 80 mol.%. Dielectric relaxation results indicates that the amplitude of the α-relaxation was suppressed significantly due to the self-alignment of the mesogenic group. The reduction of the molecular motion is beneficial to the enhancement of the temporal stability of effective second-harmonic coefficient for the polymer with a higher mesogenic group content. Moreover, the second harmonic coefficient is enhanced as the mesogenic group content increases. The self-alignment nature of liquid crystal phase is favorable for alignment of the NLO-active mesogenic group under an applied electric field and preserving such alignment after removal of the electric field. The relationship between thermal dynamic behavior and second-order nonlinear optical properties were also discussed.

Side chain liquid crystalline polymers (SCLCPs) exhibiting the desirable mechanical properties of polymer, and electro-optical properties of low molecular weight liquid crystal have been studied extensively (1-2). SCLCPs with excellent electro-optical properties have potential in electro-optical device applications, such as display, light valve, and memory devices. Moreover, SCLCPs with second-order nonlinearity have recently attracted attention because of self-alignment nature of liquid crystal (LC) phase (3-5). The self-alignment nature of LC phase is helpful for the alignment of the mesogenic group under an applied electric poling field and preserving such alignment after removal of the poling field.

SCLCPs are able to show second-order nonlinear optical (NLO) properties as their NLO-active mesogenic groups to be oriented in a non-centrosymmetric manner by an applied electric field. The NLO properties are determined by the structure of NLO-active mesogenic group (i.e. NLO chromophore), chromophore density in polymer, and the electric poling efficiency. Moreover, the poling efficiency is closely related to the ease of the reorientation of the mesogenic groups. The reorientation of the mesogenic groups is dependent on the chemical structure of SCLCPs. The SCLCPs containing a mesogenic group with large dipole moment, large intermolecular distance, and a flexible spacer are favorable for the electric poling process (4,6). Dielectric relaxation is an useful technique for studying the dynamic behavior of the polymers, since it is sensitive to the motions of the ground-state dipole moments of the NLO-active mesogenic groups. In addition, the poling efficiency was also determined by the poling conditions of an applied electric field. They are the poling temperature, poling time and strength of the poling field. The effect of poling condition on the poling efficiency have been studied for the amorphous polymers by Firestone et al. (7-8). Furthermore, the poling efficiency and temporal stability of second-order nonlinearity can be obtained by studying the relaxation behavior of the mesogenic groups. This includes the temperature dependence of the relaxation time, broadening parameter, and activation energy.

In this study, a series of copolymethacrylates containing different contents of tolane-based mesogenic groups have been synthesized. The phase behavior of this series polymers was characterized with the differential scanning calorimeter, and optical polarizing microscopy. Moreover, in order to study the relationship among the compositions, thermal dynamic behavior, and NLO properties of polymers, the relaxation behaviors of dielectric and second-harmonic (SH) coefficient were measured by using broadband dielectric relaxation spectroscopy and an in-situ second-harmonic generation (SHG) technique. Furthermore, relaxation behaviors were discussed in terms of the polymer compositions and temperature dependencies.

Experimental

The synthesis of the tolane-based mesogenic group (M6CN) has been previously reported (9). The copolymethacrylates containing tolane-based mesogenic group were prepared according to scheme 1. The free radical copolymerizations of the tolane-based monomers and methylmethacrylate were carried out in a Schlenk tube under nitrogen. The polymerization tube, which contained a chloroform solution of the monomer and initiator (AIBN), was degassed under vacuum, and finally filled with nitrogen. All polymerizations were carried out at 65 °C for 24 h. After the reaction time, the obtained polymers were precipitated in methanol, and purified by several reprecipitations from THF solutions into methanol. The content of side chain mesogenic groups was characterized with ^{1}H-NMR for the copolymethacrylates. Moreover, the molecular weights of the polymers were determined by a viscotek 200 GPC equipped with a differential refractometer and a viscometer.

The thermal transitions of polymers were determined by a differential scanning calorimeter (Seiko SSC/5200 DSC). The thermal transitions were read at the

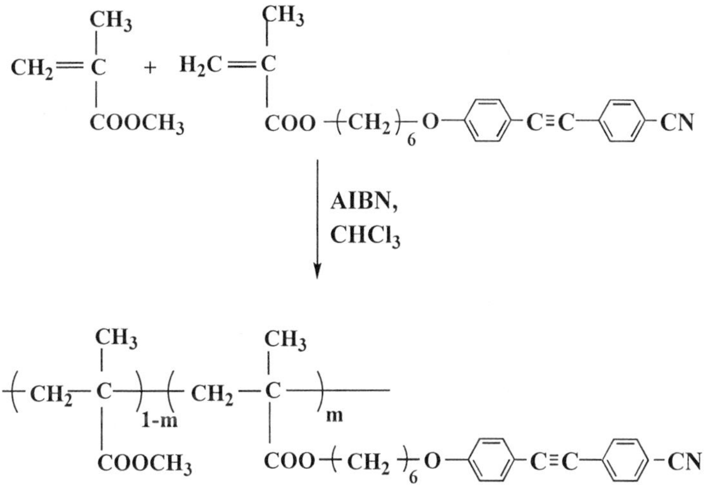

$$CH_2 = \overset{\overset{\displaystyle CH_3}{|}}{\underset{\underset{\displaystyle COOCH_3}{|}}{C}} \quad + \quad H_2C = \overset{\overset{\displaystyle CH_3}{|}}{\underset{\underset{\displaystyle COO-(CH_2)_6-O-\bigcirc-C\equiv C-\bigcirc-CN}{|}}{C}}$$

AIBN,
CHCl$_3$

$$+CH_2-\overset{\overset{\displaystyle CH_3}{|}}{\underset{\underset{\displaystyle COOCH_3}{|}}{C}}\Big)_{1-m} + CH_2-\overset{\overset{\displaystyle CH_3}{|}}{\underset{\underset{\displaystyle COO-(CH_2)_6-O-\bigcirc-C\equiv C-\bigcirc-CN}{|}}{C}}\Big)_{m} ---$$

m = 0.3 ~ 1.0

Scheme 1. Synthesis of copolymethacrylates containing tolane-based mesogenic group.

maximum of their endothermic or exothermic peaks. Glass transition temperature (T_g) was read at the middle of the change in heat capacity. Heating and cooling rates were 10 °C/min in all of these cases. The transitions were collected from the second heating and cooling scans. A Nikon Microphot-FX optical polarized microscope equipped with a Mettler FP82 hot stage and a Mettler FP80 central processor was applied toward observing anisotropic textures. Dielectric spectroscopy was determined on a Novercontrol GmbH. Measurements were performed by a Schlumberger SI 1260 impedance/gain-phase analyzer (frequency: $10^{-1} \sim 10^{6}$ Hz) and a Quator temperature controller. A nitrogen gas heating system ranged from -100 to 250 °C was used. The temperature was adjusted within the tolerance of $\pm$ 0.1 °C. The polymers were sandwiched between two parallel metal electrode plates with a spacer of 50 μm.

The polymer was dissolved in tetrahydrofuran (THF) for film preparation. Thin film was prepared by spin-coating the polymer solution onto indium tin oxide (ITO) glass substrates. The thickness and indices of refraction were measured by using a prism coupler (Metricon 2010). The poling process of the thin films was carried out using the in-situ poling technique. The details of the corona poling set-up was the same as reported earlier (10). The poling process was started at room temperature and then the temperature was increased up to the poling temperature (above T_g) with a heating rate of 10 °C/min. The corona current was maintained at 1~2 μA with a potential of 4 kV. Upon saturation of the SHG intensity, the sample was then cooled down to room temperature in the presence of the poling field. Once room temperature was reached, the poling field was terminated. Second harmonic generation measurements were carried out with a Q-switched Nd:YAG laser operating at 1064-nm. Measurement of the second harmonic coefficient, d_{33}, has been previously discussed (11), and the d_{33} values were corrected for absorption (12). Relaxation behavior of the poled sample was achieved by the in-situ SHG technique.

Results and Discussion

The compositions, molecular weights and polydispersity of the copolymers are summarized in Table I. The content of mesogenic groups in the copolymers was characterized with ^{1}H-NMR. Their molecular weights were found to be in the range of $1 \times 10^{4} \sim 1.5 \times 10^{4}$. Only amorphous phase was observed for the polymers with a lower mesogenic group content (< 65 mol. %). No LC phase was observed for the copolymers 30P, 50P, and 65P. T_g was decreased with increasing mesogenic group content, due to the plasticizer effect of the mesogenic groups (13). On the other hand, the crystalline phase was observed at temperatures below the LC phase for the polymers with a higher mesogenic group content (80P, 90P, and 100P) on the first heating scan. However, the melting point disappeared, and the glass transition was observed on the first cooling and second heating scans. Their glass transition zones were broad and the ΔC_P's (T_g) were smaller compared to those of the polymers with lower mesogenic group contents (30P, 50P, and 65P). This implies that the mobility of the polymer chains is reduced for the polymer with a higher content of the mesogenic group at temperatures close to T_g. In addition, an LC phase (smectic A)

Table I. Molecular weights, phase transitions ($^{\circ}$C) and corresponding enthalpy changes (J / g) of the copolymethacrylates.

| Polymer | m^a | GPC | | Phase transitions[b] $\left(\dfrac{\text{cooling scan}^c}{\text{heating scan}^d}\right)$ |
		$10^{-3}\,\overline{Mn}$	$\overline{Mw}\,/\,\overline{Mn}$	
30P	0.28	9.27	1.92	I 77.3 G G78.0 I
50P	0.47	14.6	2.22	I 66.5 G G 69.2 I
65P	0.63	13.0	1.94	I 64.2 G G 66.1 I
80P	0.78	10.5	1.92	* K 57.5 (3.8) S_A 81.4 (2.5) I I 80.7(2.0) S_A 57.9 G G 59.1 S_A 81.8 (2.4) I
90P	0.87	11.8	1.86	* K 54.1 (3.1) S_A 92.7 (2.9) I I 90.2(2.2) S_A 53.1 G G 54.0 S_A 93.7 (2.8) I
100P	1.00	10.1	1.61	* K 50.9 (2.6) S_A 111.4 (3.2) I I 108.1 (2.5) S_A 46.2 G G 47.0 S_A 111.8 (3.0) I

[a] m, according to the scheme.
[b] G, glass; S_A, smectic A; I, isotropic
[c] the data were obtained from the first cooling scans.
[d] the data were obtained from the second heating scans.
* the data were obtained from the first heating scans.

was observed for the copolymers 80P, 90P, and 100P. The temperature range and thermal stability of the LC phase were increased with increasing mesogenic group contents as shown in Table I. The texture of the LC phase was characterized by the optical polarized microscope.

Dielectric loss tangent (Tan δ) versus temperature and frequency for the polymers 50P, and 100P are shown in Figure 1. In Figure 1a, the α-relaxation associated with the glass transition was observed for the polymer 50P. The large amplitude of the α-relaxation was obtained due to the plasticizer effect of the mesogenic groups. Moreover, the δ-relaxation was observed at a higher temperature range, which corresponded to the rotation of mesogenic group around the polymer backbone. Similar phenomena were observed for the polymers 30P and 65P. On the other hand, a σ-relaxation was observed near the transition temperature of smectic A and isotropic phases for the polymer 100P (Figure 1b), in addition to the α- and δ-relaxations. This relaxation was possibly caused by the transition of LC and isotropic phases. Moreover, the amplitude of the α-relaxation was suppressed remarkably compared to that of the 50P. This implies that the molecular mobility of the polymer chains was reduced during glass transition for the polymer 100P. According to the literature (*14*), the α-relaxation is attributed to a combination of motions of the polymer main chain with the mesogenic group. In other words, the relaxation intensity is dependent on the dipole moment and molecular motions of the main chain backbone and side chain groups. For the polymer with mesogenic groups, the molecular motions of the main chain backbone and mesogenic groups were mutually affected via the flexible spacer (*4,15*). The self-alignment characteristic of mesogenic groups would reduce the mobility of the main chain backbone and mesogenic groups (*4,15*). Consequently, the suppression of the α-relaxation was obtained for the polymers with a higher mesogenic group content. Similar results were obtained for polymers 80P, and 90P.

When the relaxation frequencies were plotted as a function of the reciprocal temperature, a nonlinear "WLF" curve of the α-relaxation was obtained for polymers 30P, 50P, and 65P (Figure 2a). Moreover, the relaxation frequency was increased with increasing mesogenic group content, due to the increase of the plasticizer effect. The temperature dependence of the relaxation times has been described by the empirical Williams-Landel-Ferry (WLF) equation (*7-8,16*)

$$\text{Log } a_T = \text{Log } (\tau / \tau_R) = -C_1 (T - T_R) / [C_2 + (T - T_R)] \tag{1}$$

where a_T is the WLF shift factor, τ is the apparent relaxation time at temperature T, τ_R is the apparent relaxation time at the reference temperature T_R, and C_1 and C_2 are the WLF constants. The constants C_1 and C_2 can be obtained from the WLF plot of shift factors [(Log a_T) versus $1/(T-T_R)$]. The temperature dependence of the shift factor for the polymers 30P, 50P, and 65P is shown in Figure 2b. Furthermore, the apparent activation energy (ΔH_a) of the α-relaxation can be calculated as follows (*7-8,16*)

$$\Delta H_a = 2.303 \; RC_1C_2T^2 / (C_2 + T - T_R)^2 \tag{2}$$

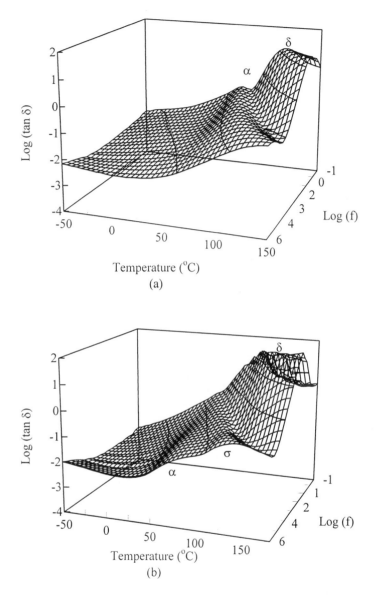

Figure 1. Dielectric loss tangent versus temperature and frequency for the polymers 50P (a), and 100P (b), respectively.

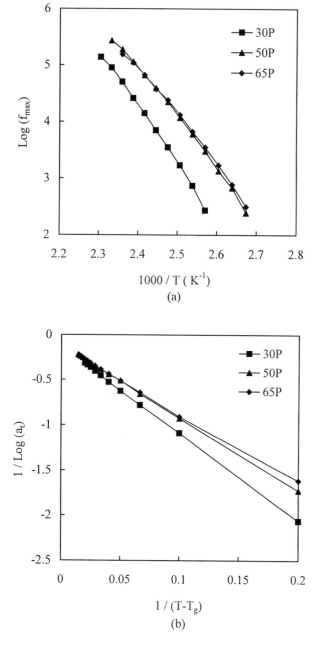

Figure 2. Temperature dependence of the relaxation frequency (f_{max}), shift factor (a_t), and activation energy (ΔH_a) of α-relaxation for polymers 30P, 50P, and 65P.

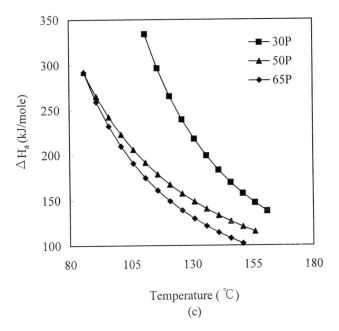

Figure 2. *Continued.*

where R is the ideal gas constant, T_R is defined as the relaxation temperature of the α-relaxation at a lower frequency (0.1 Hz). The temperature dependence of the apparent activation energy for α-relaxation is shown in Figure 2c. The result indicates that the apparent activation energy of α-relaxation was temperature-independent and decreased with increasing mesogenic group content. However, the relaxation time and activation energy were not obtained for the polymers 80P, 90P, and 100P, due to the insignificance of the α-relaxation peak.

In addition to the dielectric relaxation, thermal dynamic behavior of the polymers has been studied by an in-situ SHG technique. The time dependence of the SHG intensity during poling process for polymers 50P, and 80P are shown in Figure 3. The results suggest that the poling efficiency (maximum of SHG intensity) was increased with increasing poling temperature at the temperature range between T_g and 15 °C above T_g. The enhancement of the thermal energy was favorable for the alignment of the mesogenic group. However, the poling efficiency was reduced as the poling temperature further increased. Two factors are responsible for this. First, the high thermal energy results in a randomization in the orientation of the mesogenic group toward the electric field, and subsequently leads to the decrease of the poling efficiency (17). Secondly, the increase of the conductivity of the polymer film results in the reduction of the internal electric field. This leads to the decrease of the poling efficiency of polymers at a higher poling temperature range (18). Moreover, similar behavior of the poling process was also observed for polymers 30P, 65P, 90P, and 100P.

The temporal characteristic of effective SH coefficient after removal of the electric field at temperatures above T_g for the polymers 50P, and 80P are shown in Figure 4. The relaxation of the effective SH coefficient was dependent on the temperature. The effective SH coefficient decayed rapidly as the temperature increased. Moreover, polymer 80P exhibited a better temporal stability compared to polymer 50P. This is due to the fact that the polymer with a lower mesogenic group content possessed a larger molecular mobility at temperatures near T_g. Dielectric relaxation results indicate that the mobility of the polymer chains was suppressed remarkably during the glass transition for the polymer with LC phase i.e. high mesogenic group content, owing to the self-alignment characteristic of the mesogenic group. As a result of that, a longer relaxation time was obtained for the polymer 80P.

The temporal characteristics of SH coefficient for all of the polymers at room temperature are shown in Figure 5. For polymers 30P, 50P, 65P, the temporal stability was decreased with increasing mesogenic group contents, due to the plasticizer effect of the side chain group. Moreover, these three polymers have a better temporal stability at room temperature compared to polymers 80P, 90P, and 100P. This is due to the fact that the polymers with a lower mesogenic group content (30P, 50P, and 65P) have a higher α-relaxation temperature. For polymers 80P, 90P, and 100P, the temporal stability at room temperature was increased with increasing content of the mesogenic group despite their similar T_g's. This results indicate that the self-alignment nature of the NLO-active mesogenic group is beneficial to the enhancement of the temporal stability. In addition, the refraction indices and SH coefficient (d_{33}) of the poled polymers are summarized in Table II. The d_{33} values for

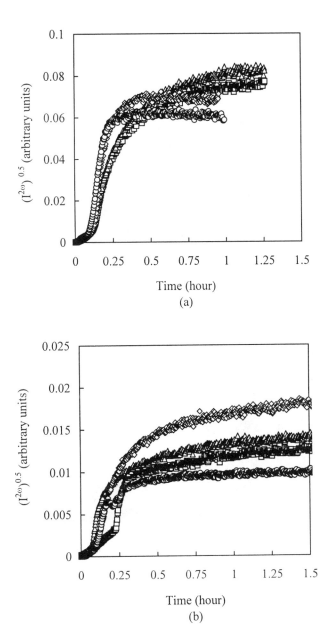

Figure 3. Time dependence of effective second harmonic coefficient during poling process for polymers 50P (a), and 80P (b), respectively (T-T$_g$ = 0 ($\square$), 5 (Δ), 10 ($\Diamond$), and 15 ℃ ($\bigcirc$)).

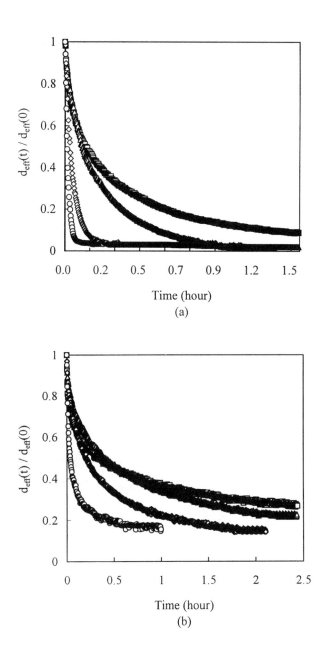

Figure 4. Temporal characteristic of effective second harmonic coefficient at temperature above T_g for polymers 50P (a), and 80P (b), respectively (T-T_g = 0 (□), 5 (Δ), 10 (◊), and 15 ℃ (○)).

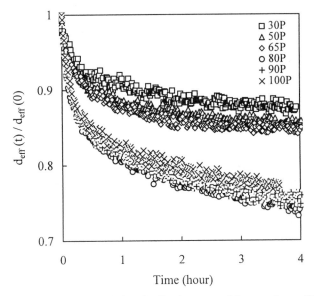

Figure 5. Temporal characteristic of effective second harmonic coefficient at room temperature for polymers 30P-100P.

TableII. The thickness, refraction indices, and second harmonic coefficient (d_{33}).

Samples	$d(\mu m)^*$	n_{532}^+	n_{1064}^+	$d_{33}(pm/V)^\#$
30P	1.08	1.37	1.30	3.7
50P	1.12	1.41	1.32	4.2
65P	1.02	1.42	1.34	4.5
80P	0.95	1.44	1.35	6.2
90P	1.08	1.45	1.36	7.4
100P	0.89	1.47	1.38	8.9

*d: thickness of the polymer films.
$^+$n: refraction indices at wavelengths 543 nm and 1064 nm.
$^\#$ d_{33}: second harmonic coefficient.

202

these polymers are in the range of 3.7 to 8.9 pm/V. The d_{33} value is enhanced as the density of the mesogenic group increases. This indicates that the steric effect among the NLO-active mesogenic groups does not occur even at high chromophore content. This is different from S'heeren's result (19) that in an amorphous NLO polymer system the SH coefficient initially increases with the increasing of the chromophore density and subsequently decreases after reaching a maximum. This is because at the high chromophore content the polymer in this work exhibits LC phase, whereas the polymer in S'heeren's work does otherwise. LC behavior indeed has a positive effect on the second-order NLO properties.

Conclusion

The effect of the mesogenic group content on the dielectric behavior and second-order nonlinearity of copolymethacrylates containing NLO-active tolane-based mesogenic groups have been studied. The self-alignment characteristic of the mesogenic group results in the reduction of the molecular mobility during the glass transition for the polymer containing a higher mesogenic group content (above 80 mol. %). As a result, a better temporal stability of SH coefficient was obtained for this type of polymers at temperatures near T_g. Moreover, the temporal stability of effective SH coefficient at room temperature was increased with increasing content of the mesogenic group for copolymers with LC phase. This indicates that the self-alignment nature of the NLO-active mesogenic group is beneficial to the enhancement of the temporal stability. In addition, the d_{33} value is enhanced as the density of the mesogenic group increases. The steric effect among the NLO-active mesogenic groups does not occur even at high chromophore contents. It is concluded that liquid crystal behavior play an important role in enhancing the second-order NLO properties.

Acknowledgment

The authors thank the National Science Council of Taiwan, ROC for financial support (Grant NSC87-2216-E007-032).

Literature Cited

1. McArdle, C. B. In *Side Chain Liquid Crystal Polymers*; McArdle, C. B., Ed.; Chapman and Hall: New York, 1989; P357.
2. Goodby, J. W.; Blinc, R.; Clark, N. A.; Lagerwall, S. T.; Osipov, M.A.; Pikin, S. A.; Sakurai, T.; Yoshino, K.; Zeks, B. In *Ferroelectric Liquid Crystals Principle, Properties and Application*; Taylor, G. W. Ed.; Gordon and Breach Science Publishers: Philadelphia, 1991.
3. Burland, D. M.; Miller, R. D.; Walsh, C. A. *Chem. Rev.* **1994**, *94*, 31.
4. Mcculloch, I. A.; Bailey, R. T. *Mol. Cryst. Liq. Cryst.* **1991**, *200*, 157.
5. Koide, N.; Ogura, S.; Aoyama, Y.; Amano, M.; Kaino, T. *Mol. Cryst. Liq., Cryst.* **1991**, *198*, 323.

. Hsiue, G. H.; Lee, R. H.; Jeng, R. J.; Chang, C. S. *J. Polym. Sci. Part B: Polym. Phys.* **1996**, *34*, 555.

. Firestone, M. A.; Ratner, M. A.; Marks, T. J.; Lin, W.; Wong, G. K. *Macromolecules* **1995**, *28*, 2260.

. Firestone, M. A.; Ratner, M. A.; Marks, T. J. *Macromolecules* **1995**, *28*, 6296.

. Hsieh, C. J.; Wu, S. H.; Hsiue, G. H.; Hsu, C. S. *J. Polym. Sci., Part A: Poly. Chem.* **1994**, *32*, 1077.

0. Mortazavi, M. A.; Knoesen; A.; Kowel, S. T.; Higgins, B. G.; Dienes, A. *J. Opt. Soc. Am.* **1989**, *B6*, 773.

1. Jeng, R. J.; Chen, Y. M.; Kumar, J.; Tripathy, S. K. *J. Macromol. Sci., Pure Appl. Chem.* **1992**, *A29*, 1115.

2. Mandal, B. K.; Chen, Y. M.; Lee, J. Y.; Kumar, J.; Tripathy, S. K. *Appl. Phys. Lett.* **1991**, *58*, 2459.

3. Dubois, J. C.; Barny, P. L.; Robin, P.; Lemoine, V.; Rajbenbach, H. *Liq. Cryst.* **1993**, *14*, 197.

4. Zental, R.; Strobl, G. R.; Ringsdorf, H. *Macromolecules* **1985**, *18*, 960.

5. Percec, V.; Pugh, C. In *Side Chain Liquid Crystal Polymers*; McArdle, C. B., Ed.; Chapman and Hall: New York, 1989, P38.

6. Ferry, J. D. In *Viscoelastic Properties of Polymers*; John Wiley & Sons, Inc.: New York, 1980, P287.

7. Bristow, J. F.; Kalika, D. S.; *Macromolecules* **1994**, *27*, 1808.

8. Inaba, R.; Sagawa, M.; Isogia, M.; Kakuta, A. *Macromolecules* **1996**, *29*, 2954.

9. S'heeren, G.; Persoons, A. *Makromol. Chem.* **1993**, *194*, 1733.

Chapter 14

Photorefractive Polymers and Polymer-Dispersed Liquid Crystals

B. Kippelen[1], A. Golemme[1,4], E. Hendrickx[1], J. F. Wang[1], S. R. Marder[2,3], and N. Peyghambarian[1]

[1]Optical Sciences Center, The University of Arizona, Tucson, AZ 85721
[2]Beckman Institute, California Institute of Technology, Pasadena, CA 91125
[3]Jet Propulsion Laboratory, California Institute of Technology, Pasadena, CA 91109

Photorefractive (PR) polymers are multifunctional field-responsive materials that combine photoconducting and electro-optic properties. In this chapter, we will review the basics of photorefractivity in polymers and liquid crystals and describe recent advances that have led to highly efficient materials. These advances in material development enable a variety of photonic applications including optical correlators for security verification.

PR materials are among the most sensitive nonlinear optical materials since they exhibit large refractive index changes when exposed to low power laser beams. The PR effect is based on the build-up of a space-charge through the photoexcitation of carriers and their transport over macroscopic distances *(1)*. Transport can occur by either diffusion of the carriers, if the excitation is nonuniform, or by drift, if an electric field is applied to the material. Charge separation results in a space-charge field that changes the optical properties of the material. These different steps are illustrated in Fig. 1. PR materials are typically illuminated by two coherent laser beams that interfere (see step *a* in Fig. 1) to produce a spatially modulated intensity distribution *I(x)* given by *(2)*:

$$I(x) = I_0 \left[1 + m\cos(2\pi x / \Lambda) \right] \tag{1}$$

Where $I_0 = I_1 + I_2$ is the sum of the intensities of the two beams, $m = 2(I_1 I_2)^{1/2} / (I_1 + I_2)$ is the fringe visibility, and Λ is the grating spacing, i.e. the distance between two light maxima. After charge separation and trapping, an internal electric field with the same spatial periodicity is formed in the material (steps *b* to *d* in Fig. 1). Finally, (step *e*) the internal electric field causes a change in refractive index leading to a phase replica of the initial light distribution (also called a grating).

[4]Current address: Dipartimento di Chimica, Università della Calabria, Rende, Italy.

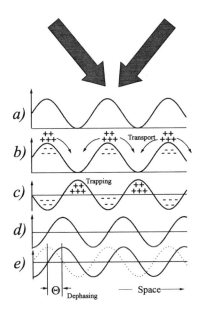

Figure 1. Illustration of the photorefractive effect. The overlap of two coherent laser beams creates an optical interference pattern *(a)*. In the high intensity regions, charge carriers are generated *(b)*. One type of carriers is transported and trapped *(c)*, creating an alternating space-charge field *(d)*. The space charge field induces a change in refractive index *(e)*. The resulting index grating is phase shifted with respect to the initial light distribution.

In traditional PR materials, this optical encoding of a grating is based on a second-order nonlinear optical process and can therefore only be observed in materials that are noncentrosymmetric *(3)*. In polymers, noncentrosymmetry and consequently electro-optic properties, can be achieved by doping the material with nonlinear chromophores that possess a permanent dipole moment and by orienting them in an electric field (the so called poling process) *(4,5)*. Initially, photorefractivity was observed in multifunctional polymers that combined photoconducting and electro-optic properties *(6)*. However, in current high-efficiency PR polymers *(7,8)*, the refractive index of the material is also modulated by a spatially periodic orientation of the chromophores *(9)*. Both electro-optic and orientational effects contribute to the overall refractive index change, but the orientational effects are dominant in the most efficient PR polymers to date *(8, 10)*. Such polymers have a glass transition temperature that is below or close to room temperature, to enable the field-responsive reorientation of the chromophores in real time. Recent advances in PR polymer development have focused on the optimization of these orientational effects *(10)*. Simultaneously, various PR liquid crystalline materials were developed in which strong orientational effects at low applied electric field can be observed *(11-13)*.

Optical media that can record elementary gratings, such as PR polymers, are in high demand. In many fields, the decomposition of a signal into a superposition of harmonic functions (Fourier analysis) is a powerful tool. In optics, Fourier analysis often provides an efficient way to implement complex operations and is at the basis of many optical systems. For instance, any arbitrary image can be decomposed into a sum of harmonic functions with different spatial frequencies and complex amplitudes as illustrated in Fig. 2. Each of these periodic functions can be considered as an elementary grating. For real-time optical recording or processing applications, the material must have a dynamic response. In other words, the light-induced gratings should be erasable or be able to accommodate any changes in the light waves that are inducing them, in real time. Materials with such optical encoding properties allow for implementation of a wide range of optical applications ranging from reconfigurable interconnects, dynamic holographic storage, to optical correlation, image recognition, image processing, and phase conjugation. Therefore, materials where the optical encoding is dynamic and based on a periodic modulation of the refractive index are the focus of intense research.

Internal Electric Field Formation

The unique properties of PR materials, including their high sensitivity to low power optical excitation, arise from the build-up and storage mechanism of the internal space-charge field. This build-up can be divided into two steps: the electron-hole generation process followed by the transport of one of the carriers over macroscopic distances. The photogeneration process is characterized by its quantum efficiency which is the ratio of the number of photogenerated free carriers that undergo transport over the number of absorbed photons. In most of inorganic PR materials, that efficiency is close to unity. In contrast, in organic PR polymers, many of the photogenerated carriers find some recombination channels. As a result, only a small

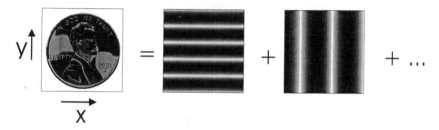

Figure 2. Illustration of the decomposition of an image into Fourier components or elementary gratings.

fraction of absorbed photons leads to free carriers for transport. The photogeneration efficiency is increased by applying an electric field and its field-dependence can be described by the geminate recombination model developed by Onsager *(14)*.

After photogeneration, free carriers migrate from the brighter regions of the light interference pattern, where they are generated, to the darker regions, where they get trapped. In contrast to inorganic PR crystals with a periodic structure, PR polymers are nearly amorphous. The local energy level of each molecule/moiety is affected by its nonuniform environment. In contrast to molecular crystals, the disorder in amorphous photoconductors leads to a distribution of localized electronic states. As a result, transport can no longer be described by band models but is attributed to intermolecular hopping of carriers between neighboring molecules or moieties. In recent years, a model developed by Bässler and co-workers has been used to describe the transport phenomena of a wide range of different materials and emerged as a solid formalism to describe the transport in amorphous organic materials *(15)*. So far, most of the predictions of this theory agree reasonably well with the experiments performed in a wide range of doped polymers, main-chain and side-chain polymers, and in molecular glasses *(16)*. In the Bässler formalism, disorder is separated into diagonal and off-diagonal components. Diagonal disorder is characterized by the standard deviation σ of the Gaussian energy distribution of the hopping site manifold (energetical disorder) and the off-diagonal component is described by the parameter Σ that describes the amount of positional disorder. Results of Monte-Carlo simulations led to the following universal law for the mobility *(15)*:

$$\tilde{\mu}(E,T) = \tilde{\mu}_0 \exp\left[-\left(\frac{2\sigma}{3kT}\right)^2\right] \exp\left\{C\left[\left(\frac{\sigma}{kT}\right)^2 - \Sigma^2\right]E^{1/2}\right\} \tag{2}$$

where $\tilde{\mu}_0$ is a mobility prefactor and C an empirical constant. Eq. (2) is valid for high electric fields (a few tens of V/μm) and for temperatures $T_g > T > T_c$, where T_g is the glass transition temperature and T_c the dispersive to nondispersive transition temperature.

The study of the transport properties of organic amorphous materials is an active field of research that is driven by the multibillion dollar industry of xerography *(17)*. Xerographic imaging systems, such as photocopiers and printers, play an important role in our daily life. In early systems, inorganic materials such as zinc oxide and amorphous silicon were used as photoconductors. However, since xerographic materials have to be prepared into very large areas and on flexible substrates, due to design and manufacturing requirements, organic photoconductors replaced their inorganic counterparts and represent today a 5 billion dollar industry. The recent development of this xerographic technology based upon organic materials, due to their low cost and ease of processing, provides strong evidence for their great potential for optical applications.

Field-Induced Refractive Index Changes

In guest/host PR polymers the macroscopic nonlinear optical properties result from the molecular constants of the dopant molecule (also called the chromophore). Chromophores for nonlinear optical properties are usually based upon aromatic π-electron systems unsymmetrically end-capped with electron donating and electron accepting groups, as shown in Fig. 3. The molecular polarization of such molecules is a nonlinear function of the electric field and can be simplified to (4,5):

$$p = \mu + \alpha E + \beta E^2 + ...$$ (3)

where μ is the permanent dipole moment of the molecule, α the linear polarizability, and β the first hyperpolarizability. For simplicity, we omit here tensorial notation. On a macroscopic level, orientation of these molecules in an applied electric field leads to electro-optic properties and to birefringence. The electro-optic properties are characterized by the second-order susceptibility tensor $\chi^{(2)}(-\omega; \omega, 0)$. In the oriented gas model, the tensor element along the direction of the poling field (Z) is given by (18):

$$\chi^{(2)}_{ZZZ}(-\omega; \omega, 0) = N\, F^{(2)}\, \beta <\cos^3 \theta > \approx N\, F^{(2)}\, \beta \frac{\mu E}{5kT}$$ (4)

where N is the density of chromophores, $F^{(2)}$ is a local field correction factor that depends on the average refractive index and the low-frequency dielectric constant of the polymer composite, θ is the angle between the poling field direction (with amplitude E), and the dipole moment of the molecule, $< >$ means averaged value using Maxwell-Boltzmann statistics, and kT is the thermal energy. A change in refractive index along the direction of the poling field leads to a birefringence that is given by (9, 19):

$$\Delta n_Z^{(1)}(\omega) = \frac{2\pi}{n} N F^{(1)} \Delta\alpha \left[<\cos^2 \theta > - 1/3\right] \approx \frac{4\pi N F^{(1)}}{45n} \Delta\alpha \left(\frac{\mu E}{kT}\right)^2$$ (5)

where n is the average refractive index, $F^{(1)} = (n^2+2)/3$ is the Lorentz –Lorenz local field correction factor, and $\Delta\alpha$ is the polarizability anisotropy of the chromophore. When cylindrical symmetry is assumed, $\Delta\alpha \approx (\alpha_{//} - \alpha_\perp)$, where $\alpha_{//}$ and $\alpha_\perp$ are the polarizability along the axis of the molecule and the polarizability in a perpendicular direction, respectively. Note that the birefringence properties have a quadratic dependence on electric field. This dependence is identical to the orientational Kerr effect, but should not be confused with the electronic electro-optic Kerr effect, which is a third-order nonlinear process described by $\chi^{(3)}(-\omega; \omega, 0, \omega)$ tensor elements. Electro-optic Kerr effects are purely electronic, have a fast response, and can be observed in all materials. Orientational effects rely on the

a)

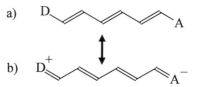

b)

Figure 3. Two limiting charge transfer resonance forms of a donor-acceptor polyene molecule: the neutral form a) and the charge separated form b).

ability of the molecule to be oriented in its matrix and can consequently be quite slow.

Chromophore Design

In low glass transition temperature (T_g) PR polymer composites, both electro-optic and orientational birefringence contribute to the total refractive index modulation. In this case, according to Eqs. (4) and (5), a figure of merit *FOM* for the design of chromophores for PR applications can be *(9, 20, 21)*:

$$FOM = A(T)\,\Delta\alpha\,\mu^2 + \beta\mu \qquad (6)$$

where $A(T) = 2/9kT$ is a numerical scaling factor. Recently, we used the BLA (Bond Length Alternation) model *(22)* to optimize the design of chromophores for PR applications *(21)*. Within that model, molecular quantities such as the dipole moment, the polarizability, as well as the hyperpolarizability can be correlated with the degree of ground-state polarization *(23,24)*. Donor-acceptor substituted molecules with a π-electron conjugation path have a ground-state structure that can be viewed as a linear combination of two limiting resonance forms: a neutral form (Fig. 3a) and a charge-separated form (Fig. 3b). The relative contribution of these two forms in the ground state can be correlated to the values of BLA or BOA (Bond Order Alternation), where BOA is the difference in the π-bond order between adjacent carbon-carbon bonds. BOA and BLA is usually varied by changing the strength of the donor and acceptor substituents, or by changing the properties of the surrounding medium such as its polarity. In model calculations, an internal field can be used to vary BLA or BOA *(23)*. The calculated values of the dipole moment μ, the linear polarizability α_{zz} along the axis of the molecule, and the first hyperpolarizability β as a function of applied field are plotted in Fig. 4 (a)-(c) for the molecule $(CH_3)_2N$-$(CH=CH)_4CHO$. The calculations indicate that that $\alpha_{zz} \gg \alpha_{xx}, \alpha_{yy}$, thus $\Delta\alpha \approx \alpha_{zz}$. From these results, the figure of merit *FOM* can be calculated as a function of BOA. Previous studies *(20)* suggested that an optimal *FOM* was obtained at the point where β vanishes and where the polarizability anisotropy is maximized. In contrast, our study *(21)* that takes into account the linear increase of μ as a function of BOA over the region considered, indicates that *FOM* is optimized for BOA where β is roughly maximized in amplitude with a negative sign as clearly shown in Fig. 4d.

Phase Stability of Guest/Host Photorefractive Polymers

Among the possible designs for low T_g polymers, the guest/host approach is the most widely used and has led to the most efficient PR polymers to date *(10)*. Because of the multifunctionality of photorefractivity, several compounds have to be mixed together to form a composite with good optical quality. This approach can lead to phase stability problems when the different constituents are not compatible. However, we would like to stress that this approach does not

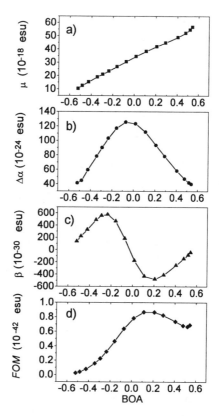

Figure 4. The dipole moment (a), polarizability anisotropy $\Delta\alpha$ (b), first hyperpolarizability β (c), and figure of merit FOM (d), as a function of bond order alternation for the molecule $(CH_3)_2 \, N(CH=CH)_4 \, CHO$ (see *(21)* and *(22)* for details).

necessarily lead to unstable materials. In many guest/host PR polymers a gradual crystallization of the chromophore limits the shelf-lifetime of the samples. For instance, the lifetime of DMNPAA:PVK:ECZ:TNF samples doped with 50 wt % of DMNPAA *(8)* was found to vary between a few hours and a few weeks depending on the starting materials and the processing conditions (DMNPAA: 2,5 dimethyl-4-*p*-nitrophenylazoanisole; PVK: poly(N-vinylcarbazole); ECZ: N-ethylcarbazole; TNF: 2,4,7-trinitrofluorenone). Encapsulation to protect the sample from moisture was found to drastically increase the lifetime to several months.

To characterize the phase stability of guest/host PR polymers we developed recently a light-scattering technique *(25)*. To follow the gradual degradation of the optical quality of the sample, we monitored the intensity of the light scattered from the sample within a fixed solid angle, using the system shown in Fig. 5. The samples were illuminated with an expanded beam from a He-Ne laser or a laser diode (690 nm). The transmitted light was focused by the first lens on a small silver dot (diameter 200 μm) deposited on a transparent glass slide, and reflected. Since the light scattered from crystallites forming in the sample propagates in different directions and is not focused onto the silver spot, it passes through the glass slide. This scattered light was then imaged onto a detector and its intensity was recorded as a function of time. The scattered light intensity measured as a function of time in a sample of DMNPAA:PVK:ECZ (39:41:20 wt. %) held at 55 °C is shown in Fig. 6. Crystallization does not start immediately upon heating: an induction period for the formation of nucleation sites that scatter light is observed and followed by a period of accelerated crystallization during which the nuclei grow in radius. This behavior is characteristic of phase separations occurring through the mechanism of nucleation and growth. To quantify the shelf lifetime of guest/host PR polymers we define a reference time t_{ref} that corresponds to the intercept of the linear extrapolations of the induction and growth phases. We investigated the crystallization onset in PVK-based samples doped with different chromophores. All these measurements were carried out at temperatures that are well above room temperature, i.e., well above the glass transition temperature, but well below the melting point of the chromophores.

NPADVBB-Based Polymers

To improve the phase stability of PVK-based PR polymers we replaced the chromophore DMNPAA with isomeric mixtures of the chromophore NPADVBB (4-(4'-nitrophenylazo)1,3-di[(3"or 4"-vinyl)benzyloxy]-benzene) (see Table I). In NPADVBB, the vinyl groups on both phenyl rings can be either in the 3" or 4"-position. Hence, NPADVBB is a mixture of four isomers, with the vinyl groups in positions (3",3"), (3",4"), (4",3") and (4",4"). The shelf lifetime and crystallization behavior of PVK:ECZ samples doped with DMNPAA or NPADVBB were investigated by light scattering experiments as a function of temperature and sample composition *(25)*. The PR properties of the samples were tested by four-wave mixing and two-beam coupling experiments in the tilted geometry described previously *(25,26)*. The laser source was a HeNe laser (633 nm). The two writing beams were s-polarized, had a power of 0.8 mW each, and were focused to spot sizes of ~ 450 μm in the sample. The angle between the beams outside the sample was $2\theta = 20.5°$. The tilt angle was $\psi = 60°$. The probe beam was p-polarized, had a power of 2.2 μW, and was focused to a spot size of ~450 μm. The steady-state diffraction efficiencies

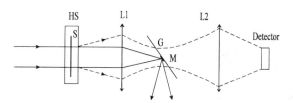

Figure 5.   Light scattering system used for the characterization of the shelf lifetime of guest/host photorefractive polymers. S: sample; HS: hot-stage; G: glass slide; M: mirror; L1 and L2: lenses.

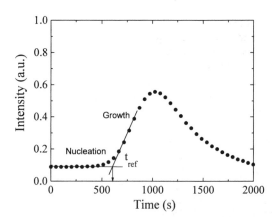

Figure 6. Intensity of the scattered light as a function of time for a photorefractive polymer composite of DMNPAA:PVK:ECZ with composition 39:41:20 wt. % held at 55°C.

Table I. Chemical structure, name, melting point T_M, and reference time of several chromophores doped in a PVK/ECZ matrix measured by light scattering experiments at 85 °C.

Chemical structure	Name	T_M (°C)	Polymer composition (wt. %)	t_{ref} (s)
	DMNPAA	161	DMNPAA:PVK:ECZ (39:41:20)	60
	NPADVBB	127	NPADVBB:PVK:ECZ (39:41:20)	6600
	DHADC-MPN	95	DHADC-MPN:PVK:ECZ (39:41:20)	13000

as a function of applied field in samples with composition DMNPAA:PVK:ECZ:TNF (40:39:19:2 wt. %) and NPADVBB:PVK: ECZ:TNF (40:39:19:2 wt. %) are shown in Fig. 7. For both samples, maximum diffraction efficiency is observed at an applied field of nearly 65 V/μm. The PR origin of both signals was confirmed by two-beam coupling experiments *(27)*. The dynamics of the build-up of the PR grating was similar in both samples and sub-second. The results of phase stability measurements for different chromophores in a PVK/ECZ matrix are summarized in Table I. These results demonstrate that the substitution of DMNPAA with isomeric mixtures of NPADVBB increases the shelf lifetime of the samples by two orders of magnitude without any loss in diffraction efficiency. The shelf lifetime at room temperature of these new composites is therefore several years and is no longer a limitation.

Efficient Infrared Photorefractive Polymers

Early dopant molecules such as DMNPAA were incorporated into polymer composites because of their electro-optic properties. With the evidence of strong orientational birefringence effects in low T_g PR polymers *(8,9)*, new design rationales for PR chromophore development have emerged *(20, 21)*. To explore our chromophore design rationale *(21)* discussed above (see Eq. 6), we have focused our studies on linear molecules such as polyenes, rather than on chromophores that contain benzene rings such as DMNPAA or NPADVBB. This is because polyenes exhibit a considerable charge transfer that is confined along the quasi one-dimensional π-conjugated bridge, providing a large $\Delta\alpha$. In addition, polyenes can have an important charge separation in the ground state that provides large dipole moment μ. To comply with the practical requirements of a well-performing PR polymer, we synthesized the polyene molecule DHADC-MPN (2-N,N-dihexylamino-7-dicyanomethylidenyl-3,4,5,6,10-pentahydronaphthalene) (Table I). The hexyl groups help impart solubility to this highly dipolar molecule. The incorporation of the polyene into the fused ring systems enhances thermal and photochemical stability. The molecule was used as a dopant molecule in mixtures of PVK and ECZ. Sensitivity in the visible (633 nm) was provided by (TNF). By using the sensitizer (2,4,7-trinitro-9-fluorenylidene)malonitrile (TNFDM), the spectral response of the photosensitivity could be extended to the near infrared (λ = 830 nm). The spectral response of the new sensitizer in PVK/ECZ is shown in Fig. 8. The PR properties, in particular the dynamic range, or Δn, were tested by four-wave mixing experiments in the tilted geometry. The thickness of all the samples was 105 μm and the polymer was sandwiched between two transparent indium-tin oxide (ITO) electrodes. Two-beam coupling, i.e., energy exchange between the two interfering laser beams, was observed in all the PR composites presented here and confirmed the PR nature of the optical encoding. The normalized diffraction efficiency (corrected for absorption and reflection losses and for small electro-absorption effects) of a composite DHADC-MPN:PVK:ECZ:TNF (40:39:19:2 wt. %) as a function of applied field is shown in Fig. 9a. For comparison, we also plotted the normalized diffraction efficiency of a composite DMNPAA:PVK:ECZ:TNF (40:39:19:2 wt. %). In both samples the oscillatory behavior of the diffraction efficiency η is in agreement with Kogelnik's coupled-wave theory *(28)*:

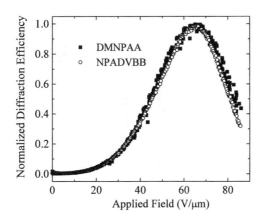

Figure 7. Diffraction efficiency as a function of applied field for samples with compositions DMNPAA:PVK:ECZ:TNF (40:39:19:2 wt. %) and NPADVBB:PVK:ECZ:TNF (40:39:19:2 wt. %) measured at 633 nm.

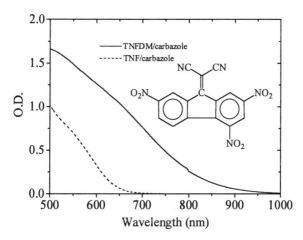

Figure 8. Linear optical absorption spectrum of the charge transfer complexes of TNF/carbazole in PMMA doped with ECZ, and TNFDM/carbazole in PVK. Inset: chemical structure of TNFDM.

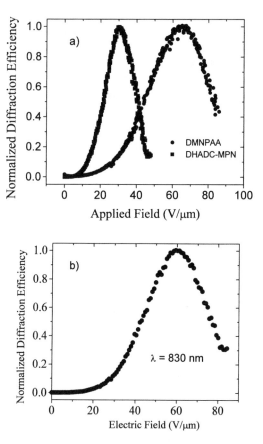

Figure 9. (a) Normalized diffraction efficiency versus field measured at 633 nm in a composite DMNPAA:PVK:ECZ:TNF (40:39:19:2 wt. %) (circles) and in a composite DHADC-MPN:PVK:ECZ:TNF (40:39:19:2 wt. %) (squares). (b) Normalized diffraction efficiency versus field measured at 830 nm in a composite DHADC-MPN:PVK:ECZ:TNFDM (25:49:24:2 wt. %).

$$\eta \propto \sin^2 \left[\frac{G \pi \Delta n d}{\lambda} \right] \tag{7}$$

where G is a geometrical factor that depends on the polarization of the beams and the experimental geometry, Δn is the refractive index modulation amplitude, d the thickness of the grating, and λ the wavelength of the light. The composite based on DHADC-MPN exhibits a first maximum at an applied field $E_{\pi/2}$ = 30 V/µm, to be compared with $E_{\pi/2}$ = 65 V/µm for the DMNPAA-based composite (Fig. 9a). At this first maximum of the diffraction efficiency, the value of the argument of the *sin* function in Kogelnik's expression for the diffraction efficiency (Eq. 7) is $\pi/2$. This reduction in $E_{\pi/2}$ by a factor larger than 2 is indicative of the large efficiency of DHADC-MPN based polymers. Calculations of Δn using Kogelnik's expression (Eq. 7) show that Δn in the DHADC-MPN composite is over four times higher than in a DMNPAA composite with the same doping level and in the 0 to 50 V/µm field range. At fields > 50V/µm, beam-fanning effects are observed in the DHADC-MPN-based polymers and the analysis of the four-wave mixing data is more complicated. When using TNFDM as a sensitizer, the spectral sensitivity is extended to the near infrared. In the normalized diffraction efficiency of a composite of DHADC-MPN:PVK:ECZ:TNFDM (25:49:24:2 wt. %) (Fig. 9b) maximum diffraction is observed at $E_{\pi/2}$ = 59 V/µm at 830 nm. The real diffraction efficiency η_{max} at $E_{\pi/2}$ is 30% for that composite. This value can be further optimized by reducing the sensitizer concentration, that is reducing the absorption of the sample at 830 nm, and can reach η_{max} = 74% in a composite of DHADC-MPN:PVK:ECZ:TNFDM (25:49:25:1 wt. %).

These materials are, to the best of our knowledge, the first efficient PR polymers in the near infrared. These results suggest that our proposed design rationale, that is to synthesize molecules that combine high μ and high $\Delta \alpha$, provides an efficient route to optimize the performance of low glass transition temperature PR polymers. The PR polymer composites with high Δn and spectral sensitivity in the near infrared developed here offer new opportunities for numerous photonic applications. As a sensitive holographic recording medium with high dynamic range, we used them for instance, for imaging through scattering media *(29)*. Of particular importance is their compatibility with the emission of high quality GaAs semiconductor laser diodes and commercial solid-state femtosecond lasers, such as Ti:Sapphire lasers. More importantly, their near infrared spectral response is compatible with the transparency of biological tissues (700-900 nm) and therefore the imaging through scattering media technique shown in *(29)* could be extended to medical imaging.

Photorefractive Polymer-Dispersed Liquid Crystals

As shown in previous sections, PR polymers with a low T_g, exhibit strong electric-field induced orientational effects that are responsible for their high efficiency. However, a drawback of these materials is the high electric field (30-100 V/µm) that needs to be applied in order to orient these molecules efficiently. Another obvious way to induce a molecular reorientation with an electric field involves the use of materials with a dielectric anisotropy, such as liquid crystals. In contrast to dipoles dispersed in amorphous polymers, liquid crystal molecules in mesophases can be reoriented with much lower electric fields (few V/cm). However, due to the coherence length in bulk nematic liquid crystals, the resolution of these materials is limited and their PR performance is low for small values (< 5 μm) of the grating spacing *(12,13)*.

To combine the high resolution of PR polymers, and the high refractive index changes associated with field-induced reorientation of nematic liquid crystals, we developed PR polymer-dispersed liquid crystals *(30)*. Such materials are prepared by dispersing liquid crystal domains of almost spherical shape in a photoconducting solid organic polymer matrix. The sensitizing, photoconducting, and trapping properties necessary for the build-up of a space-charge field are provided by the polymer matrix and the refractive index changes are due to the liquid crystal droplets as illustrated in Fig. 10. In addition to the PR properties, these new materials present also the characteristic field-dependent transmission properties of PDLCs *(31)*. For our experiments we used poly-methylmethacrylate (PMMA) for the polymer matrix, doped with ECZ for hole transport, TNF for charge generation, and the eutectic nematic mixture E49 (purchased from Merck) for the orientational properties. For this study, we prepared samples with the following composition: PMMA:E49:ECZ:TNF (45:33:21:1 wt. %). For an applied field of 22 V/μm, an external diffraction efficiency of 8% was observed in 53 μm-thick samples for a grating spacing of 4.5 μm. The corresponding internal diffraction efficiency defined as the ratio between the diffracted and transmitted beam intensities was 40%. This corresponds to a refractive index modulation amplitude of $\Delta n = 2 \times 10^{-3}$. That value is five times higher than the value of $\Delta n = 3.8 \times 10^{-4}$ measured with the same applied field in the highly efficient PR polymers DMNPAA:PVK:ECZ:TNF *(8)* and illustrates the high efficiency of these new materials. The speed of these materials is rather slow at this stage (minute) due to the non optimized transport properties of the PMMA matrix doped with ECZ. However, due to the structural flexibility of organic photoconductors and liquid crystals, materials with further optimized properties can be expected. This new class of materials combines PR properties with the field-dependent transmission properties of PDLCs, which makes them intriguing materials for new optical applications.

An All-Optical, All Polymeric Optical Correlator for Security Applications

To demonstrate the technological potential of high efficiency PR polymers in device geometry, we developed an all-optical, all-polymeric pattern recognition system for security verification where the PR polymer was used as the real-time optical recording and processing medium *(32)*. The low cost security verification system we proposed is based on the optical encoding of documents with pseudo-randomly generated phase masks. These masks are attached onto documents and their inspection is performed by comparing them optically to a master pattern. To check if the phase mask on the document is identical to the master, an optical correlation is performed in the PR polymer that acts as the recording medium of the correlator. The principle of the correlator is the following: a laser beam is encoded

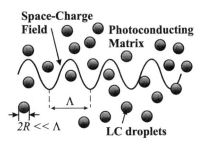

Figure 10. Design principle of photorefractive polymer-dispersed liquid crystals.

with the phase information attached to the document and is focused, together with a reference laser beam, into the PR polymer, leading to the recording of a Fourier hologram. A third laser beam is encoded with the phase information of the master and is diffracted on the hologram stored in the PR polymer. If the phase information on the document and that of the master are strictly identical, the diffraction in the PR polymer leads to a strong intensity peak that is detected by a photodiode. If the phase information does not match, only a weak background signal is generated. Thus, the detection of that strong intensity peak provides a rapid way to check the authenticity of the document that is tested.

The recording medium is a key element in this type of all-optical architecture. The limited performance and/or the high cost of existing nonlinear optical materials has severely limited the technological potential of all-optical correlators: inorganic PR crystals have been investigated but their processing and high cost has limited their use in wide-spread applications. Due to limited optical material performance, other correlator designs have been proposed over the years: nonlinear joint-transform correlators, for instance, show good performance for pattern recognition and are capable of real-time operation (33). However, because these systems use either sophisticated liquid crystal light valves (34), CCD detectors, and/or a computer to perform Fourier transforms, they do not meet the low cost requirement.

This security verification system has the following features that make it practical for wide-spread applications: the use of a highly efficient PR polymer and its compatibility with semiconductor laser diodes keep the overall manufacturing cost to levels that are significantly lower than that of any previously proposed optical correlator. The system is fast because the processing is implemented optically in parallel. Furthermore, the high resolution of the PR polymers allows the use of shorter focal length lenses in the 4f correlator, thus, making its design more compact compared with one using liquid crystal light valves. Fig. 11 shows a photograph of a compact version of the system for security verification. The correlator was built on a 11" x 8.5" breadboard.

Summary

In summary, we have developed highly efficient PR polymers based on isomeric mixtures of chromophores and have improved the shelf lifetime by two orders of magnitude. These new composites are stable at room temperature for several years. By following a new design rationale, that is by using chromophores such as DHADC-MPN that combine high dipole moment and high polarizability anisotropy, we have been able to further improve the refractive index modulation amplitude by a factor of four. By using TNFDM as a sensitizer, and DHADC-MPN as a chromophore, we obtained PR composites that showed total diffraction in the near infrared at 830 nm. We have developed a new class of organic PR materials: PR polymer-dispersed liquid crystals. They show high refractive index changes at lower fields compared with polymers and are a promising new class of materials. Finally, we have illustrated the strong potential of PR polymers for photonic applications, by demonstrating an all-optical, all-polymeric optical correlator for security verification.

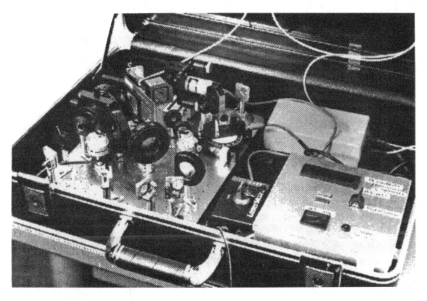

Figure 11. Photograph of an optical correlator for security verification that uses a photorefractive polymer as active medium.

224

Acknowledgments
This work was supported by the US Office of Naval Research (ONR) through the MURI Center CAMP, by the US National Science Foundation (NSF), by an international CNRS/NSF travel grant, by the USAF Office of Scientific Research (AFOSR), and by a NATO travel grant. E. H. is a postdoctoral fellow of the Fund for Scientific Research-Flanders (Belgium). We acknowledge collaboration with D. D. Steele, J. L. Maldonado, and B. L. Volodin from the University of Arizona, and with Prof. B. Javidi from the University of Connecticut for the work on the optical correlator.

Literature cited
[1]   P. Günter and J.-P. Huignard, *Photorefractive Materials and Their Applications I*,   Vol. 61 (Springer-Verlag, Berlin, 1988).
[2]   H. J. Eichler, P. Günter, and D. W. Pohl, *Laser-induced dynamic gratings*, Vol. 50 (Springer-Verlag, Berlin, 1986).
[3]   A. Ashkin, G. D. Boyd, J. M. Dziedic, R. G. Smith, A. A. Ballmann, and K. Nassau, *Appl. Phys. Lett.* **9**, 72 (1966).
[4]   J. Zyss, *Molecular Nonlinear Optics, Materials, Physics, and Devices* (Academic Press, Inc., San Diego, 1994).
[5]   P. N. Prasad and D. J. Williams, *Nonlinear Optical Effects in Molecules and Polymers* (John Wiley & Sons, Inc., New York, 1991).
[6]   S. Ducharme, J. C. Scott, R. J. Twieg, and W. E. Moerner, *Phys. Rev. Lett.* **66**, 1846 (1991).
[7]   B. Kippelen, Sandalphon, N. Peyghambarian, S. R. Lyon, A. B. Padias, and H. K. Hall Jr., *Electron. Lett.* **29**, 1873 (1993).
[8]   K. Meerholz, B. L. Volodin, Sandalphon, B. Kippelen, and N. Peyghambarian, *Nature* **371**, 497 (1994).
[9]   B. Kippelen, S. R. Marder, E. Hendrickx, J. L. Maldonado, G. Guillemet, B. L. Volodin, D. D. Steele, Y. Enami, Sandalphon, Y. J. Yao, J. F. Wang, H. Röckel, L. Erskine, N. Peyghambarian, Science **279**, 54 (1998).
[10] W. E. Moerner, S. M. Silence, F. Hache, and G. C. Bjorklund, *J. Opt. Soc. Am. B* **11**, 320 (1994).
[11] E. V. Rudenko and A. V. Sukhov, *JETP Lett.* **59**, 142 (1994).
[12] C. Khoo, H. Li, and Y. Liang, *Opt. Lett.* **19**, 1723 (1994).
[13] G. P. Wiederrecht, B. A. Yoon, and M. R. Wasielewski, *Science* **270**, 1794 (1995).
[14] L. Onsager, *Phys. Rev.* **54**, 554 (1938).
[15] H. Bässler, *Adv. Mat.* **5**, 662 (1993).
[16] P. M. Borsenberger, E. H. Magin, M. Van der Auweraer, and F. C. De Schryver, *Phys. Stat. Sol. (a)* **140**, 9 (1993).
[17] P. M. Borsenberger and D. S. Weiss, *Organic photoreceptors for imaging systems* (Marcel Dekker, Inc., New York, 1993).
[18] K. D. Singer, M. G. Kuzyk, and J. E. Sohn, *J. Opt. Soc. Am. B* **4**, 968 (1987).
[19] J. W. Wu, *J. Opt. Soc. Am. B* **8**, 142 (1991).
[20] R. Wortmann, C. Poga, R. J. Twieg, C. Geletneky, C. R. Moylan, P. M. Lundquist, R. G. DeVoe, P. M. Cotts, H. Horn, J. E. Rice, and D. M. Burland, *J. Chem. Phys.* **105**, 10637 (1996).

[21] B. Kippelen, F. Meyers, N. Peyghambarian, and S. R. Marder, *J. Am. Chem. Soc.* **119**, 4559 (1997).

[22] S. R. Marder, D. N. Beratan, and L.-T. Cheng, *Science* **252**, 103-106 (1991).

[23] F. Meyers, S. R. Marder, B. M. Pierce, and J. L. Brédas, *J. Am. Chem. Soc.* **116**, 10703 (1994).

[24] S. R. Marder, B. Kippelen, A. K.-Y. Jen, and N. Peyghambarian, *Nature* **388**, 845 (1997).

[25] E. Hendrickx, B. L. Volodin, D. D. Steele, J. L. Maldonado, J. F. Wang, B. Kippelen, and N. Peyghambarian, *Appl. Phys. Lett.* **71**, 1159 (1997).

[26] Sandalphon, J. F. Wang, B. Kippelen, and N. Peyghambarian, *Appl. Phys. Lett.* **71**, 873 (1997).

[27] E. Hendrickx, J. F. Wang, J. L. Maldonado, B. L. Volodin, Sandalphon, E. A. Mash, A. Persoons, B. Kippelen, and N. Peyghambarian, *Appl. Phys. Lett.*, in press (1998).

[28] H. Kogelnik, *Bell. Syst. Tech. J.* **48**, 2909 (1969).

[29] D. D. Steele, B. L. Volodin, O. Savina, B. Kippelen, N. Peyghambarian, H. Röckel, and S. R. Marder, *Opt. Lett.* **23**, 153 (1998).

[30] A. Golemme, B. L. Volodin, B. Kippelen, and N. Peyghambarian, *Opt. Lett.* **22**, 1226 (1997).

[31] P. S. Drzaic, *Liquid crystal dispersions* (World Scientific, Singapore, 1995).

[32] B. L. Volodin, B. Kippelen, K. Meerholz, B. Javidi, and N. Peyghambarian, *Nature* **383**, 58 (1996).

[33] B. Javidi, J. Wang, and Q. Tang, *Pattern Recogn.* **27**, 523 (1994).

[34] B. Javidi, G. Zhang, A. H. Fazlollahi, and U. Efron, *Appl. Opt.* **33**, 2834 (1994).

Chapter 15

Novel Photorefractive Materials Based on Multifunctional Organic Glasses

Qing Wang, Nikko Quevada, Ali Gharavi, and Luping Yu[1]

Department of Chemistry and The James Franck Institute, The University of Chicago, 5735 South Ellis Avenue, Chicago, IL 60637

A small molecular system based on carbazole moiety was developed for photorefractive applications. One of the molecules has been shown to form stable and amorphous films and exhibit good photorefractive performance. A net optical gain of 11 cm^{-1} was observed. An attractive feature of this molecule is that its photorefractive response is fast. Because of its structural versatility, this molecular system is worth further exploration.

In this paper, we describe the synthesis and characterization of a multifunctional, small molecular system based upon carbazole moiety for photorefractive applications. It is known that photorefractive effect causes refractive index change in a noncentrosymmetric material due to photoinduced space-charge field and electro-optic effect. (1) Thus, to manifest photorefractive effect, a material must possess photoconductivity and electro-optic response. This is one of the design principles for the synthesis of the above mentioned molecules. Carbazole and its derivatives have long been recognized as good photoconductors and extensive work on composite photorefractive materials have revealed that carbazole containing systems exhibit an efficient photorefractive effect. (2-5) Carbazole also has the flexibility for structure modification. By attaching a nonlinear optical (NLO) chromophore to the carbazole, this molecule will exhibit dual functions necessary for the photorefractive effect: photoconductivity and second-order NLO activity. Furthermore, in our approach, the photocharge transport and the electro-optic (EO) functions are performed by two separate constituents. This provides us the flexibility of maximizing both charge transport and NLO activity separately and fine-tuning the photorefractive performance. A small amount (<1 wt%) of charge-generating sensitizer, such as 2,4,7-trinitro-9-fluorenone (TNF), is added to the compound to enhance photoconductivity. Another attractive feature is that we can choose different sensitizers to match the laser wavelength for the application desired since photosensitizers are available throughout the visible to near-infrared.

Results and Discussion

Synthesis and Structural Characterization. Scheme 1 describes the

[1]Corresponding author.

Scheme 1. Synthesis of Compound 5 and 7

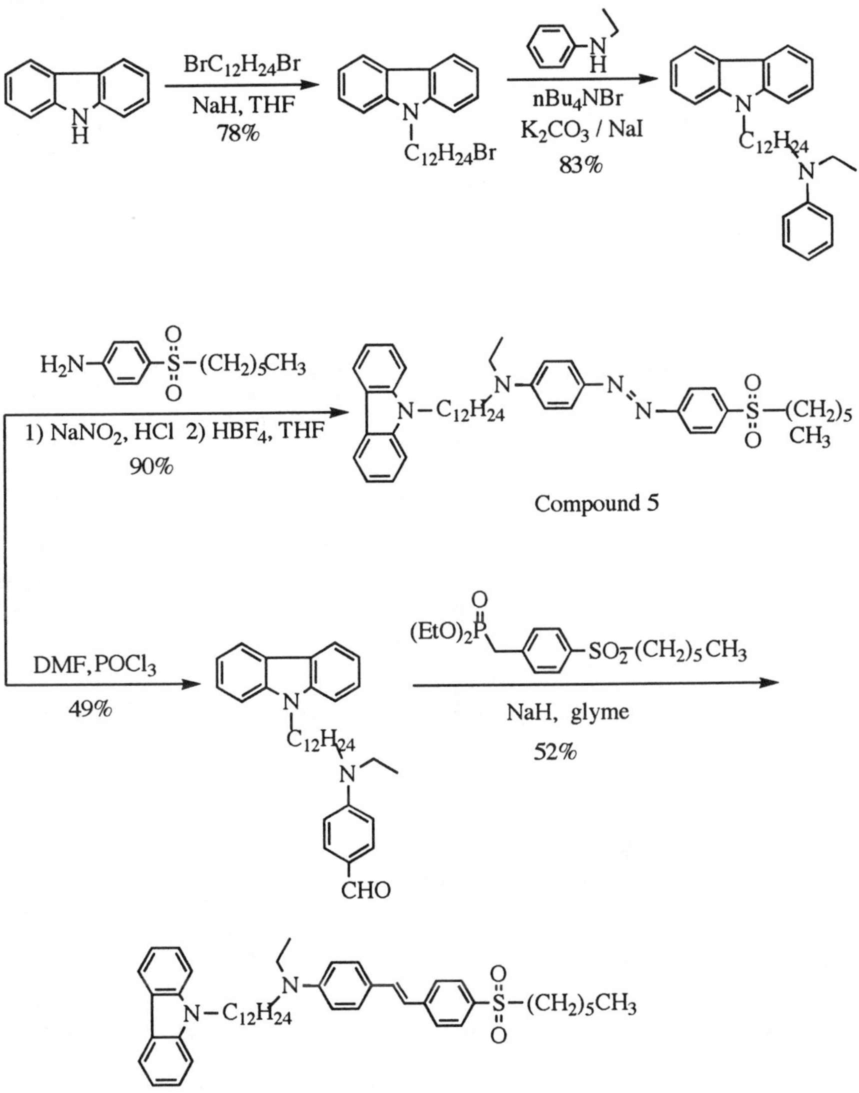

Compound 5

Compound 7

synthetic approach for compound **5** and **7** which are stable glasses at room temperature. This synthetic approach is also versatile and allows the syntheses of many other multifunctional molecules. Thus far, we have synthesized a number of carbazole based molecules (Table I). In these molecules, the carbazole groups are linked to different kinds of NLO chromophores by long aliphatic spacers. These long aliphatic chain serves the purpose of introducing disorder, preventing crystallization, and lowering the glass transition temperature (Tg) of the compounds. All compounds have been characterized by ^{1}H NMR, ^{13}C NMR, IR, UV-visible, differential scanning calorimetry (DSC) and elemental analysis. As shown in Table I, the relationship between the structure and the thermal behavior of this system is complex. The majority of these compounds will crystallize under certain condition except for compound **5** and **7**. Further studies are now in progress to examine the glass-forming properties, morphological changes of these molecules, and their effect on photorefractive responses .

Here we report the results of the experiments performed on molecule **5**. At room temperature, this molecule forms a stable amorphous glassy material which exhibits a low Tg (-35 $^\circ$C) and no crystallization based on our DSC experiments. Consequently, the films prepared from CH_2Cl_2 solutions containing 0.9 wt% TNF are transparent and of excellent optical quality. No changes in optical quality of the film was observed for more than 12 months, indicating an excellent composition stability. The UV/vis spectrum of the thin film showed an absorption maximum of the chromophore (N-ethyl-4-p-hexylsulfonylazo-aniline) at 456 nm, a shoulder of 348 nm due to the carbazole, and an absorption tail extending beyond 600 nm due to the charge transfer complex formed by carbazole and TNF (Fig. 1).

Photoconductivity and Electrooptic Activity. Molecule **5** has been shown to be both photoconductive and second-order NLO active. The samples for physical property measurement were prepared in the form of films. Uniform films of ~80μm thickness were fabricated by sandwiching the materials between two indium tin oxide (ITO) coated glass substrates, the electric field thus being applied perpendicular to the sample surface. The photoconductivity was studied at a wavelength of 632.8 nm using a photocurrent method. (6) Figure 2 shows that the photocurrent is strongly dependent on the applied electric field. A photoconductivity of 1.2×10^{-11} $\Omega^{-1}cm^{-1}/(W/cm^2)$ and a quantum yield of 5.63×10^{-6} at an applied field of 370 kV/cm were obtained. Second harmonic generation (SHG) experiments were carried out in order to probe the orientation of the NLO chromophores at room temperature. The sample was irradiated with infrared light (1064 nm) from a Q-switched Nd:YAG laser and the intensity of the second harmonic signal was monitored. As shown in Figure 3, after switching on the electric field, the value of d_{33} increased as the electric field is increased.

Steady State Properties of the Photorefractive Grating. The most important feature of the photorefractive effect, which distinguishes itself from many other mechanisms that may result in refractive index grating, is the finite phase shift between the refractive index grating and the interference pattern in a two-beam coupling (2BC) experiment. This finite phase shift causes an asymmetric energy transfer between the two writing beams. Therefore, the observation of energy transfer in 2BC experiments is usually considered a proof of the photorefractive nature of the recorded grating. (1) We performed the 2BC experiment at 632.8 nm (He-Ne laser, 30mW, p polarized). Two writing beams with equal intensities were

Table I. Chemical structures and thermal properties of the carbazole-based molecules

Compound 1: $X=\overset{(CH_2)_{12}CH_3}{\underset{S}{\langle\!\langle\rangle}}$ R= $C_{12}H_{24}$ NLO= B

Compound 2: X= H   R= $C_{12}H_{24}$ NLO= C

Compound 3: $X=\overset{(CH_2)_{12}CH_3}{\underset{S}{\langle\!\langle\rangle}}$ R= $C_{12}H_{24}$ NLO= C

Compound 4: $X=\overset{(CH_2)_{12}CH_3}{\underset{S}{\langle\!\langle\rangle}}$ R= C_6H_{12} NLO= A

Compound 5: X= H   R= $C_{12}H_{24}$ NLO= D

Compound 6: X= H   R= $C_{12}H_{24}$ NLO= E

Compound 7: X= H   R= $C_{12}H_{24}$ NLO= F

Compound	T_g (^{0}C)	T_c (^{0}C)	T_m (^{0}C)	T_d (^{0}C)
1	-11		98	345
2		65	114	>400
3		37	89	>400
4	33	60	74	324
5	-35		85	340
6	-2	12	72	>400
7	-29		58	>400

* The glass transition temperature (T_g), the crystallization temperature (T_c), the melting point (T_m), and the decomposition temperature (T_d) were measured by using the DSC-10 system. (10 ^{0} C/min)

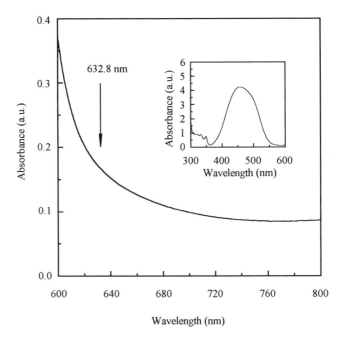

Figure 1. UV/vis spectrum of the thin film of 99.1 wt% compound 5 and 0.9 wt% TNF.

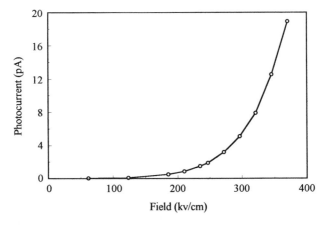

Figure 2. The electric field dependence of the photocurrent of compound 5/ TNF.

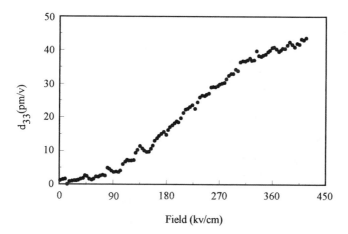

Figure 3. The d33 value of compound 5 as a function of the external electric field.

focused and intersected in the sample at incident angle of $23°$ with respect to the bisector of the two incident beams, thus producing interference fringes with $\Lambda=0.8$ μm. The normal of the sample was rotated $30°$ with respect to the bisector of the writing beams in order to provide a projection of the grating wave vector along the applied field. The experiment was performed by blocking one of the incident beams and monitoring the transmitted intensity of the other beam. As the field was increased, the intensity of one beam increased, while that of the second beam decreased. When the electric field was turned off, the intensities returned to their original levels and no energy change was observed. These observations confirm that the grating is due to an orientation photorefractivity and not to thermal or absorption grating. (7) Figure 4 shows the 2BC coupling gain constant, Γ, and the absorption coefficient, α, as a function of the applied field. The gain coefficient Γ was calculated according to the following equation: (8)

$$\Gamma = \frac{1}{L} \ln \frac{1+\alpha'}{1-\beta\alpha'}$$

where L is the optical path for the beam, α' is the ratio of the intensity modulation of the signal beam ($\Delta I_s/I_s$) and β is the intensity ratio of the two incident laser beams (I_s/I_q). The optical gain coefficient increases exponentially with the external field. Such a dependence is typical for amorphous photorefractive organic materials, as both photoconductivity and orientation of second-order NLO chromophore depend on the external electric field. Beyond the field strength of 460 kV/cm, a net gain occurs, where the gain coefficient, Γ, exceeds the absorption coefficient, α, of the material. A net gain of 11 cm^{-1} was obtained at the applied electric field of 576 kV/cm.

The index grating recorded in the material was further tested by the four-wave mixing (FWM) experiment. In this experiment, the two p-polarized beams (632.8 nm, 40 mW) were used as writing beams. The grating was read with a weak p-polarized beam (780 nm, 4 mW) counterpropagating to one of the writing beams. The diffraction efficiency, η, defined as the ratio of the diffracted to incident reading beam power, is recorded. The value of η increased rapidly with the external field E due to an increase in both the photorefractive space-charge field E_{sc} and the EO response $n^3 r_{eff}$. As shown in Figure 5, the diffraction efficiency reaches a maximum of 56% at E= 415 kV/cm and a value of 1.76×10^{-3} for the refractive index modulation (Δn) was achieved.

Kinetics of the Photorefractive Grating. The dynamics of holographic grating formations were studied by measuring the time constants of the grating formation and their electric field dependence in the FWM experiment. In the Figure 6, a typical write-erase cycle is shown. The arrows mark the position at which the second writing beam was turned on and subsequently turned off. Quantitative information about the grating growth can be obtained using the single-carrier model of photorefractivity in the limit where the decay of the diffraction efficiency is related to the space charge field decay. In this case, the diffraction efficiencies obeys the equation: (9)

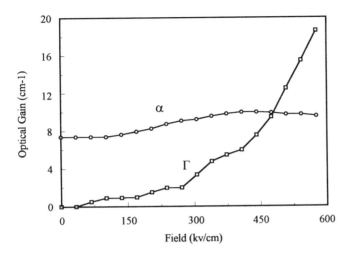

Figure 4. Dependence of the gain coefficient (Γ) and the absorption coefficient (α) on the applied field for a 60-μm-thick film of 99.1 wt% compound 5 and 0.9 wt% TNF.

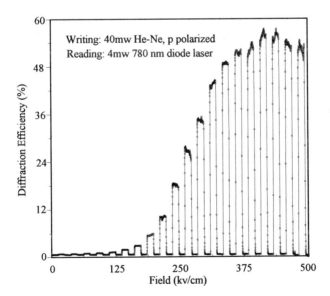

Figure 5. Dependence of the diffraction efficiency on the applied field for a 71-μm-thick film of 99.1 wt% compound 5 and 0.9 wt% TNF.

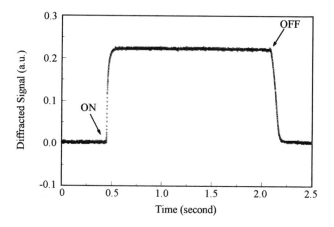

Figure 6. Typical write-erase cycle for the film of 99.1 wt% compound 5 and 0.9 wt% TNF. "ON" and "OFF" marked arrows denotes the moments of switching both writing beams "on" and switching one of the writing beams "off", respectively.

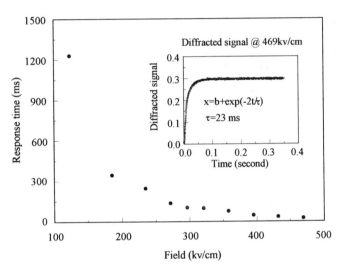

Figure 7. Electric field dependence of the time constants of grating formation for a 99.1 wt% compound 5; 0.9 wt% TNF sample of thickness 81 μm.

$$\eta(t) = \eta(0)\exp[-2t/\tau]$$

which τ is the characteristic response time and $\eta(0)$ is the steady state diffraction efficiency. For the space charge field formation, charges have to be generated and transported to dark regions. Consequently, the response time is determined by the efficiency of the photocharge generation and their drift mobility. As the results show, the grating writing time constant at an electric field of 469 kV/cm is around 23 ms, which is so far the fastest response time reported for organic photorefractive materials to our knowledge. This may be due to the large charge mobility, efficient charge transporting property of carbazole, and "small" size of the molecule as compared with photorefractive polymers. Such a fast response time makes this type of material promising for real-time holographic applications. Figure 7 shows the dependence of the response time on the applied field.

Conclusions

A carbazole-based multifunctional glassy molecule has been shown to exhibit large photorefractive effect. This system also shows very fast kinetics of photorefractive grating formation, occurring in the millisecond time scale. This is an attractive feature from the point of view of real time applications. It was also found that the photorefractive properties were strongly dependent upon the applied field due to the low Tg of the material. The synthetic flexibility and simplicity of this system allows modification of the structures to further improve the photorefractive properties. Work on similar type of materials is currently in progress.

Acknowledgments

This work was supported by the U.S. Air Force Office of Scientific Research and National Science Foundation. Support from the National Science Foundation Young Investigator program is gratefully acknowledged.

Literature Cited

1. *Photorefractive Materials and their Applications* ; Gunter, P.; Huignard, J. P., Eds.; Springer-Verlag: Berlin, 1988; Vol. 1 and 2.
2. Moerner, W. E.; Silence, S. M. *Chem. Rev.*, **1994**, *94*, 127.
3. Zhang, Y.; Burzynski, R.; Ghosal, S.; Casstevens, M. K. *Adv. Mater.*, **1996**, *8*, 111.
4. Meerholz, K.; Volodin, B. L.; Sandalphon; Kippelen, B.; Peyghambarian, N. *Nature*, **1994**, *371*, 497.
5. Grunnet-Jepsen, A.; Thompson, C. L.; Moerner, W. E. *Science*, **1997**, *277*, 549.
6. Li, L.; Lee, J. Y.; Yang, Y.; Kumer, J.; Tripathy, S. *Appl. Phys.*, **1991**, *279*, 1353.
7. Wiederrecht, G. P.; Yoon, B. A.; Wasielewski, M. R. *Science*, **1995**, *270*, 1794.
8. Yu, L. P.; Chen, Y. M.; Chan, W. K. *J. Phy. Chem.*, **1995**, *99*, 2797.
9. Orczyk, M. E.; Swedek, B.; Zieba, J.; Prasad, P. N. *J. Appl. Phys.*, **1994**, *76*, 4990.

Chapter 16

An Oligo(3-alkylthiophene) Containing Material Showing High Photorefractivity

Wenjie Li, Alireza Gharavi, Qing Wang, and Luping Yu[1]

Department of Chemistry and The James Franck Institute, The University of Chicago, 5735 South Ellis Avenue, Chicago, IL 60637

A multifunctional photorefractive material was designed and synthesized by incorporation of an oligo(3-alkylthiophene) moiety with a nonlinear optical (NLO) chromophore. The conjugated oligo(3-alkylthiophene) segment played a dual role of the photocharge generator and the charge transporter. Large net optical gain coefficient and diffraction efficiency, as well as a fast response time, were achieved in this small molecule system. The photorefractive performance of this molecule showed a strong dependence on the applied electric field due to the orientational enhancement effect as well as the linear electro-optic effect. These properties make it an attractive material candidate for molecular optical devices.

The photorefractive (PR) effect is based on a combination of photoconductivity and electro-optic (E-O) effect (*1-2*). It is defined as the spatial modulation of the refractive index due to charge redistribution in an optically nonlinear and photoconducting material. This effect arises when photogenerated charge carriers separate by drift and/or diffusion processes and become trapped to produce a nonuniform charge density distribution. The displacement of charges in turn creates an internal space-charge field which modulates the refractive index of the material through the electro-optic effect. The refractive index grating can then be detected by utilizing holographic techniques such as two-beam coupling and four-wave mixing experiments.

The PR effect was originally found in ferroelectric single crystals of LiNbO₃ and LiTaO₃ (*3-6*). Since then, numerous prototype devices based on the PR materials have been proposed and some of them have been demonstrated, such as reversible optical holography, image correlation, amplification, dynamic novelty filtering, and other optical signal processing techniques (*1-2,7-12*). Recently, the PR effect was also observed in polymeric materials which exhibit both electro-optic effect and photoconductivity (*7-9*). A relatively convenient approach to prepare PR polymers is to mix all of the necessary molecular components into a polymer matrix forming a composite (*10-12*). Extremely large photorefractive effects were observed in some systems when specific compositions were used (*13-16*). A

[1]Corresponding author.

© 1999 American Chemical Society

general observation is that only those composite materials which have low glass transition (T_g) temperatures (lower than room temperature) show large net optical gain. It was suggested that the nonlinear optical (NLO) chromophores in these low T_g materials could be aligned not only by the externally applied electric field, but also in situ by the sinusoidally varying space-charge field during the grating formation. As a result, the total poling field that orients the molecules is the superposition of the uniform external field and the spatially modulated internal space-charge field. This effect was originally referred to as orientational enhancement effect (17). However, one major drawback for these composite materials is their phase separation, caused by the incompatibility between small molecules and the polymer host. In some cases, the phase separation may completely ruin their optical properties.

To utilize the advantage of the orientational enhancement effect, and at the same time to minimize the phase separation, we designed and synthesized a simple, small molecular system which contains a 3-alkyl-substituted oligothiophene molecule covalently connected to a NLO chromophore. The design concept of this molecule is inherited from our previous work on conjugated photorefractive polymers (18-21). The molecule contains all the functionalities necessary to show the PR effect. Polythiophenes and their derivatives are known for their interesting electrical and optical properties (22-23). The oligo(3-alkylthiophene) is photoconductive in the visible light region. The E-O component could be generated by aligning the dipole of the NLO chromophore under an applied electric field at room temperature. Amorphous films of this compound can easily be prepared by slightly heating the material and then sandwiching it with two indium tin oxide (ITO) coated glass slides. The films thus made are transparent and electric field can be applied through the conductive ITO coatings.

Experimental Section

Synthesis. Synthesis of compound **3** (Scheme 1) was described in our previous publication (24). Compound **5** was prepared according to the literature procedures (25) All of the other chemicals were purchased from the Aldrich Chemical Company and used as received unless otherwise noted.

Compound **2**: To a stirred solution of methyl 4-methyl benzoate (4.5 g, 30 mmol) and NBS (5.34 g, 30 mmol) in 25 ml carbon tetrachloride, benzoyl peroxide (36 mg, 0.15 mmol) was added. The mixture was refluxed for 4 h before it was filtered by suction filtration when the solution was still hot. The filtrate was concentrated by rotary evaporation and was mixed with triethyl phosphite (12.5 g, 75 mmol) in a 50 ml round-bottom flask. The resulting mixture was then stirred under reflux for 20 h. The excess triethyl phosphite was removed by vacuum distillation. The residue was purified by flash chromatography (silica gel, ethyl acetate / methanol (100 / 2)) to give 7.3 g colorless liquid (85% yield): ^{1}H NMR (CDCl$_3$, ppm) δ 7.92 (d, 2 H), 7.32 (d, 2 H), 3.99 (m, 4 H), 3.88 (s, 3 H), 3.18 (d, 2 H), 1.23 (t, 6 H).

Compound **4**: A solution of sodium hydride (42 mg, 1.7 mmol) and compound **2** (490 mg, 1.7 mmol) in 2 ml ethylene glycol dimethyl ether (DME) was stirred at room temperature for 0.5 h. About one-tenth of the above mixture was transferred to a solution of compound **3** (280 mg, 0.17 mmol) in 5 ml DME. The resulting mixture was stirred under refluxing for 15 h. The reaction solution was poured into an ice-water mixture. The product was extracted by ether three times. The combined organic layers were concentrated by rotary evaporation. The residue was then dissolved in 2 ml benzene, and a solution of potassium hydroxide

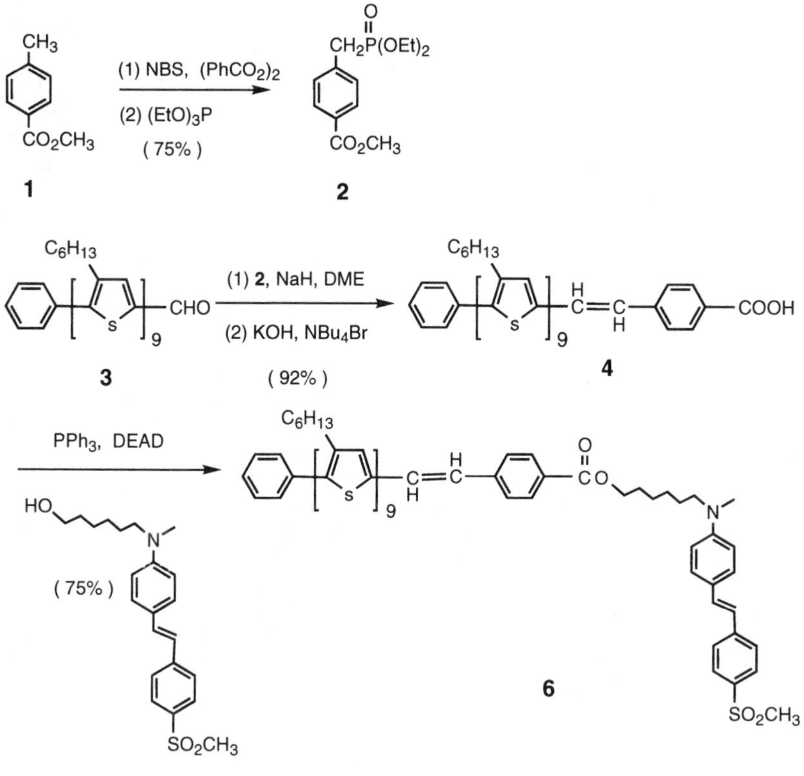

Scheme 1. Synthesis of oligothiophene containing photorefractive material.

(480 mg, 8.6 mmol) in 2 ml H_2O / DMSO (1/1) was added. Tetrabutylammonium bromide (10 mg, 0.03 mmol) was also added. The resulting reaction mixture was stirred at 85 °C for 2 h. The solution was poured into water. Acetic acid was used to neutralize the solution. The product was extracted by ether and dried over $MgSO_4$. The solvent was removed by rotary evaporation, and the crude product was purified by flash chromatography (silica gel, hexane / ethyl acetate (100 / 60))to give 0.27 g red solid (92% yield). ^{1}H NMR (CDCl$_3$, ppm) δ 8.01 (d, 2 H), 7.48 (d, 2 H), 7.41 (d, 2 H), 7.38 (d, 2 H), 7.30 (m, 1 H), 7.21 (d, 1 H), 6.99 (s, 1 H), 6.92 (m, 8 H), 6.84 (d, 1 H), 2.78 (t, 16 H), 2.64 (t, 2 H), 1.69 (m, 18 H), 1.33 (broad, 54 H), 0.90 (m, 27 H).

Compound 6: Diethyl azodicarboxylate (DEAD) (40 mg, 0.23 mmol) was added to a solution of compound 4 (280 mg, 0.16 mmol), compound 5 (81 mg, 0.21 mmol) and triphenylphosphine (60 mg, 0.23 mmol) in 3 ml THF. The mixture was stirred at room temperature for 3 h. After the solvent was removed with a rotary evaporator, the crude product was purified by a silica gel chromatography column using hexane / ethyl acetate (2 / 1) as the eluent to give 0.25 g red solid (75% yield). ^{1}H NMR (CDCl$_3$, ppm) δ 8.01 (d, 2 H), 7.86 (d, 2 H), 7.60 (d, 2 H), 7.50 (d, 2 H), 7.42 (m, 6 H), 7.36 (m, 1 H), 7.24 (d, 1 H), 7.18 (d, 1 H), 7.03 (s, 1 H), 6.98 (m, 8 H), 6.89 (d, 1 H), 6.88 (d, 1 H), 6.68 (d, 2 H), 4.32 (t, 2 H), 3.37 (t, 2 H), 3.04 (s, 3 H), 2.99 (s, 3 H), 2.80 (t, 16 H) 2.67 (t, 2 H), 1.68 (m, 20 H), 1.34 (m, 60 H), 0.91 (m, 27 H). Anal. Calcd for $C_{127}H_{165}NS_{10}O_4$: C, 72.97; H, 7.96. Found: C, 73.05; H, 7.99.

Characterization. Nuclear magnetic resonance (NMR) spectra were obtained using a GE QE 300 MHz spectrometer. Infrared (IR) spectra were recorded on a Nicolet 20 SXB FTIR spectrometer. A Shimadzu UV-2401PC UV-Vis recording spectrophotometer was used to record the absorption spectra.

Sample Preparation for Physical Property Measurements. The samples for physical property measurements were prepared in the form of films sandwiched between two ITO coated glass substrates for application of the dc field. The voltage was applied across the films so that the dc field is perpendicular to the sample surface. A filtered solution of compound 6 in CH_2Cl_2 was cast manually onto an ITO glass substrate and dried under vacuum overnight to remove the solvents. The sample was then heated to about 80 °C and sandwiched with another ITO glass substrate. A typical film thickness ranged from 60 μm to 75 μm.

Results and Discussion

Synthesis. The oligothiophene containing photorefractive compound was synthesized according to Scheme 1. In our previous publication (24), we have described the synthesis of oligothiophene aldehyde 3. The Wittig reaction between aldehyde 3 and phosphonate 2 in refluxing ethylene glycol dimethyl ether, followed by the hydrolysis in a base condition, afforded the carboxylate acid 4. The target molecule (compound 6) was synthesized in 75% yield by utilizing the Mitsunobu reaction between compound 4 and the NLO chromophore 5. Compound 6 is a red solid at room temperature and is quite soluble in most organic solvents.

Structural Characterization. The structure of compound 6 was verified by ^{1}H NMR, ^{13}C NMR and elemental analysis. In the FTIR spectrum of compound 6, the absorption band at 1713 cm^{-1} is ascribed to the stretching vibration of the ester C=O bond. The asymmetric stretching vibration bands of the ester -C(=O)-O and O-C- occur at 1276 cm^{-1} and 1112 cm^{-1} receptively. Another strong absorption band was observed at 957 cm^{-1}, which is a typical absorption of the out-of-plane bending vibration of the trans-substituted vinylene groups. These results further support the structure of compound **6**.

The UV/vis spectrum of compound **6** in CHCl$_3$ shows a red shift of 24 nm in the absorption peak compared to the absorption spectrum of the NLO chromophore **5**. The solid-state UV/vis spectrum of compound **6** exhibits the absorption maximum at 440 nm and a tail extending beyond 600 nm (Figure 1). This red-shifted absorption enables us to excite the compound by utilizing a He-Ne laser (λ = 632.8 nm).

Physical Property Measurements. The films prepared from the CH$_2$Cl$_2$ solution are transparent and photoconductive as studied at the working laser wavelength of 632.8 nm. The photoconductivity was determined at by measuring the voltage which resulted from the passing of the photocurrent through the thin film and a 1-MΩ resistor. A lock-in amplifier was used to measure the photovoltage. The photocurrent was found to be electric field-dependent as shown in Figure 2; as the electric field increases, the photocurrent increases superlinearly, a phenomenon similar to most of the PR polymers. A photoconductivity of ca. 1.59 $\times$ 10^{-9} Ω^{-1} cm^{-1} /(W/cm^2) was obtained for compound **6** under an external electric field of 462 kv/cm at the wavelength of 632.8 nm. The dark conductivity of this sample also increases with the applied electric field but remains at least one magnitude lower than the photoconductivity. The quantum yield of photocharge generation, ϕ, defined as the number of free electrons and holes formed per absorbed light quantum, is usually used to characterize the photogeneration process. It can be determined by the following equation (26):

$$\phi = \frac{I_{ph}}{eP[1-\exp(-\alpha L)]}$$

where I$_{ph}$ is the photocurrent, P is the power of the incident light, α is the absorption coefficient of the material, L is the distance between the electrodes. From the photoconductivity results, the maximum quantum yield for photocharge generation was estimated to be 1.05 $\times$ 10^{-2} at 462 kv/cm.

The dipole of the NLO chromophore in this small molecule system was readily aligned by the applied electric field at room temperature, as indicated by the value of the second harmonic coefficient, d$_{33}$, as a function of the electric field at room temperature (Figure 3). The second harmonic generation (SHG) experiments were performed by using a fundamental frequency (1064 nm) of a model-locked Nd:YAG laser. The SHG signal generated by the sample was detected by a photomultiplier tube (PMT), then amplified, and averaged in a boxcar integrator. The d$_{33}$ value shows a trend of leveling off as the field further increases, possibly due to the saturation of the electric field induced chromophore alignment.

The photorefractive properties of this material were examined by the two-beam coupling, four-wave mixing, and holographic image formation experiments at different applied electric fields. In the two-beam coupling experiment, two coherent laser beams (632.8 nm, 30 mW, p-polarized) intersect in the PR sample to generate

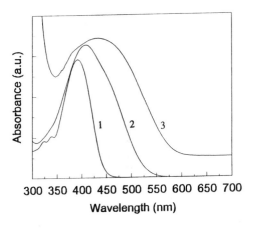

Figure 1. UV/vis spectra of the NLO compound **5** in CHCl₃(1), compound **6** in CHCl₃(2) and solid-state(3).

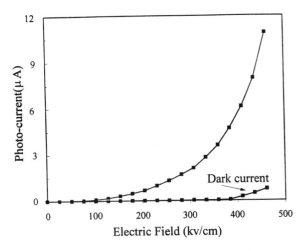

Figure 2. The photocurrent and dark current response of compound **6** as a function of the external electric field.

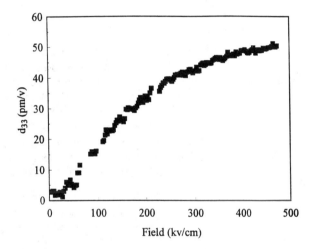

Figure 3. The second harmonic generation coefficient d_{33} of compound **6** as a function of the external electric field.

the refractive index grating. The grating component that is 90° out of phase with the light intensity pattern leads to the asymmetric energy exchange between the two beams. As a result, one beam gains energy, the other one losses energy. The two writing laser beams used in our experiment have comparable intensities. The normal of the sample was tilted by 30° with respect to the bisector of the two writing beams in order to obtain a non-zero projection of the E-O coefficient. The experiment was carried out by chopping one of the two incident beams while monitoring the transmitted intensity of the other.

After the electric field was switched on, the asymmetric energy transfer between the two beams was observed (inset of Figure 4), which verifies the photorefractive nature of the sample. The optical gain coefficient Γ could be calculated from the intensities of the two beams by the following equation (1-2):

$$\Gamma = \frac{1}{L} \ln(\frac{1+\alpha}{1-\beta\alpha})$$

where L is the optical path length for the beam with gain, α is the ratio of the intensity modulation of the signal beam $\Delta I_s/I_s$, and β is the intensity ratio of the two writing beams I_s/I_q. The gain coefficient Γ also showed a strong dependence on the applied electric field (Figure 4). This strong field dependence of the optical gain was anticipated because of the fact that both the SHG signal and the quantum yield of the photogeneration of charge carriers are strongly dependent on the external field. A Γ value of 102 cm^{-1} was obtained at E = 706 kv/cm. This gain exceeds the absorption coefficient (19 cm^{-1} at 632.8 nm) of the sample, giving a net optical gain of 83 cm^{-1}. When the external field was switched off, the gain and loss signals disappeared immediately, eliminating the possibility of beam coupling due to thermal grating (15).

The index grating formed can be probed by the four wave mixing (FWM) experiment where the diffraction efficiency of a probe laser beam ($\lambda = 780$ nm, 3 mW, p-polarized) from the photorefractive grating was measured. According to the Bragg diffraction condition, the probe beam was diffracted by the index grating and measured by a lock-in amplifier. The diffraction efficiency η is calculated as the ratio of the intensity of the diffracted beam measured after the sample and the intensity of the probe beam before the sample. As shown in Figure 5, the diffraction efficiency also has a strong dependence on the electric field and reaches almost 40% at E = 706 kv/cm. From this value, a refractive index modulation of $\Delta n = 2.55 \times 10^{-3}$ was calculated according to the following expression (16):

$$\eta = \sin^2[\frac{\pi d \Delta n \cos(\theta_1 - \theta_2)}{\lambda\sqrt{\cos\theta_1\cos\theta_2}}]$$

where d is the sample thickness, λ is the probe laser wavelength, and θ_1 and θ_2 are the internal angles of incidence of the two writing beams.

The time constant for the grating formation, τ, can also be measured in the FWM experiment. The initial growth of the diffracted signal was recorded by a oscilloscope and fit with the function $\eta(t) \sim \exp[-2t/\tau]$ for the single-carrier model of refractive grating kinetics (14, 27). A fast response time of 42 ms was obtained at

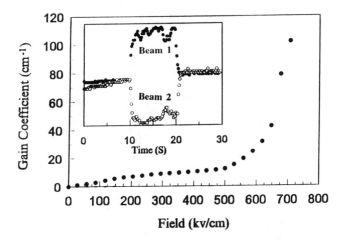

Figure 4. The dependence of the two-beam coupling gain coefficient on the applied external electric field. The inset shows the two beam coupling signal.

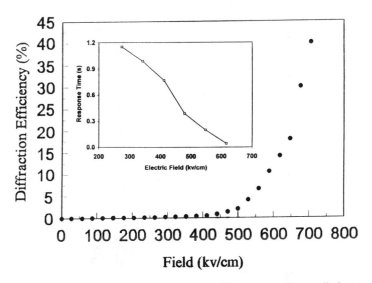

Figure 5. The dependence of the diffraction efficiency on the applied external electric field. The inset shows the field dependence of the response time.

246

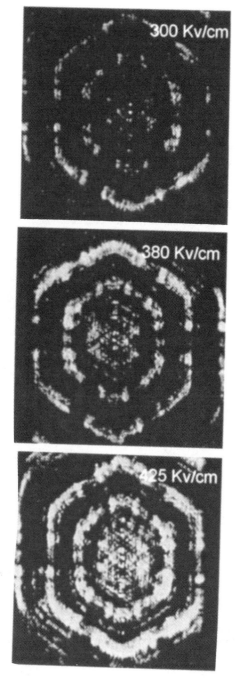

Figure 6. Holographic image formation at different applied field. The object is shown in the bottom right of the series.

247

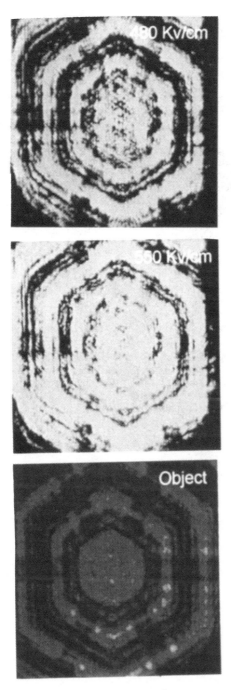

Figure 6. *Continued.*

an applied field of 616 kv/cm. The speed for the PR grating is determined by the photogeneration rate of charge carriers and by their migration over distances in the range of several times the wavelength of the laser source. Excess trapping sites could slow down the migration rate of the charge carriers. In organic materials, normally the trapping sites are impurities and structural defects. The fast response time constant could be attributed to the relatively higher purity and less defects in our material than the polymeric counterparts. The grating formation time constant was found to increase almost twenty-six times when the electric field was decreased from 616 kv/cm to the lowest value for which a grating was detectable (inset of Figure 5).

By incorporating a mask into one of the wring beams in the FWM experiment, a photorefractive hologram can be written in the sample. The image was carried by an expanded beam and then was focused on the sample. The hologram was then detected by the probe reading beam under the Bragg condition as the diffracted signal and was recorded by a CCD camera. Figure 6 shows a series of holograms recorded under successively increasing electric field. Erasure of the holograms occurred quickly upon blocking the writing beams, consistent with the fast grating formation process.

Conclusions

In conclusion, we have designed and synthesized a multifunctional photorefractive molecule by combining the well-known photoconductor oligothiophene and a nonlinear optical (NLO) chromophore. In this single molecular system, the conjugated oligothiophene part acts as both the photocharge generator and the charge transporter. This material showed a greatly improved photorefractive performance. A net optical gain of 83 cm^{-1} and a diffraction efficiency of nearly 40% were obtained at an applied field of 706 kv/cm. The system also has a fast response time to the applied electric field. The time constant for the index grating formation at an applied field of 616 kv/cm was measured to be 42 ms. The high photorefractivity and the fast dynamic process make this material very attractive for further exploration.

Acknowledgments

This work was supported by the U.S. National Science Foundation and Air Force Office of Scientific Research. Support from the National Science Foundation Young Investigator program is gratefully acknowledged.

Literature Cited

1. Gunter, P.; Huignard, J. P.; Eds. *Photorefrative Materials and their Applications*; Springer-Verlag: Berlin, **1988**; Vols 1 and 2.
2. Yariv, A.; *Optical Electronics*, 4th ed.; Harcourt Brace Jovanovich: Orlando, FL, **1991**.
3. Ashkin, A.; Boyd, G. D.; Dziedzic, J. M.; Smith, R. G.; Ballmann, A. A.; Nassau, K.; *Appl. Phys. Lett.* **1966**, *9*, 72.
4. Chen, F. S.; *J. Appl. Phys.* **1967**, *38*, 3418.
5. Chen, F. S.; LaMacchia, J. T.; Fraser, D. B.; *Appl. Phys. Lett.* **1968**, *13*, 223.
6. Chen, F. S.; *J. Appl. Phys.* **1969**, *40*, 3389.
7. Sunter, K.; Hulliger, J.; Gunter, P.; *Solid State Commun.* **1990**, *74*, 867.

8. Sunter, K.; Gunter, P.; *J. Opt. Soc. Am. B* **1990**, *7*, 2274.
9. Ducharme, S.; Scott, J. C.; Tweig, R. J.; Moerner, W. E.; *Phys. Rev. Lett.* **1991**, *66*, 1846.
10. Moerner, W. E.; Silence, S. M.; *Chem. Rev.* **1994**, *94*, 127.
11. Yu, L.P.; Chan, W. K.; Peng, Z.; Gharavi, A.; *Acc. Chem. Res.* **1996**, *29*, 13.
12. Zhang, Y.; Burzynski, R.; Ghosal, S.; Casstevens, M. K.; *Adv. Mater.* **1996**, *8*, 111.
13. Meerholz, K.; Volodln, B. L.; Sandalphon, Klppelen, B.; Peyghambarian, N.; *Nature* **1994**, *371*, 497.
14. Orczyk, M. E.; Swedek, B.; Zieba, J.; Prasad, P.N.; *J. Appl. Phys.* **1994**, *76*, 4990.
15. Wiederrecht, G. P.; Yoon, B. A.; Wasielewski, M. R.; *Science* **1995**, *270*, 1794.
16. Lundquist, R. M.; Wortmann, R.; Geletneky, C.; Twieg, R. J.; Jurich, M.; Lee, V. Y.; Moylan, C. R.; Burland, D. M.; *Science* **1996**, *274*, 1182.
17. Moerner, W. E.; Silence, S. M.; Hache, F.; Bjorklund, G. C.; *J. Op. Soc. Am. B.* **1994**, *11*, 320.
18. Yu, L. P.; Chan, W. K.; Bao, Z.; Cao, S. X. F.; *Macromolecules* **1993**, 26, 2216.
19. Chan, W. K.; Chen, Y.; Peng, Z.; Yu, L. P.; *J. Am. Chem. Soc.* **1993**, *115*, 11735.
20. Peng, Z.; Bao, Z.; Yu, L. P.; *J. Am. Chem. Soc.* **1994**, *116*, 6003.
21. Peng, Z.; Gharavi, A.; Yu, L. P.; *J. Am. Chem. Soc.* **1997**, *119*, 4623.
22. Tourillion, G.; in *Handbook of Conducting Polymers*; Vol.1 (Ed: T. A. Skotheim), Marcel Dekker, New York, **1986**, pp. 294-351.
23. Roncali, J.; *Chem. Rev.* **1992**. *92*, 711.
24. Li, W.; Maddux, T.; Yu, L.P.; *Macromolecules*, **1996**, *29*, 7329.
25. Rebello, D. R.; Dao, P.T.; Phelan, J.; Revelli, J.; Schildkraut, J. S.; Scozzafava, M.; Ulman, A.; Willand, C. S.; *Chem. Mater.* **1992**, *4*, 425.
26. Yu, L.P.; Chan, W.K.; Peng, Z.; Li, W.; Gharavi, A.; "Photorefractive Polymers", in *Handbook of Organic Conductive Molecules and Polymers: Vol.4, Conducting Polymers: Transport, Photophysics, and Applications* (Ed: H. S. Nalwa), John Wiley & Sons, Washington DC, **1997**, pp. 123-132.
27. Kukhtarev, N.; Markov, V.; Odulov, S.; *Opt. Commun.*, **1977**, *23*, 338.

Chapter 17

Methodology for the Synthesis of New Multifunctional Polymers for Photorefractive Applications

K. D. Belfield, C. Chinna, O. Najjar, S. Sriram, and K. J. Schafer

Department of Chemistry and Biochemistry, University of Detroit Mercy, P.O. Box 19900, Detroit, MI 48219-0900

Well-defined multifunctional polymers, bearing charge transporting (CT) and nonlinear optical (NLO) functionality covalently attached to each repeat unit, were synthesized and characterized. A series of polysiloxanes were prepared that possessed either diarylamine, carbazole, or triarylamine CT moieties. High T_g polyurethanes, comprised of alternating CT and NLO units, were synthesized from NLO functionalized diols and N,N-bis(4-isocyanatophenyl)aniline. Polyimides, exhibiting high thermal stabilities, were obtained from a fluorinated dianhydride and a multifunctional (CT and NLO) diamine. Polymers were prepared containing two NLO chromophores per repeat unit, thereby doubling the NLO chromophore density over conventional NLO materials.

Photorefractivity, first reported in 1966 as a detrimental effect in lithium niobate, is now considered one of the most promising mechanisms for producing holograms (1). Since the first report of an organic photorefractive crystal in 1990, substantial efforts have been expended in the development of photorefractive organic crystals, glasses, and polymers. Photorefractivity holds great potential in holographic optical data storage, optical computing and switching, and integrated optics (1). Photorefractive materials can, in principle, execute such integrated optoelectronic operations as switching, modulation, threshholding, and parallel processing for image processing and display.

The photorefractive (PR) effect is defined as a spatial modulation of the refractive index due to charge redistribution in an optically nonlinear material. The effect involves photogeneration of charge carriers, by a spatially modulated light intensity, which separate, become trapped, and produce a non uniform space-charge field. This internal space-charge field can modulate the refractive index to create a phase grating or hologram (2). In order to be photorefractive, a material has to combine photosensitivity and photoconductivity, and possess an electric field dependent refractive index. Uniform illumination eliminates the space-charge distribution and erases the hologram. The materials are suitable for dynamic (i.e., reversible) and real time holography.

Organic Crystals and Polymeric Photorefractive Materials

In 1990, the first observation of the PR effect in an organic material utilized a carefully grown nonlinear optical organic crystal, 2-(cyclooctylamino)-5-nitropyridine, doped with 7,7,8,8-tetracyanoquinodimethane (3). The growth of high quality doped organic crystals, however, is a very difficult process because most dopants are expelled during crystal preparation. Polymeric materials on the other hand, can be doped with various molecules of quite different sizes with relative ease. Further, polymers may be formed into a variety of thin films and waveguide configurations as required by the application. The noncentrosymmetry requisite for second order nonlinearity of polymers containing NLO chromophores can be produced by poling, whereas in crystals it depends on the formation of crystals with noncentrosymmetric crystal structures, a formidable limitation. Photorefractive polymers also promise many of the traditional advantages associated with polymers, such as good thermal stability, low dielectric constant, geometric flexibility, and ease of processing (2).

The PR effect can occur in certain multifunctional materials which both photoconduct and show a dependence of index of refraction upon the application of an electric field. In order to manifest the photorefractive effect, it is thought that these polymers must contain photocharge generating (CG) and transporting (CT) functionality, charge trapping sites, and nonlinear optical (NLO) chromophores. Finally, the index of refraction must depend on the space charge field. The mere physical presence of these functionalities does not guarantee that any diffraction grating produced by optical illumination arises from the PR effect. Hence, a polymeric material can potentially be made photorefractive by incorporating these properties directly into the polymer or by doping guest molecules into the polymer to produce these properties. It is usually desirable to covalently incorporate some of the elements into the polymeric host itself to minimize the amount of inert volume in the material.

Photorefractive Polymer Composites

A number of composite systems have been reported in which a polymer possessing one of the requisite functionalities, e.g., covalently attached NLO chromophores, is doped with the others, e.g., CG and CT dopants, or a photoconductive polymer, e.g., poly(N-vinylcarbazole), is doped with the sensitizer and NLO dopants (1, 4-7). The high dopant loading levels necessary (up to 50 wt%) result in severe limitations of doped systems, including diffusion, volatilization, and/or phase separation (crystallization) of the dopants. In addition, plasticizers and compatibilizers are often used to lower the glass transition temperature (T_g) of the polymer and increase solubility of the dopants in the host polymer, respectively. This, in turn, dilutes the effective concentration of CT, CG, and NLO moieties, diminishing the efficiency and sensitivity of the photorefractive polymer composite.

Single Component PR Polymers

It naturally followed that polymers were developed in which all the necessary moieties were covalently attached, particularly CT and NLO. Several "fully functionalized" polymeric systems have been reported although not all have unambiguously demonstrated photorefractivity. Among the reports of this class of polymer are a methacrylate polymer containing carbazole (CT) and tricyanovinylcarbazole groups (NLO and CG) linked to the polymer by an alkyl spacer (8), a random copolymer of a 4-amino-4'-nitrostilbene-functionalized methacrylate, carbazole-derived methacrylate, and a long chain alkyl methacrylate in which the long alkyl chain served to plasticize and solubilize the polymer (9), a 1:1 random copolymer of a 4-amino-4'-nitrostilbene-

252

functionalized methacrylate and methyl methacrylate (DANS/MMA) was studied for PR properties (10), a fully functionalized PR polymer in which the NLO chromophore, charge transporter, and the charge generator were all attached to a polyurethane backbone (11), a conjugated poly(phenylene vinylene) copolymers (12,13), a polyimide was reported that contained NLO chromophores and porphyrin electron-acceptor moieties (14), random and block polyesteramide copolymers in which a charge transporting (diphenylhydrazone-containing) diol and NLO-functionalized diamine was reacted with a diacid chloride (15), and ring-forming, molybdenum-catalyzed olefin metathesis polymerization between NLO and CT-functionalized diacetylene monomers (16). However, in all of these approaches, shortcomings included random distribution and spacing of charge transport and NLO moieties, factors that may lead to inhomogeneities and inefficient charge transport.

Our research aims to address both stability and synthetic efficacy challenges in creating well-defined, single-component PR polymers. Herein, we report the synthesis of multifunctional polymers having CT and NLO functionality attached to each repeat unit of the polymer. In fact, some of our systems contain two NLO chromophores per repeat unit, thereby doubling the NLO chromophore density over conventional NLO polymers. Two paradigms are introduced, one in which the NLO chromphores are incorporated into the CT moiety, the second involves formation of a perfectly alternating arrangment of CT and NLO units (Figure 1).

Siloxane polymers, recently prepared in our laboratories based on the former strategy, are illustrated in Figures 2 and 3 (17). The syntheses were quite efficient. Each reaction on the polymer, i.e., hydrosilylation, regiospecific bromination, and Heck reaction, went to hig conversion, confirmed spectroscopically and by elemental analysis. These were the first reports of the use of phosphonate groups as electron-withdrawing functionalities in NLO-chromophore-containing polymers (17,18).

In fact, we recently reported the first synthesis of and electro-optic characterization of a 4-amino-4'-phosphorylated stilbene derivative (19). This, indeed, exhibited second-order NLO properties.

The synthetic methodology we developed is characterized by "high fidelity" and is adaptable to allow the preparation of a "catalog" of polymers. This latter aspect is potentially important for systematic structure-property investigations in which the CT and/or NLO functionalities can quite easily be systematically varied. These polymers, when doped with CG dopants, are expected to exhibit photorefractive behavior.

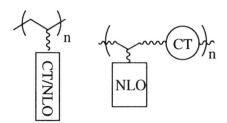

Figure 1. Paradigms for well-defined CT and NLO-containing single component-type PR polymers.

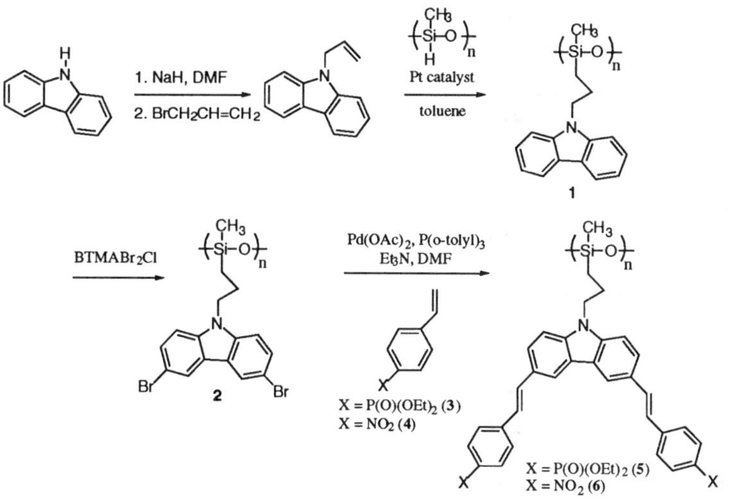

Figure 2. Synthesis of carbazole-based PR polysiloxanes.

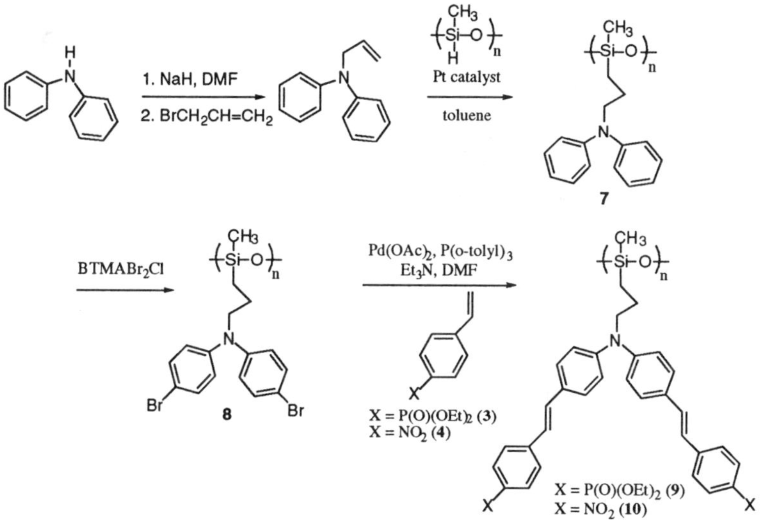

Figure 3. Synthesis of diphenylamine-based PR polysiloxanes.

Experimental

Synthesis of poly[4-(N,N-diphenylamino)-1-(3-propoxymethylbenzene) methylsiloxane] 12. Poly(hydrogen methylsiloxane) (1.6 g, 26.37 mmol) was added to a solution of 4-(N,N-diphenylamino)-1-(2-propenoxy)methylbenzene (10.8 g, 34.28 mmol) in 70 mL toluene. After the addition of dichloro(dicyclopentadiene)platinum (1.05 mg, 0.0026 mmol), the mixture was heated to 60 °C for 5 days. The disappearance of the Si-H peak (2150 cm^{-1}) was monitored by FT-IR. The mixture was then poured in 600 mL cold hexane. The solid obtained was dissolved in THF and reprecipitated twice in hexane (65% yield). ^{1}H NMR (300 MHz, CDCl$_3$): δ 0-0.25 (bm, 3H, SiCH$_3$), 0.45-0.65 (bs, 2H, SiCH$_2$), 1.55-1.75 (bs, 2H, CR$_2$CH$_2$CR$_2$), 3.30-3.50 (bs, 2H, CR$_2$CH$_2$O), 4.25-4.45 (bs, 2H, ArCH$_2$), 6.80-7.20 (bm, 14H, ArH).

Synthesis of poly[4-(N,N-bis(4-bromophenyl)amino)-1-(3-propoxymethylbenzene) methylsiloxane] 13. To a solution of **12** (4 g, 10.66 mmol) in 280 mL DMF was added a solution of N-bromosuccinimide (3.736 g, 20.99 mmol) in 60 mL DMF. The reaction mixture was stirred at room temperature overnight. The mixture was then poured into cold water and the solid obtained was precipitated from DMF. ^{1}HNMR (300 MHz, CDCl$_3$): δ0-0.25 (bm, 3H, SiCH$_3$), 0.45-0.60 (bs, 2H, SiCH$_2$), 1.50-1.80 (bs, 2H, CR$_2$CH$_2$CR$_2$), 3.30-3.50 (bs, 2H, CR$_2$CH$_2$O), 4.20-4.50 (bs, 2H, ArCH$_2$), 6.70-7.0 (bm, 6H, ArH, *ortho* to Br and CH$_2$), 7.0-7.7.40 (bm, 6H, ArH, *ortho* to N). Anal. Calcd for C$_{23}$H$_{23}$Br$_2$NO$_3$Si: C, 50.29; H, 4.22; N, 2.55. Found: C, 51.58; H, 4.28; N, 2.47.

Synthesis of triarylaminonitrostilbene siloxane polymer 14. To a solution of **13** (1 g, 1.88 mmol) in 15 mL DMF, Pd(OAc)$_2$ (0.004 g, 0.018 mmol), tri-*o*-tolyphosphine (0.023 g, 0.075 mmol), Et$_3$N (0.47 g, 4.6 mmol) and 4-nitrostyrene (0.8 g, 5.46 mmol) were added. The dark reaction mixture was stirred at 90 °C for two days. The mixture was poured in 500 mL cold water. The polymer obtained was dissolved in DMF, reprecipitated in water, and washed with cold hexane and diethyl ether. The polymer was then redissolved in CHCl$_3$, washed with water and saturated NaHCO$_3$. The solvent was removed under reduced pressure and the polymer was dried *in vacuo*, affording a red product. ^{1}H NMR (200 MHz, CDCl$_3$): δ 0.00-0.25 (bs, SiCH$_3$), 0.45-0.70 (bs, SiCH$_2$), 1.5-1.8 (bm, CH$_2$CR$_2$Si), 3.3-3.4 (bm, 2H, CR$_2$CH$_2$-), 4.25-4.50 (bs, ArCH$_2$), 6.65-7.70 (bm, ArH), 8.0-8.2 (bd, ArH *ortho* to NO$_2$); UV-vis: λ_{max}= 434 nm (240 - 570 nm).

Synthesis of N,N-diethanol-4'-amino-4-stilbenephosphonic acid diethyl ester (15). A mixture of 4-bromo-N,N-diethanolaniline (1.5 g, 7.0 mmol), **3** (1.51 g, 0.31 mmol), Pd(OAc)$_2$ (0.15 g, 0.7 mmol), tri-*o*-tolyl phosphine (0.429 g, 1.4 mmol) and 1.1 mL of Et$_3$N in 30 mL of CH$_3$CN was heated in a screw capped tube purged with argon at 90 °C for 48 h. The reaction mixture was cooled, filtered, poured into cold water, and extracted with EtOAc. The organic extract was dried over magnesium sulfate and the solvent was removed *in vacuo*. Stilbene **15** was further purified by flash chromatography through silica gel, affording 1.21 g of fluorescent yellow solid (65%). UV (10^{-5} M) λ_{max} = 256 nm (240-460 nm); ^{1}H NMR (CDCl$_3$): δ 1.30 (t, 6H), 3.60 (t, 4H), 3.85 (t, 4H), 4.05 (m, 4H), 4.55 (s, 2H), 6.70 (d, 2H), 6.70 (d,2H), 6.85 (d, 1H), 7.1 (d, 1H), 7.35 (d, 2H), 7.45 (dd, 2H), 7.60 (dd, 2H). HRMS (FAB), [M+Na$^+$] calc: 442.1759, actual: 442.1769 amu.

Synthesis of N,N-diethanol-4'-amino-4-nitrostilbene (16). A mixture of of 4-bromo-N,N-diethanolaniline (0.872 g, 3.35 mmol), 4-nitrostyrene (0.500 g, 3.35 mmol), Pd(OAc)$_2$ (0.0752 g, 0.335 mmol), tri-*o*-tolyl phosphine (0.204 g, 0.670 mmol), and Et$_3$N (0.513 mL, 3.685 mmol) in DMF was heated in a screw tapped tube purged with argon at 90 °C for 48 h. The reaction mixture was cooled, filtered, and poured into ice cold water. The organic layer was extracted with EtOAc. The organic extract was dried over MgSO$_4$ and solvent removed *in vacuo*. Stilbene

1 6 was further purified by flash column chromatography through silica gel, affording a red solid in 68% yield. ^{1}H NMR (DMF-d$_7$): δ 3.6-3.8 (bm, OH, OCH$_2$), 3.9 (t, NCH$_2$), 6.7 (d, ArH o- to N), 6.9-7.1 (2d, =CH stilbene), 7.2-7.8 (ArH), 8.2 (d, ArH o- to NO$_2$).

Synthesis of N,N-bis(4-isocyanatophenyl)aniline (17). A solution of bis(tricholoromethyl)carbonate (0.719 g, 2.42 mmol) in dry toluene was added dropwise into a stirred solution of 4, 4'-diaminotriphenylamine (1.0 g, 3.633) in dry toluene at room temperature under N$_2$. The reaction mixture was stirred at 100 °C for 2 h. The reaction mixture was filtered and the solvent evaporated, resulting in 1.01 g of a brown solid (85%). ^{1}H NMR (benzene-d$_6$): δ 6.45 (d, 4H, ArH $ortho$ to N), 6.70 (d, 4H, ArH $ortho$ to NCO), 6.90 (m, 3H, ArH of Ph), 7.05 (m, 2H, ArH $ortho$ to N in Ph). HRMS (EI, 70 ev) calc: [M$^+$] : 327.1008, actual : 327.1007 amu.

Synthesis of phosphorylated polyurethane 1 8. A 40 mL heavy walled glass vial was charged with diol **1 5** (0.3558 g, 0.848 mmol) and 5 mL of dry DMF. Diisocyanate **1 7** (0.3618 g, 1.102 mmol) was added, along with another 5 mL of DMF. Two drops of Et$_3$N was added to facilitate the polymerization. The polymerization mixture was stirred at 90 °C, yielding a dark green viscous solution. The reaction was stopped after 36 h and then added dropwise into cold methanol. Polymer **1 8** precipitated out as a dark green solid. It was reprecipitated again to further purify it, resulting in 0.3066 g of dark green solid. ^{1}H NMR (DMF-d$_7$): δ 1.2 (t, OCH$_2$CH$_3$), 3.4-3.8 (bm OCH$_2$CH$_2$N), 4.1 (q, OCH$_2$CR$_3$), 6.9-7.1 (2d, ArH), 6.7, 7.1 (2d, =CH stilbene), 7.2 (m, ArH), 7.5-7.6 (bd, ArH), 7.6-7.8 (bm, ArH), 7.9-8.1 (bm, ArH). FT-IR neat film: 3435 cm^{-1}, 3247 cm^{-1} (v N-H of urethane), 2935cm^{-1}(v$_{as}$ CH$_2$), 2878 cm^{-1} (v$_{as}$ CH$_2$), 1659 cm^{-1} (v$_{as}$ C=O of urethane), 1439-1416 cm^{-1} (v bending CH$_2$), 1391 cm^{-1} (v CH$_3$ stretching), 1255 cm^{-1} (v P=O phosphonate ester), 1104, 1062 cm^{-1} (v P-O-C stretch).

Synthesis of nitrostilbene polyurethane 1 9. A 20 mL heavy walled glass vial was charged with diol **1 6** (0.1889 g, 0.575 mmol) and 5 mL of dry DMF. Diisocyanate **17** (0.2449 g, 0.748 mmol) in 5 mL of DMF was added. Two drops of Et$_3$N was added to facilitate the polymerization. The polymerization mixture was stirred at 90 °C, yielding an orange-red viscous solution. The reaction was stopped after 48 h and added dropwise into cold methanol. Polymer **1 9** precipitated out as an orange-red solid powder. It was reprecipitated and washed with methanol. UV: λ$_{max}$ = 316 nm (range 260-548 nm). FT-IR (neat film) cm^{-1}: 3430 (v N-H of urethane), 2961 (v$_{as}$ CH$_2$ stretch), 1651 (v C=O of urethane, broad), 1504 (v$_{as}$ NO$_2$ stretching), 1310-1220 (v$_s$ NO$_2$ stretch), 927 (v trans C=C stretch), 870 (v C-N stretch).

Synthesis of brominated polyimide 2 0. A 20 mL vial was charged with 4-bromo-4',4''-diaminotriphenylamine (0.500 g, 1.412 mmol) and 2 g of dry NMP. The mixture was stirred until complete dissolution of the diamine was achieved, and then (hexafluoroisopropylidene)diphthalic anhydride (6-FDA) (0.627 g, 1.412 mmol) was added and washed in with another 2 g of NMP. The polymerization mixture turned dark red in color. It was stirred for a period of 48 h to yield a highly viscous polyamic acid solution **PAA-1**.

To PAA-2 was added dry pyridine (0.451 g, 5.70 mmol) and acetic anhydride (0.582 g, 5.70 mmol). The reaction mixture was stirred for approximately 16 h at ambient temperature, and finally at 70 °C for an additional 6 h. The resulting polyimide solution was then precipitated in water with vigorous stirring and washed thoroughly with methanol. The dark green powder was dried for 48 h under vacuum, resulting in 0.904 g of **20**. ^{1}H NMR (DMF-d$_7$): δ 6.9 (d, ArH), 7.0 (d, ArH), 7.1 (bt, ArH), 7.3 (d, ArH), 7.4-7.6 (bm, ArH anhydride), 7.9 (s, ArH anhydride), 8.1-8.2 (bm, ArH *ortho* and *meta* to imide N). Elem. Anal.: theoretical: %C = 58.29, %H = 2.38, %N = 5.51; actual: %C = 55.85, %H = 2.52, %N = 5.17. T$_g$ = 137 °C.

Synthesis of phosphorylated polyimide 21 via Heck reaction of 20. A 20 mL heavy walled vial was charged with **20** (0.400 g, 0.525 mmol), **3** (0.151 g, 0.63 mmol), Et$_3$N (0.011 g, 0.063 mmol), and 5 mL freshly distilled DMF. The reaction mixture was heated to 85 °C and stirred vigorously. The color of the reaction mixture turned dark red from green after 32 h. The reaction mixture was heated for a total of 64 h, yielding a dark red viscous liquid. The resulting solution was then precipitated in distilled water with vigorous stirring and washed thoroughly with methanol. The black colored powder was dried under vacuum, affording 200 mg of polyimide **21**. UV-vis : λ$_{max}$ = 294 nm (range 258-576 nm). FT-IR (neat film) cm^{-1}: 2963 (ν sp^2 C-H stretch), 1661 (ν CON stretch), 1505 (ν phenyl C=C stretching), 1257 (ν P=O of phosphonate ester), 1102, 1063 (ν POC stretch), 967 (ν *trans* C=C stretch). T$_g$ = 212 °C.

Synthesis of N,N-bis(4-aminophenyl)-4-stilbenephosphonic acid diethyl ester 22. In a screw cap vial was taken 4-bromo-4',4"-diaminotriphenylamine (1.0035 g, 2.85 mmol), **3** (0.6163 g, 2.565 mmol), Pd(OAc)$_2$ (0.064 g, 0.285 mmol), of tri-*o*-tolylphosphine (0.173 g, 0.570 mmol), and Et$_3$N (0.32 g, 3.135 mmol) in 30 mL CH$_3$CN. The reaction mixture was heated while stirring at 100 °C in an oil bath. After complete consumption of **3** (by TLC), the reaction mixture was passed through a celite bed and poured into ice cold water, yielding a dark brown precipitate. The solid was filtered and dried under vacuum. The product was further purified by column chromatography (CH$_2$Cl$_2$/EtOAc), resulting in 1.0104 g of **22**. ^{1}H NMR (CDCl$_3$): δ 1.3-1.4 (t,H, OCR$_2$CH$_3$), 3.6 (bs, 4H, NH$_2$), 4.1 (m, 4H, OCH$_2$CR$_3$), 6.6 (d, 4H, *ortho* to NH$_2$), 6.8 (d, 2H, *ortho* to 3°N), 6.9 and 7.1 (m, 2H, stilbene =CH), 7.3 (d, 2H, *meta* to 3° N), 7.5 (dd, 2H, *meta* to P(O)(OR)$_2$), 7.7 (dd, 2H, ArH *ortho* to P(O)(OR)$_2$). UV-vis : λ$_{max}$ = 406 nm (range 270-522 nm).

Synthesis of phosphorylated polyimide 23. A 20 mL vial was charged with diaminostilbene **22** (0.3025 g, 0.589 mmol) and 1 g of dry NMP. The mixture was stirred until complete dissolution of diamine was achieved, followed by addition of 6-FDA (0.2616 g, 0.589 mmol) and 1 g of NMP. The mixture turned dark red in color, it was stirred for 48 h, yielding a dark red viscous polyamic acid solution **PAA-2**.

To **PAA-2** was added dry pyridine (0.188 g, 2.378 mmol) and acetic anhydride (0.243 g, 2.378 mmol). The reaction mixture was stirred for 16 h at ambient temperature and finally at 70 °C for an additional 6 h. The resulting polyimide solution was then precipitated into water with vigorous stirring and washed thoroughly with methanol. The dark colored powder was dried for 48 h under vacuum, affording 0.304 g of polyimide **23**. ^{1}H NMR (DMF-d$_7$): δ 1.3 (t, 6H, OCR$_2$CH$_3$), 4.1 (m, 4H, OCH$_2$CR$_3$), 7.1-8.2 (24H, ArH). Elemental analysis: theoretical: %C = 63.85, %H = 3.72, %N = 4.56; actual: %C = 62.25, %H = 3.69, %N = 4.42. T$_g$ = 167 °C. UV-vis : λ$_{max}$ =298 nm (260 - 476 nm).

Results and Discussion

Polysiloxanes

A triarylamine-containing polysiloxane was synthesized bearing nitrostilbene moieties (Figure 4). Vilsmeier reaction of triphenylamine, followed by reduction and allylation, afforded alltriphenylamine **11**. This was subjected to Pt-catalyzed hydrosilylation (6) with poly(hydrogen methylsiloxane). The reaction was conveniently monitored by FT-IR, by observing disappearance of the Si-H stretch at 2150 cm^{-1}, and by ^{1}H NMR, by observing disappearance of the Si-H proton resonance at 4.7 ppm. Regiospecific bromination (20) of **12** was accomplished with NBS in DMF at room temperature, yielding brominated polysiloxane **13**. Elemental analysis confirmed near quantitative dibromination of each triaryl unit, while ^{1}H NMR confirmed the *para*-regiospecificity of the reaction. The highly functionalized nitrostilbene-containing polysiloxane **14** was secured through efficient Pd-catalyzed Heck coupling of **13** with 4-nitrostyrene. A bright red powdery solid polymer was obtained with λ_{max} = 434 nm and solubility in THF, DMF and CHCl$_3$. A high fidelity, atom economical synthesis resulted in formation of a triaryl-based polysiloxane bearing one CT and two NLO moieties covalently attached to each repeat unit. Though the presence of the styrenic groups on the aryl rings may affect the CT of the arylamine, photoconductivty was recently demonstrated for polymer **6** in Figure 2 (18), an analogous material.

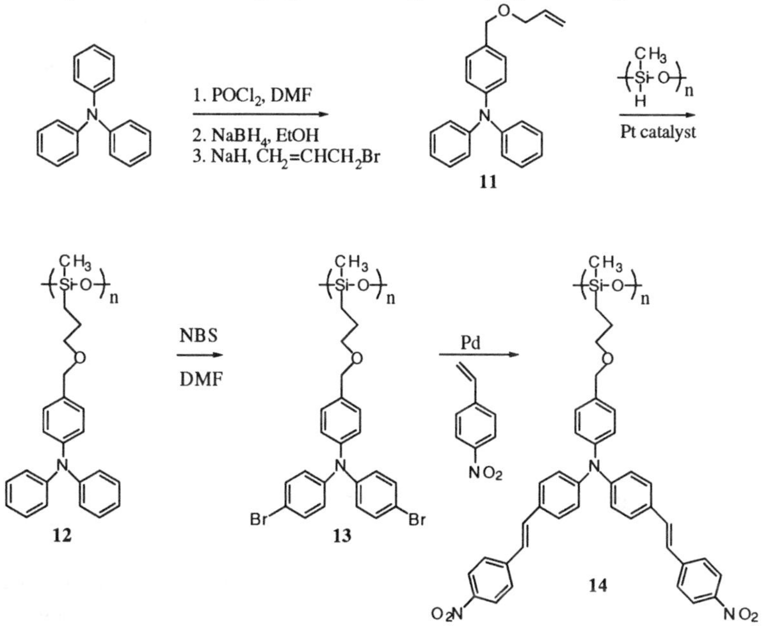

Figure 4. Synthesis of a triphenylamine-based PR polysiloxane.

Polyurethanes

Our aim was to utilize the highly efficient and atom economical Pd-catalyzed Heck reaction to prepare stilbene-based NLO chromophoric diols. The diols would then be condensed with a charge transporting diisocyanate. The synthesis of (N,N-diethanol)-4'-amino-4-stilbenephosphonic acid diethyl ester **15** was conducted as outlined in

Figure 5. The first step involved rapid regioselective bromination of N,N-diethanolaniline using benzytrimethylammonium chlorobromate (20). The ^{1}H NMR indicated the symmetrical nature of the molecule. The next step involved the Heck reaction between 4-bromo-N,N-diethanolaniline and styrene phosphonate 3. This was accomplished using 10 mol% palladium (II) acetate and 20 mol% tri-o-tolylphosphine. To increase the efficiency of the reaction by lowering the volume of activation, the reaction was performed in a heavy walled screw cap vial at 90 °C. Good yield of phosphorylated diol 15 was realized, whose structure was confirmed by both ^{1}H NMR and high resolution FAB MS. (N,N-Diethanol)-4'-amino-4-stilbenephosphonic acid diethyl ester 15 was bright yellow in color with λ_{max} = 256 nm. (N,N-Diethanol)-4'-amino-4-nitrostilbene 16 was synthesized in an analogous manner (Figure 5). Thus, Pd-catalyzed Heck reaction between 4-bromo-N,N-diethanolaniline and 4-nitrostyrene provided nitro diol 16 in good yield as a red solid.

It has been established previously that triarylamines possess photoconductive and hole transporting properties (21). We set out to prepare a charge trnasporting triarylamine-based diisocyanate. This was accomplished by reaction of 4,4'-diaminotriphenylamine with triphosgene, affording the 4,4'-diisocyanatotriphenylamine 17 as a brown solid whose sturcture was confirmed by ^{1}H NMR and high resolution MS (observed 327.1007, calc. 327.1008 amu). Further substantiation of 17 was garnered by observation of a strong NCO absorption at 2261 cm^{-1} by FT-IR.

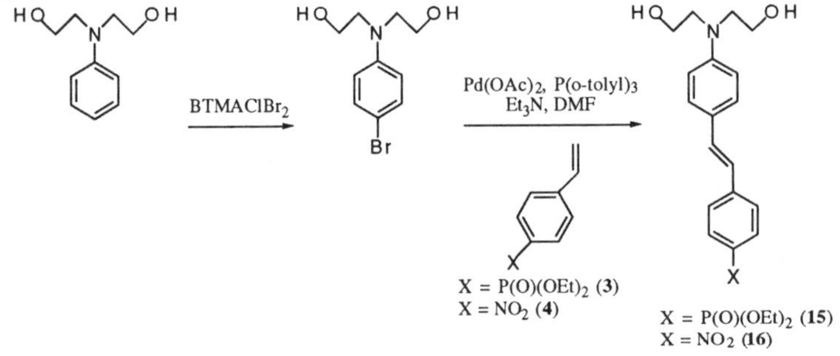

Figure 5. Synthesis of NLO-containing diols.

A phosphorylated stilbene containing polyurethane 18 was synthesised by triethylamine-catalyzed reaction of N,N-diethanol-4'-amino-4-stilbenephosphonic acid diethyl ester 15 with the charge transporting diisocyanate 17, as shown in Figure 6. A greenish polyurethane with perfectly alternating CT and NLO moieties was obtained, having λ_{max} = 316 nm and absorption extending to 438 nm. The relatively long absorbance maximum is suggestive of some charge transfer interaction between the NLO chromophore and the CT unit. The TGA analysis showed that this polymer is highly thermally stable, it decomposes only after 246 °C. The T_g of the material was found to be 175 °C. These results indicate that the material sould have sufficient stability of poled order.

A nitro analog to polyurethane 19 was prepared by triethylamine-catalyzed polycondensation of (N,N-diethanol)-4'- amino-4-nitrostilbene 16 and the charge transporting diisocyanate 17 (Figure 6). The λ_{max} for 19 was also 316 nm but its absorption extended out to 548 nm. The ^{1}H NMR spectrum was consistent with the polyurethane structure. The TGA analysis revealed high thermal stability of the polymer, with decomposition beginning at 247 °C. The T_g of polyurethane 19 was

167 °C, indicating the material should exhibit a highly stable poled order and would possess good thermal stability. Thus, it was demonstrated that this methodology is versatile, allowing for ready preparation of well-defined polyurethanes with alternating CT and NLO moieties.

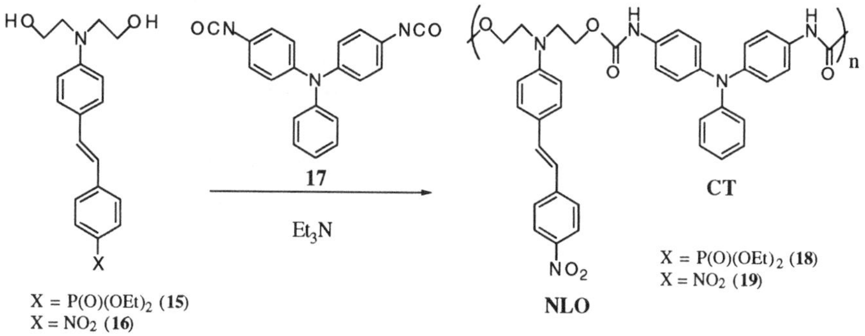

Figure 6. Synthesis of polyurethanes with alternating CT and NLO units.

Polyimides

In order to realize high T_g and high thermal stability, the synthesis of well-defined, single component polyimides was undertaken. The fluorinated dianhydride 6-FDA was selected to achieve both good solubility and high thermal stability. The aromatic polyimide moiety has also been postulated to assist in charge transport (22). Two approaches were investigated. The first approach involved preparing a multifunctional diamine bearing CT and NLO functionality, and conducting a condensation polymerization between the diamine and 6-FDA. The second strategy called for the preparation of an arylhalide-containing polyimide, followed by Heck reaction with appropriately derivatived styrene. Both methods are described below and result in the formation of polyimides with diamines units comprised of CT triarylamine-based NLO chromophores.

Brominated polyimide **20** was prepared with a view of making multifunctional PR polyimides with different NLO chromophores. N-(4-bromophenyl)-N,N-bis(4-aminophenyl)amine (a charge transporting diamine) was subjected to polycondensation with 6-FDA, as shown in Figure 7. The intermediate polyamic acid was chemically imidized. Polyimide **20** was also obtained as a green powder, soluble in THF and DMF. The absorption spectrum revealed a maxima at 318 nm and absorption ranging from 246 - 414 nm. Polyimide **20** was highly thermally stable and displayed a T_g of 137 °C. It was stable until a temperature of 216 °C, after which only a 1.5% weight loss occurred between 216 °C and 500 °C. NMR and elemental analysis clearly established formation of the product. IR was a useful diagnostic tool to identify the formation the polyimide through observation of characteristic imide and Ar-Br bond stretching absorptions.

To demonstrate the usefulness of **20** as a building block to create single component PR polymers, it was treated with styrene phosphonate **3** in a Pd-catalyzed Heck reaction (Figure 8). A deep red, THF and DMF soluble powder was obtained having $\lambda_{max} = 294$ nm and an absorption range of 258 - 576 nm. The T_g for phosphorylated polyimide **21** was relatively high at 212 °C and no weight loss was detected up to 230 °C. ^{1}H NMR and IR spectroscopic analysis were consistent with the structure.

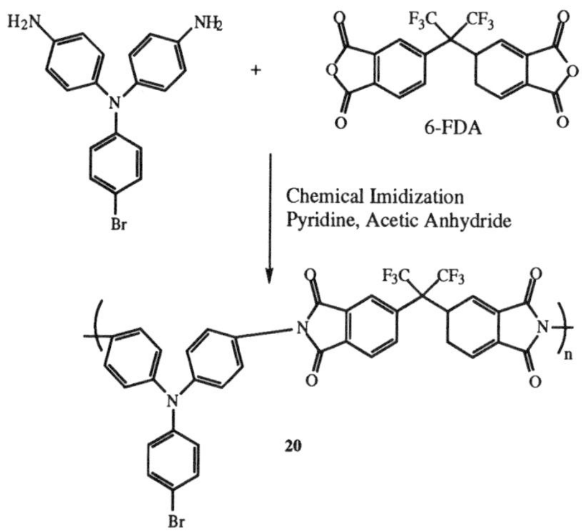

6-FDA

Chemical Imidization
Pyridine, Acetic Anhydride

20

Figure 7. Synthesis of brominated polyimide **2 0**.

The second approach to create PR polyimides involved preparation of a diamine possessing both CT and NLO moieties. N,N-Bis(4-aminophenyl)-4-stilbenephosphonic acid diethyl ester **2 2** was designed with the view of making a single component system wherein the charge transporter and the NLO chromophore are incorporated as one unit. The synthesis of this chromophore highlights the effectiveness and generality of the Heck reaction, as shown in Figure 9.

N,N-Bis(4-nitrophenyl)-4-bromoaniline was generated by the reaction of 4-fluoronitrobenzene and 4-bromoaniline in the presence of CsF in DMSO (23). The resulting dinitrotriphenylamine was then reduced with either 5 wt% of 5% Pt/C or 4 wt% of 5% Pd/C and hydrogen, affording good conversion to N,N-bis(4-aminophenyl)-4-bromoaniline. Controlling the wt% of Pt or Pd was essential in avoiding reduction (displacement) of bromine, higher weight percents lead to a mixture of products resulting from reduction of nitro and bromine. The final step in this sequence involved the Heck reaction between the diaminobromotriphenylamine derivative and styrene phosphonate **3**. ^{1}H NMR indicated the presence of the *trans* stilbene olefinic protons. FT-IR clearly indicated the presence of a *trans* stilbene double bond from the stretches at 159 cm^{-1} and 965 cm^{-1}. The P=O stretch at 1274 cm^{-1} and the P-O-C stretch at 1128 cm^{-1} and 1070 cm^{-1} were clear indication of formation of diamine **2 2**. Phosphorylated diamine **2 2** exhibited UV-visible absorbance from 270 - 522 nm with λ_{max} = 406 nm.

In the final step of this strategy, the CT/NLO diamine **2 2** underwent polycondensation with 6-FDA, resulting in formation of phosphorylated polyimide **2 3** (Figure 10). The reaction was carefully monitored for color and viscosity changes, and chemical imidization was utilized to complete the imide bond formation. Polyimide **2 3** was obtained as dark brown solid, soluble in THF and DMF. The polymer had a broad absorption band with the maxima at 298 nm, ranging from 260 - 476 nm. Polyimide **2 3** was very stable up to 249 °C, with a T_g observed at 167 °C. This clearly indicates the possibility of long term stability and stable poled order. ^{1}H NMR and elemental analysis unambiguously indicated formation of the desired polymer. In fact, other than the T_g, polyimides **2 1** and **2 3** were very similar.

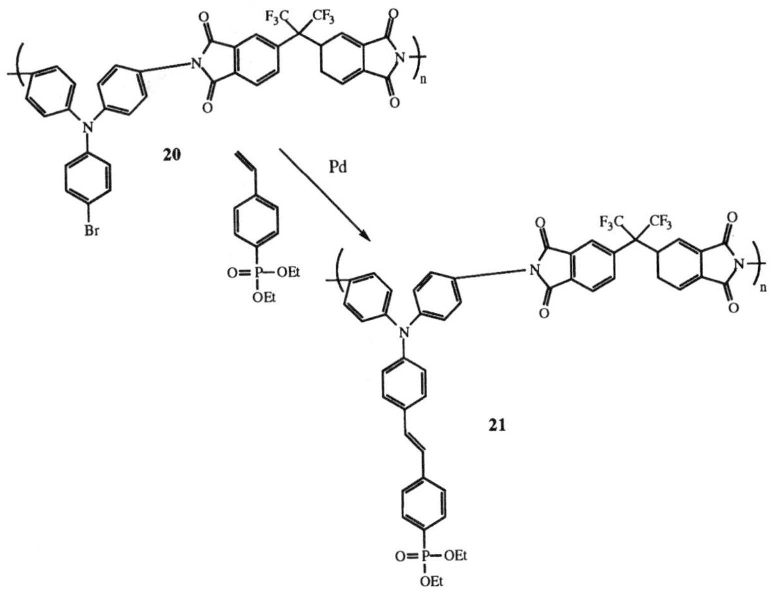

Figure 8. Synthesis of phosphorylated polyimide **21** via Heck reaction of **20**.

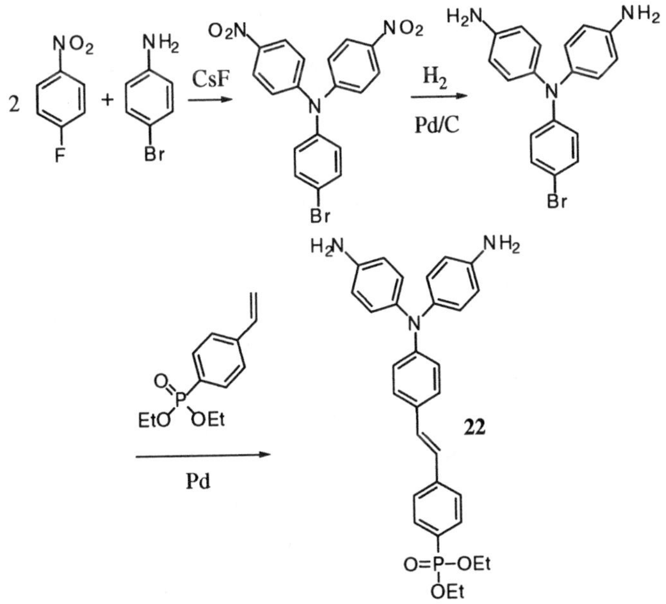

Figure 9. Preparation of N,N-bis(4-aminophenyl)-4-stilbenephosphonic acid diethyl ester **22**.

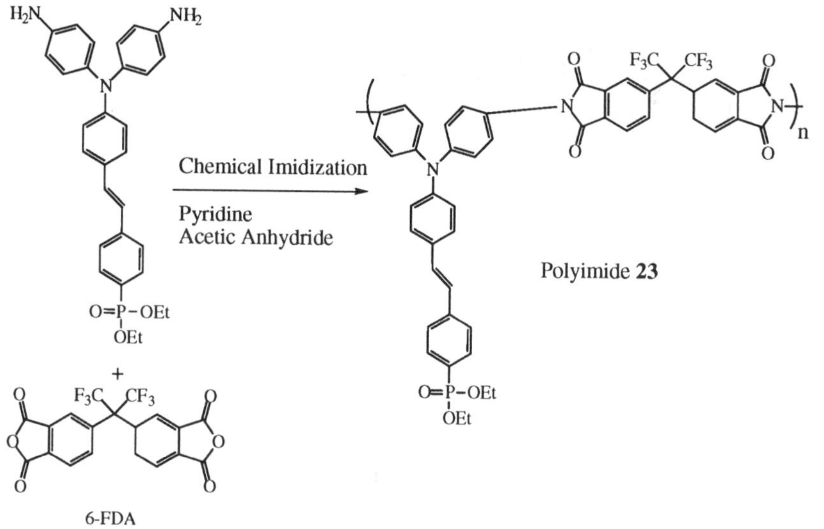

Figure 1 0. Preparation of phosphorylated polyimide **2 2** via polycondensation of **2 1** and 6-FDA.

Conclusions

The palladium-catalyzed Heck reaction proved to be a very efficient and atom economical approach to synthesize stilbene derivatives on polymers as disparate as polysiloxanes and polyimides. A variety of different donor-acceptor substituted stilbenes have been synthesized. The regiospecific bromination and Heck reactions clearly show the utility and the generality of these processes to controllably and efficiently modify polymers.

CT and NLO-containing polyimides were successfully prepared by two distinct strategies. These polymers displayed very high thermal stability (up to 250 °C) and had high T_gs. Through Pd-catalyzed coupling reactions, the brominated polyimide should prove useful as a substrate for creation of an array of PR polyimides possessing different NLO moieties.

Two different polyurethanes with well-defined, alternating NLO and CT units were prepared through a rationally designed synthesis. The UV-visible spectrum revealed some charge transfer interaction between the NLO chromophore and the CT unit. The polyurethanes synthesized have high thermal stability up to 240 °C, with relatively high T_gs, a clear indication that good long term stability can be expected from these materials. This represents the first report of the use of a phosphonate ester group as an electron-withdrawing group in the synthesis of stilbenoid photorefractive polyurethane and polyimide-based polymers.

Acknowledgments

Financial support from the NSF (DUE-9550885 and DUE 9650923) is acknowledged. The authors thank Dr. Andrew Tyler of Harvard University for mass spectrometry assistance. The Harvard University Mass Spectrometry Facility is supported by grants from NSF (CHE-900043) and NIH (SIO-RR06716).

Literature Cited

1. (a) Moerner,W. E. *Nature* **1994**, *371*, 475. (b) *Photorefractive Materials and Their Application*; Günter, P.; Huignard, J. P., Eds.; Springer-Verlag: Berlin, 1988, Vol. 1 and 2.
2. Moerner, W. E.; Silence, S. M. *Chem. Rev.* **1994**, *94*; 127.
3. (a) Sutter, K.; Günter, P.*J. Opt. Soc. Am. B.* **1990**, *7*, 2274. (b) Sutter, K.; Hullinger, J.; Schlesser, R.; Günter, P. *Opt. Lett.* **1993**, *18*, 778.
4. (a) Ducharme, S.; Jones, B.; Takacs, J. M.; Zhang, L. *Opt. Lett.* **1993**, *18*, 152. (b) Jones, B. E.; Ducharme, S.; Liphardt, M.; Goonesekera, A.; Takacs, J. M.; Zhang, L.; Athalye, R. *J. Opt. Soc. Am. B* **1994**, *11*, 1064.
5. Meerholz, K.; Volodin, B. L.; Sandalphon; Kippelen, B.; Peyghambarian, N. *Nature* **1994**, *371*, 497.
6. Zobel, O.; Eckl, M.; Strohriegl, P.; Haarer, D. *Adv. Mater.* **1995**, *7*, 911.
7. Cox, A. M.; Blackburn, R. D.; West, D. P.; King, T. A.; Wade, F. A.; Leigh, D. A. *Appl. Phys. Lett.* **1996**, *68*, 2801.
8. Tamura, K.; Padias, A. B.; Hall, H.K, Jr.; Peyghambarian, N. *Appl. Phys. Lett.* **1992**, *60*; 1803.
9. Zhao, C.; Park, C.; Prasad, P. N.; Zhang, Y.; Ghosal, S.; Burzynski, R. *Chem. Mater.* **1995**, *7*, 1237.
10. Sansone, M. J.; Teng, C.C.; East, A. J.; Kwiatek, M. S. *Opt. Lett.* **1993**, *18*; 1400.
11. Yu, L.; Chan, W.; Bao, Z.; Cao, S. X. F. *Macromolecules* **1993**, *26*, 2216.
12. Bao, Z. N.; Chan, W. K.; Yu, L. P. *Chem. Mater.* **1993**; *6*; 2.
13. Bao, Z. N.; Chen, Y.; Cai, R.; Yu, L. P. *Macromolecules* **1993**, *26*; 5281.
14. Peng, Z. H.; Bao, Z. N.; Chen, Y. M.; Yu, L. P. *J. Am. Chem. Soc.* **1994**, *116*; 6003.
15. Dobler, M.; Weder, C.; Ahumada, O.; Neuenschwander, P.; Suter, U. W.; Follonier, S.; Bosshard, C.; Günter, P. *Polym. Mater. Sci. Eng.* **1997**, *76*, 308.
16. Lee, J.-H.; Moon, I. K.; Kim, H. K.; Choi, S.-K. *Polym. Mater. Sci. Eng.* **1997**, *76*, 314.
17. Belfield, K. D.; Chinna, C.; Najjar, O.; Sriram, S. *Polym. Preprints* **1997**, *38(1)*, 203.
18. Belfield, K. D.; Chinna, C.; Najjar, O.; Sriram, S.; Schafer, K. J. *Proc. SPIE - Int. Soc. Opt. Eng.* **1997**, *3144*, 154.
19. Belfield, K. D.; Chinna, C.; Schafer, K. J. *Tet. Lett.* **1997**, *38*, 6131.
20. Kajigaishi, S.; Kakinami, T.; Yamasaki, H.; Fujisaki, S.; Okamoto, T. *Bull. Chem. Soc. Jpn.* **1988**, *61*; 2681.
21. Stolka, M.; Pai, D. M.; Renfer, D. S.; Yanus, J. F. *J. Polym. Sci. Polym. Chem. Ed.* **1983**, *21*; 969.
22. Iida, K.; Nohara, T.; Nakamura, S.; Sawa, G. *Jpn. Appl. Phys.* **1989**, *28*; 1390.
23. Oishi, Y.; Takado, H.; Yoneyama, M.; Kakimoto, M. A.; Imai, Y. *J. Polym. Sci. Polym. Chem. Ed.* **1990**, *28*; 1763.

RESPONSIVE POLYMERS IN CHEMISTRY AND BIOLOGY

Chapter 18

Stimuli-Responsive Behavior of *N,N*-Dimethylaminoethyl Methacrylate Polymers and Their Hydrogels

Fu-Mian Li, Shuang-Ji Chen, Fu-Sheng Du, Zhi-Qiang Wu, and Zi-Chen Li

Department of Chemistry, Peking University, Beijing 100871, China

In this chapter, we report on the thermo- and pH-responsive behavior of poly(N,N-dimethylaminoethyl methacrylate) (poly(DMAEMA)) in aqueous solution as well as those of the hydrogels prepared from the slightly crosslinked homopolymer and copolymers. The nature of the thermo-responsive behavior is the balance of hydrophobility and hydrophilicity of poly(DMAEMA), while the pH-sensitive behavior is due to the existence of tertiary amino-group, which becomes protonated with the decrease of pH of the aqueous medium. These properties of poly(DMAEMA) were used to develop hydrogels with good comprehensive properties including mechanical properties through copolymerization of DMAEMA with other (meth)acrylate derivatives, among which the copolymers and hydrogels prepared from DMAEMA and butyl methacrylate (BMA) were studied in detail. By using a unique method and photoredox system to initiate the copolymerization of DMAEMA with BMA, a kind of strongly stuck stable asymmetric bilayer sheets which show reversible thermo- and pH-responsive behavior was developed as an intelligent soft material.

Polymers and hydrogels which are sensitive to environmental stimuli, such as temperature, pH, light, ionic strength, magnetic field, electric field, have promising potential applications in the field of drug delivery systems, separation, artificial muscle, chemo-mechanical systems, sensors, and so on (*1*). One of the hot areas of research of stimuli-responsive polymers and hydrogels focuses on the thermal-responding behavior of polymers and hydrogels of N-isopropylacrylamide (NIPAAm) (*2,3*). The polymers of acrylate derivatives having aliphatic tertiary amino-group effectively respond to the change of pH of the environment, and we have reported the combined thermo- and pH-responsive behavior of polyacrylates having morpholinyl moieties (*4,5*). In this chapter, the thermo- and pH-responsive behavior

266

of N,N-dimethylaminoethyl methacrylate (DMAEMA) polymer, its copolymers and hydrogels have been investigated aiming at the preparation of a hydrogel not only possessing stimuli-responsive behavior but also with good mechanical properties. More important, by using a unique photopolymerization system, we prepared a strongly stuck asymmetric bilayer sheet which shows reversible thermo- and pH-responsive behavior with good mechanical properties.

Thermo-responsive Behavior of Poly(DMAEMA) in Aqueous Solution

With the increase of temperature, the poly(DMAEMA) solutions abruptly become opaque. This temperature-sensitive behavior is caused by the dehydration of the hydrated DMAEMA molecules in the polymer chain above transition temperature. Depending on the thermo-responsive polymer and its concentration in aqueous media, the critical point of phase separation or the cloud point (C.P.) can be quite different. This was examined in our case by monitoring the temperature-dependent transmittance changes at 550nm of poly(DMAEMA) in deionized water as shown in Figure 1. For a given concentration of poly(DMAEMA), the solution shows a distinctive reversible transition in a few seconds within a narrow temperature range. With the increase of the concentration of poly(DMAEMA), the C.P. shifts to lower temperature. For example, the C.P. of 1.0 wt% polymer solution is 52 ℃, whereas that of 10 wt% is 40 ℃. This can be explained by the polymer chains association in concentrated aqueous solution which is easier than that in diluted one. The thermo-responsive property of poly(DMAEMA) in solution has been used by Okubo as a thermo-sensitive coagulant and by Ito for a thermo-sensitive light shield (*6, 7*).

Additionally, monomeric DMAEMA and its model compound, N,N-dimethyl-aminoethyl *iso*-butyrate (DMAEiBA), show similar temperature-sensitive behavior in aqueous solution as poly(DMAEMA), but the C.P. of DMAEMA solution is much lower than that of its polymer solution at the same monomeric unit concentration; the C.P. shifts to higher temperature with the increase of monomer concentration, which is just opposite to the trend for poly(DMAEMA) aqueous solution.

Thermo-responsive Behavior of DMAEMA Hydrogels

DMAEMA hydrogels were prepared by bulk (0.5 wt% 2,2-azobisisobutyronitile, AIBN, as an initiator, 60 ℃, 24hr.) or aqueous polymerization (0.5 wt% ammonium persulfate, APS, as an initiator, 25 ℃, 48hr.) in the presence of various amounts of ethylene glycol dimethacrylate (EGDMA) as a crosslinker. The crosslinked DMA-EMA hydrogel swells and shrinks in response to temperature change reversibly as the other well-known thermo-responsive hydrogels. Shown in Figure 2 is the temperature-dependent change of swelling ratio (S.R.) for poly(DMAEMA) hydrogel, which was prepared by initiation with AIBN in bulk and crosslinking with 1.0 wt% EGDMA. The hydrogel shrinks instantaneously when the temperature is higher than 45 ℃. This temperature is usually referred as phase separation temperature which

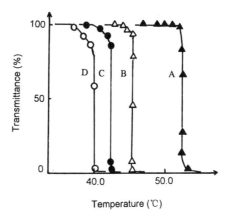

Figure 1. Temperature-dependent transmittance changes of poly(DMAEMA) aqueous solutions. Conc.(wt%): A. 1.0; B. 2.5; C. 5.0; D. 10.0.

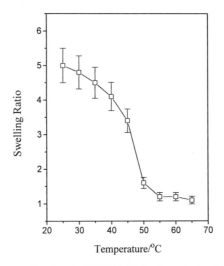

Figure 2. Swelling ratio of poly(DMAEMA) hydrogels in deionized water as a function of temperature. Hydrogels were prepared by bulk polymerization with 1.0wt% EGDMA as a crosslinker.

reflects the thermo-responsive behavior of hydrogels. The physical properties of this thermo-responsive poly(DMAEMA) hydrogel, including equilibrium swelling ratio and the phase separation temperature, were studied, and it was found that they are very much dependent on the preparative procedure. For example, poly(DMAEMA) hydrogels prepared by bulk polymerization show better mechanical properties, smaller swelling ratio and less sensitive thermo-responsive behavior. On the contrary, the hydrogels from aqueous medium by using APS as an initiator show poor mechanical properties even though they have larger swelling ratios and sensitive thermo-responsive behavior (see Table II).

DSC measurement is an effective means to understand the thermal behavior of hydrogels whether caused by physical or chemical factors (8,9). Figure 3 shows the DSC traces of poly(DMAEMA) hydrogel in dry state and swollen state (S.R=2.5). It can be seen that in swollen state, the sharp transition near -2 ℃ and broad transition near 3 ℃ were due to the melting endotherms of freezable bound water and free water, respectively (10). The transition at around 48 ℃ is the characteristic endotherm of phase separation resulted from the dissociation of bound water and polymer chains. This temperature corresponds well with that obtained from the temperature-dependent swelling ratio change as discussed above and supports the conclusion that poly(DMAEMA) hydrogel is a thermo-responsive hydrogel similar to poly(NIPAAm) hydrogel as reported by Otake et al. (11). In dry state, poly(DMAEMA) hydrogel displays its glass transition temperature at about 30 ℃ as shown in Figure 3A, while in the swollen state, this temperature shifted downward to about 20 ℃.

pH-responsive behavior of poly(DMAEMA) in aqueous medium and poly(DMAEMA) hydrogel

Since DMAEMA has an aliphatic tertiary amino-moiety, pH value of the medium will very much affect its thermo-responsive behavior. Both pH-dependent transmittance changes in poly(DMAEMA) aqueous solution or pH-dependent swelling ratio change of poly(DMAEMA) hydrogels indicate that a discontinuous change near pH=7 was observed, which is similar to the results reported by Siegel et al. (12). In aqueous solution and at room temperature, when the pH value was higher than 7, the clear solution of linear poly(DMAEMA) would become opaque and the poly-(DMAEMA) hydrogel would shrink. At a certain pH value, both DMAEMA linear polymer and its hydrogel show phase separation with the change of temperature and this phase reparation temperature increases with the decrease of pH of the medium. In a strongly acidic medium, such as pH=1.0, there is no phase transition point within the temperature range from 10 to 80 ℃ for both linear poly(DMAEMA) and poly(DMAEMA) hydrogels. Changing pH value should influence the degree of protonation of tertiary amino group of DMAEMA, which would change the polymer hydrophilicity. The highly protonated tertiary amino group in acidic medium will increase the polymer hydrophilicity as well as electron-statically repel polymer chains, thus resulting in the increase of C.P..

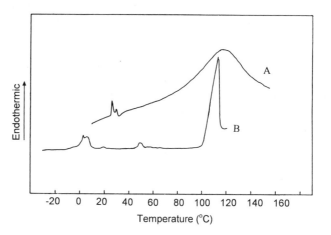

Figure 3. DSC thermograms of poly(DMAEMA) hydrogel in (A) dry state and (B) swollen state. (S.R.=2.5).

Properties of Copolymers of DMAEMA with Other Acrylate Derivatives

Though poly(DMAEMA) in aqueous solution and poly(DMAEMA) hydrogel show reversible thermo-responsive behavior, the lack of mechanical strength would limit their application. In order to solve this problem and also to vary the swelling ratio and phase separation temperature of poly(DMAEMA) hydrogel, the copolymers of DMAEMA were prepared with different acrylate derivatives in the presence and absence of crosslinker. The comonomers used were methyl methacrylate (MMA), methyl acrylate (MA), butyl methacrylate (BMA), butyl acrylate (BA), *iso*butyl methacrylate (*i*-BMA), *iso*butyl acrylate (*i*-BA), 2-hydroxyethyl methacrylate (HEMA) and 2-hydroxyethyl acrylate (HEA). The properties of hydrogels prepared from 83 mol% of DMAEMA and 17 mol% of comonomer with 1.0 wt% EGDMA as a cross-linker are summarized in Table I. Both the S.R and phase separation temperature of

Table I. Properties of hydrogels prepared from copolymers of DMAEMA and comonomers (feed molar ratio: DMAEMA/comonomer=83/17)*

	MMA	BMA	*i*-BMA	HEMA	MA	BA	*i*-BA	HEA
S.R.	2.3	1.6	1.7	2.4	3.6	1.8	1.9	3.5
T(°C) [a]	30-37	24-29	25-29	32-43	31-39	26-30	28-31	31-39

* Prepared in bulk, 1.0 wt% EGDMA as a crosslinker.

[a] Phase separation temperature as detected by temperature-dependent swelling ratio measurement.

the copolymer hydrogels are lower than those of the pure poly(DMAEMA) hydro-gels. Also, those for the hydrogels prepared from copolymers of DMAEMA and acrylates are higher than those for the corresponding methacrylates.

Some properties of these copolymers have been reported (*12,13*). The mechanical properties including tensile strength and tenacity of the hydrogels from the above copolymers were tested here. They showed improved mechanical proper-ties compared to those of the poly(DMAEMA) hydrogels. Especially, the hydrogel prepared by the copolymerization of DMAEMA and BMA demonstrated substan-tially improved mechanical properties and this promoted us to study in more detail the thermo- and pH-responsive behavior of hydrogels from DMAEMA-BMA copolymer aiming at the development of an useful intelligent soft material.

Table II shows the influence of BMA content on the phase separation temperature, swelling ratio and mechanical properties including tensile strength and elongation of DMAEMA-BMA copolymer hydrogels. It can be seen that with the increase of BMA content in copolymers, both the phase separation temperature and the swelling ratio decrease while the tensile strength and elongation increase. This can be attributed to the decrease of hydrophilicity of the copolymers with the increase of

Table II. Effect of BMA content on the properties of DMAEMA-BMA copolymer hydrogels*

BMA/ DMAEMA[a]	Phase separation temperature (°C)	S.R.	Tensile strength (Kg/cm²)	Percent of Elongation (%)
0[b]	47-49	35.2	<0.001	-
0	47-50	5.1	0.008	1.3
9/91	34-37	3.9	1.2	1.6
17/83	31-34	3.2	4.5	5.0
23/77	29-32	2.4	8.0	7.4
29/71	28-31	2.0	8.7	8.0
33/67	19-23	1.8	13.6	7.8

* Prepared in bulk, 0.5 wt% EGDMA as a crosslinker.
[a] Feed molar ratio.
[b] Prepared in aqueous solution, 1.0 wt% EGDMA as a crosslinker.

BMA content as a comonomer. When the feed ratio of BMA was increased to 33 mol%, the phase separation temperature shifted from 50°C of the pure DMAEMA hydrogels to about 20°C of the DMAEMA-BMA copolymer hydrogels. In addition, increase in the amount of crosslinker used decreased the swelling ratio and increased the tensile strength of the copolymer hydrogels, while the elongation seemed unaffected. Thus, hydrogels based on DMAEMA with satisfied comprehensive properties can be obtained through carefully selecting the experimental conditions.

pH-responsive Behavior of Bilayer Sheet

An intelligent soft material with satisfied mechanical properties constructed by BMA homopolymer layer and DMAEMA-BMA copolymer layer was prepared via a two step photopolymerization procedure as developed by us as outlined in Figure 4. Here **Layer A** is totally hydrophobic with only a very small amount of DMAEMA to form a redox system with benzophenone (Bp). The composition of **Layer B** can be adjusted by varying the monomer ratio. The thickness of the two layers can also be easily adjusted. The merit of this two step photopolymerization procedure is that these two layers can be stuck tightly and will not strip even when immersed into water for a long time, which is due to the penetration of the interface between the two layers. The photopolymerization was conducted by using photoredox initiation system. One of the well-known photoredox initiation system of photopolymerization for vinyl monomers consists of Bp and an aliphatic tertiary amine, such as triethylamine (TEA). The advantage of Bp-TEA photoredox initiation system is that the polymerization can be performed in the environment exposed to air at room temperature, and the oxygen does not inhibit the polymerization, but accelerates the polymerization instead. The mechanism of this photo-redox system involves:

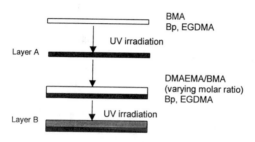

Figure 4. Schematic preparation of asymmetrically stimuli-responsive sheet.

$$CH_3CHN(CH_2CH_3)_2 \ + \ O_2 \ \longrightarrow \ CH_3\overset{\overset{O-O\cdot}{|}}{C}HN(CH_2CH_3)_2$$

$$CH_3\overset{\overset{O-O\cdot}{|}}{C}HN(CH_2CH_3)_2 \ + \ TEA \ \longrightarrow \ CH_3\overset{\overset{O-OH}{|}}{C}HN(CH_2CH_3)_2$$

$$+$$

$$CH_3\overset{\cdot}{C}HN(CH_2CH_3)_2$$

Since DMAEMA is an aliphatic tertiary amine, the N,N-dimethylaminoethyl moiety of DMAEMA can form a photoredox system with Bp to initiate the photopolymerization of itself. The photopolymerization can be carried out at very mild conditions. If the Bp is polymerizable, for example, 4-methacryloyl benzophenone, a polymerizable photoredox initiation system will be formed (*14*). Figure 5 illustrates the thermo- and pH-responsive shape change of asymmetric bilayer sheet made from DMAEMA-BMA copolymer. Since **Layer B,** in this case, is more hydrophobic than **Layer A**, the bilayer sheet exhibits an asymmetrical shape change against the pH or temperature change of the environment.

Table III. Curling time (min.) of bilayer sheets from DMAEMA-BMA copolymer hydrogels of different composition at different pH values*

pH	DMAEMA/BMA (molar ratio, layer B)			
	1:9	2:8	3:7	1:2
1.0	45	24	15	16
3.0	47	20	15	10
5.0	60	35	27	17

Preparation of bilayer sheet: **Layer A**: DMAEMA/BMA=1/1(molar ratio), 1.0 wt% Bp, 2.0 wt% EGDMA, UV irradiation for 2 hrs, then **Layer B** of different composition with 1.0 wt% Bp and 2.0 wt% EGDMA was irradiated by UV light for 10hrs, Size: 40mm x 5mm x 4mm (thickness of each layer is 2mm)..

Other types of bilayer sheet made from DMAEMA and BMA can be prepared by using the same method and changing the two layer compositions. The pH-responsive behavior of one kind of bilayer sheet (**Layer A**: DMABMA\BMA=1/1; **Layer B**: DMABMA\BMA=1/4, both are in molar ratio) was examined in detail by placing the sheets in deionized water as shown in Figure 6. It can be seen that the bilayer began to curl very quickly. Since **Layer A** is more hydrophilic than **Layer B**, the sheet bent gradually until a circle was formed. The curling time, the time when

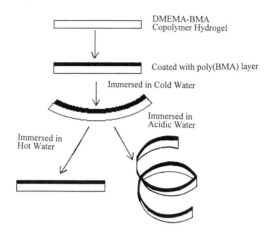

Figure 5. Schematic illustration of thermo-and pH-responsive change of asymmetric bilayer sheet made from DMAEMA-BMA copolymer.

Figure 6. Time-dependent change of the asymmetric bilayer sheet made from DMAEMA-BMA copolymer in deionized water of pH 5.0. (inner layer: DMAEMA/BMA=1/4, exterior: DMAEMA/BMA=1/1, by mole).

the straight sheet changed to circle, was used to describe this behavior. Depending on the compositions of the two layers and the pH value of the medium, the curling time can be adjusted as desired, which was shown in Table III. Based on these results, an intelligent soft material with controllable mechanical properties can be designed and prepared for practical purpose.

Acknowledgments

This project (No. 59633020) was supported by the National Natural Science Foundation of China

References

1. Hoffman A.S.; Ratner, B.D.; *ACS Symposium Series*, **1976**, 31, 1
2. DeRossi, D.; Kajiwara, K.; Osada Y. and Yamauchi, A.; "*Polymer Gels*", Plenum, New York, **1991**.
3. Bae,Y.H.; Okano,T.; Kim,S.W.; *J.Contr.Rel.*, **1989**, 9, 271
4. Zhang, Z.Y.; Chen, T.M.; Li, F.M.; *Acta Polymeric Sinica*, **1993**, No 2, 237
5. Chen, S.J.; Xue, M.; Li, F.M.; *Acta Polymerica Sinica*, **1995**, No 3, 373
6. Matsumoto, T.; Nakamei, K.; Okubo M., et al.; *Kobunshi Ronbunshu*, **1974**, 31, 669
7. Ito H.; Japan Kokai 85-235638
8. Lee, H.B.; Jhon, M.S.; Andrande, J.D.; *J. Colloid Interface Sci.*, **1975**, 51, 225
9. Higuchi A.; Iijima,T.; *Polymer*, **1985**, 26, 1833
10. Maeda Y., Kitano H.; *Spectrochimica Acta Part A*, **1995**, 51, 2433
11. Otake K.,Inomata H.; Konno, M.; Saitao S.; *Macromolecules*, **1990**, 23, 283
12. Siegel, R.A.; Firestone, B.A.; *Macromolecules,* **1988**, 21, 3254
13. Baines, F.L.; Billingham, N.C.; Armes, S.P.; *Macromolecules*, **1996**, 29, 3416
14. Du, F.S.; Zhang, P.; Li, F.M.; *J. Appl. Polym. Sci.*, **1994**, 51, 2139

Chapter 19

Thermoreversible Gelation of Syndiotactic Poly(methyl methacrylate) Based Block Copolymers in *o*-Xylene

J. M. Yu[1] and R. Jerome

CERM, University of Liege, B6 Sart-Tilman, 4000 Liege, Belgium

Thermoreversible gelation has been studied in *o*-xylene for syndiotactic poly(methyl methacrylate) based block copolymers. The dynamic properties of solutions and gels have been analyzed and discussed on the basis of scaling assumptions. At the gel point, where the loss angle tan $\delta_c = G''/G'$ is independent of the probing frequency, the samples obey the typical power law $G'(\omega) \sim G''(\omega) \sim \omega^{\Delta}$. The scaling exponent Δ is found in the 0.65-0.75 range, independently of the experimental conditions, the copolymers composition and the nature of the midblock. Modulus-frequency master curves have been built up by using appropriate reaction time dependent renormalisation factors for the individual frequency and modulus data. The scaling of these factors with reaction time has allowed to calculate the static scaling exponents for the increase observed in both modulus and viscosity.

The study of polymer gels has stimulated considerable interest from both theoretical and experimental points of view. In the last decade, much attention has been paid to modifications in the structure and viscoelastic properties of systems going through a sol-gel transition. As a rule, gelation may be of physical or chemical origins, depending on the structure of the cross-links. In physical gels, the cross-linking is reversible and the cross-linking sites can be of a large size and of a high functionality. In contrast, chemical gels are permanently cross-linked by covalent bonds and the branching point has a well-defined functionality, i.e. that one of the cross-linker.

Chemical gelation has been extensively investigated by sophisticated experiments (*1*) and accounted for by different theories from the original mean-field theory of Flory (*2*) to the concept of fractal geometry and the connectivity transition

[1]Current address: ICI Polyurethanes, Everslaan 45-B-3078 Everberg, Belgium.

model of percolation ($1,3,4$). A special attention has been paid to the viscoelastic behavior of near-critical gels (5-14). Analysis of the dynamics near the critical gelation point has led to predictions for the frequency dependence of shear storage and loss moduli (eq.1)

$$G^*(\omega)=G'(\omega)+iG''(\omega) \qquad (1)$$

where $G^*(\omega)$, $G'(\omega)$ and $G''(\omega)$ are the complex, storage and loss moduli, respectively, and ω the angular frequency. At the gel point, these moduli are predicted (5-7) and observed($5,8,9$) to scale with frequency(f) according to eq.2

$$G'(\omega){\sim}G''(\omega){\sim}\omega^\Delta \qquad (2)$$

where $\omega=2\pi f$, f is the frequency and Δ is the scaling exponent.

As a rule, the loss angle at the gel point (δ_C) which is a measure of the phase difference between G' and G'' ($G''/G'=\tan \delta_C$), has an universal value at least at low frequencies(8)

$$\delta_C=\Delta(\pi/2) \qquad (3)$$

Analysis of dynamics at the gel point and theory of viscoelasticity provide a method to determine the static scaling exponents. Indeed, scaling arguments allow to show that the viscoelastic functions, G' and G'', at different stages of the network formation, can be superimposed into a master curve, provided that frequency and complex modulus are renormalized by appropriate reaction time (t_r) dependent factors. The theory shows that the renormalisation factors for the frequency and the complex modulus are the longest relaxation time (τ_Z) and the steady-state creep compliance (J_e^0) (15), respectively, which at the gel point scale with $\varepsilon = (|t_r\text{-}t_g|)/t_g$, according to eqs.4 and 5

$$\tau_Z{\sim}\varepsilon^{s\text{-}t} \qquad (4)$$

$$J_e^0 \sim \varepsilon^{\text{-}t} \qquad (5)$$

In these expressions, t_g is the reaction time at gel point, s and t are the static scaling exponents which describe the divergence of the static viscosity, $\eta_0{\sim} \varepsilon^{\text{-}s}$, at $t_r{<}t_g$ and the static elastic modulus, $G_0 \sim \varepsilon^{\text{-}t}$, at $t_g{<}t_r$. The gelation mechanism has been discussed on the basis of several models based on the percolation theory (for review, see ref. 16), that provide power laws for the divergence of the static viscosity and the elastic moduli. Characteristic values for the s, t and Δ exponents are predicted by each of these models (Table I).

In contrast to chemical gelation, physical gelation is not so well understood. The transient nature of the physical network junctions makes it difficult to study these systems near the gel point. According to de Gennes (*1*), the physical gelation can either fit the universal law for "strong gelation" or be comparable to a "glass transition" which is then referred to as "weak gelation". Although Δ=0.7 is reported for chemical gels (*8,9*), Δ is observed in the range of 0.1 to 0.8 for physical gels (*10-12*).

This paper is a discussion of the main results we have on the thermoreversible gelation of block copolymers of the MXM type in *o*-xylene, where M is syndiotactic poly(methyl methacrylate) (sPMMA) and X is either polybutadiene (PBD), hydrogenated PBD (PEB), poly(styrene-b-butadiene-b-styrene) (SBS) triblock or the hydrogenated version of this triblock (SEBS) (*25-28*).

Table I. Predicted and experimental exponents for the evolution of structure and viscoelasticity near the gel point

	t	s	D	Ref.
Theories				
Mean Field	3	-	1	2, 3
Percolation Electrical Analogy	1.94	0.75	0.72	1, 17
Percolation Rouse approx.	2.7	1.32	0.67	6, 16
Experimental				
Chemical Gels				
Tetraethoxy silane	2.2-2.6	0.79-0.99	0.70-0.72	18
Epoxy	-	1.4	0.70	19
Polyurethane	-	-	0.69	20, 44
Polyester	-	-	0.69	7
Polysaccharide	-	-	0.70	21
Physical Gels	-	-		
10% PVC			0.80	12
Crystallizing polypropylene	-	-	0.13	10
Bacterial Elastomer	-	-	0.11	22
Gelatin	1.82	1.48	-	13
Gelatin			0.62	23
Gelatin, *i*-carrugeenan, xanthan-carob	-	-	0.50-0.65	11
Pectin	1.93	0.82	0.70	24

Materials and sample preparation

The MXM block copolymers were prepared by sequential anionic polymerization of butadiene, styrene and methyl methacrylate with the diadduct of *t*-BuLi onto *m*-diisopropenylbenzene (*m*-DIB) as a difunctional initiator. The MB diblock copolymer was prepared in the same way except for a monofunctional initiator. Details on the synthesis (*29*) and subsequent hydrogenation (*30*) of the polybutadiene blocks were reported elsewhere. Chemical composition, molecular weight and chain microstructure of the MXM copolymers are listed in Table II.

sPMMA was prepared by anionic polymerization in THF at -78°C by using the reaction product of *s*-BuLi with one equivalent of 1,1-diphenylethylene (DPE) as initiator (*31*). iPMMA was synthesized by anionic polymerization of MMA initiated by *t*-butylMgBr in toluene at -78°C (*32*). The molecular characteristics of sPMMA and iPMMA are reported in Table II.

Homogeneous solutions of homo sPMMA, MB and MXM copolymers, added with sPMMA or not, were prepared in *o*-xylene at 80°C. Solutions of blends of sPMMA - polybutadiene (PBD) - sPMMA, or MBM triblock copolymers, with iPMMA were prepared in *o*-xylene at 130°C.

Table II. Characteristics of homo PMMA and MXM block copolymers
(M being syndiotactic PMMA).

samples	midblock (X) [a]	Mn $(x10^{-3})^{b}$ (MXM)	Mw/Mn	syndio(%)[c] in PMMA	1,2-units [c] in PBD (%)
M1	/	30	1.15	80	
M2	/	78	1.10	80	
iPMMA	/	30	1.10	d	
MB	/	40-80	1.10	78	44
MBM1	B	25-80-25	1.10	80	41
MBM2	B	35-36-35	1.10	77	47
MBM3	B	46-36-46	1.10	77	43
MBM4	B	51-77-51	1.10	79	46
MBM5	B	51-100-51	1.10	80	45
MBM6	B	12-50-12	1.10	78	15
MBM7	B	13-61-13	1.10	77	68
MEBM1	EB	25-80-25	1.10	80	
MSBSM	SBS	19-18-79-18-19	1.10	79	44
MSEBSM	SEBS	19-18-79-18-19	1.15	79	

[a] B=polybutadiene, EB=hydrogenated polybutadiene, SBS=poly(styrene-*b*-butadiene-*b*-styrene), SEBS=hydrogenated SBS, I=isoprene; [b] measured by SEC and ^{1}H NMR; [c] measured by ^{1}H NMR; [d] iso(%)=90

Investigation techniques

The final structure of polymer solutions in *o*-xylene was investigated by infrared (IR). DSC measurements were carried out under nitrogen atmosphere in sealed pans at a heating rate of 20° C/min.

The dynamic mechanical properties were measured in the linear regime with a Bohlin CS apparatus equipped with coaxial cylinders (d = 25 mm). The polymer solution, preheated at 80°C (or 130°C), was rapidly added between the coaxial cylinders thermostated at the requested temperature. Solutions were cooled from 80°C or 130°C down to the requested temperature within ca. 1 minute (as measured with a thermocouple) and the measurements were then immediately started.

Thermoreversible gelation

On cooling a MXM block copolymer solutions of a high enough concentration in o-xylene, a transparent gel was formed at temperature lower than 25°C. Conversely, the subsequent heating of this gel resulted in a transparent solution. This sol-gel transition is reversible. At room temperature, the gel is fragile at low concentration (<5wt%) and it becomes stronger at higher concentration (>10wt%). MBM/iPMMA blends in o-xylene form stronger gels than MBM.

Mechanism of the gelation

It is well-known that syndiotactic poly(methyl methacrylate) (sPMMA) could crystallize in the presence of some solvents (33, 34). This phenomenon is strongly dependent on the PMMA stereoregularity and the solvent used. The aggregation takes place in some solvents, such as o-dichlorobenzene, butyl acetate and o-xylene, even in dilute solutions as result of interactions between long parallel syndiotactic sequences.

Solutions of sPMMA homopolymer in o-xylene have been extensively investigated. It has been shown (35, 36) that extended helices are formed in oriented swollen films. The solvent participates to the intermolecular association of the sPMMA chains and it is incorporated into the ordered structure. Spevacek et al. (37, 38) have observed a regular chain conformation by IR and particularly the transition from a predominantly *trans gauche* (TG) to an *all trans* (TT) conformation when sPMMA is dissolved in o-xylene, as indicated by an increase in the intensity of the -CH_2- rocking vibration at 860 cm^{-1} at the expense of the intensity of the vibration at 843 cm^{-1}. Recently, Berghmams et al. have investigated the mechanism for the thermoreversible gelation of homo-sPMMA solutions in o-xylene (34) and in toluene (39). They have proposed a two-step mechanism, the first step being a fast intramolecular conformational change, followed by a time dependent intermolecular association into a tridimensional network. This mechanism has been extended to MBM triblock copolymers (Scheme 1).

When iPMMA is added to MBM block copolymers, stereocomplexation between iPMMA and sPMMA chains occurs. A melting point as high as 185°C has been observed for toluene cast iPMMA/MBM blends (40).

IR analysis. The mechanism of the thermoreversible gelation has been investigated by the IR analysis of the parent sol and gel. Indeed, the self-association of sPMMA in o-xylene has been is accompanied by a change in the chain conformation, as observed in

282

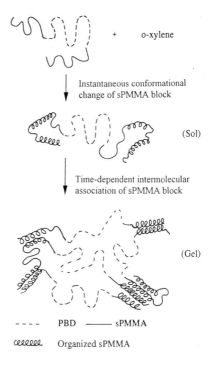

+ o-xylene

Instantaneous conformational
change of sPMMA block

(Sol)

Time-dependent intermolecular
association of sPMMA block

(Gel)

- - - - PBD ——— sPMMA

ℓℓℓℓℓℓ Organized sPMMA

Scheme 1. Gelation of MBM triblock copolymer in *o*-xylene (Reproduced with permission from Ref.25, Copyright 1996 Elsevier Science) the frequency range for the carbonyl stretching vibration (1700 -1750 cm^{-1}) and for the -CH$_2$- rocking vibration at 843 and 860 cm^{-1} typical of TG and TT conformations, respectively. Figure 1 compares the IR absorption of a 10wt% MBM solution in CHCl$_3$ (Figure 1a) and in o-xylene (Figure 1b, c). In CHCl$_3$ the sPMMA self-association does not occur (33) and the C = O absorption is observed at 1731 cm^{-1} (Figure 1a). In contrast, a gel is formed in o-xylene, and an additional absorption is observed at a higher wave number (1738 cm^{-1}) at room temperature (Figure 1b), which is characteristic for intermolecular interactions (34). When this gel is heated up to 80°C, the absorption at 1738 cm^{-1} disappears and a symmetrical absorption is again observed for the C = 0 stretching vibration at 1733 cm^{-1} (Figure 1c), consistently with the dissociation of the tridimensional sPMMA network. Figure 1 also shows the 800-870 cm^{-1} absorption range with a well-defined band at 843 cm^{-1} and a shoulder at 860 cm^{-1} in CHCl$_3$ (Figure 1a). The intensity of the absorption at 860 cm^{-1} is dramatically increased at the expense of the peak at 843 cm^{-1} when a gel is formed in o-xylene (Figure 1b). This modification is characteristic of a transition from a random coil into a regular *all-trans* conformation for the sPMMA chains (34). This absorption pattern remains unmodified when the gel is melted by heating at 80°C (Figure 1c) except for a decrease in relative intensity. This observation indicates that the regular helix conformation of sPMMA persists upon the chain dissociation. These observations for solutions of MBM in o-xylene confirm that the two-step mechanism proposed for the self-association of homo sPMMA (34, 35) can be extended to the MBM triblock copolymers.

Solutions of homo sPMMA M1, MB diblock and MBM1 triblock copolymers (Table II) in o-xylene (10wt%) have been studied at 25°C (Figure 2). The regular *all-trans* conformation at 860 cm^{-1} does not dominate in the solution of homo sPMMA (Figure 2a). This situation persists even at concentrations as high as 30 wt%. The *all-trans* conformation dominates the random coil conformation for the MB diblock copolymer (Figure 2b), and this effect is more pronounced when sPMMA is part of the MBM triblock copolymer (Figure 2c). It is clear from Figure 2 that the sPMMA conformation in o-xylene at 25°C is strongly dependent on the constraints resulting from the chemical bonding to an immiscible partner such as polybutadiene. The absorption due to carbonyl vibration (1700-1750 cm^{-1}) is also shown in Figure 2. A symmetrical absorption for the sPMMA homopolymer is observed at 1731 cm^{-1}, whereas this peak is slightly shifted toward higher wave numbers in case of the MB diblock copolymer. An additional absorption becomes very pronounced at higher wave numbers for the MBM triblock solution. This observation is in agreement with the macroscopic properties of these solutions. Indeed, solution of the homo sPMMA solution does not form a gel even at 0°C for one week. Gelation of the MB diblock solution is rather slow (one day at 0°C), in contrast to gelation of the MBM triblock solution which is very fast (a few minutes at 10°C).

Calorimetric study. Thermal transitions have been investigated by DSC for gels of MBM and MBM/iPMMA blends in o-xylene (10wt%) aged at 0°C for one week. Figure 3 shows the DSC thermograms typical of three gels, i.e. the MBM1 gel, the MBM1/iPMMA gel with a 10/1 s/i mixing ratio and the same formed at a 2/1 mixing

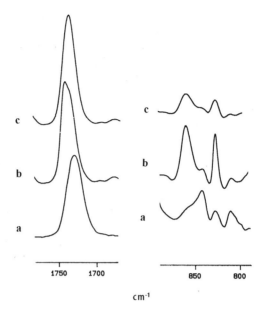

Figure 1. FTIR spectrum of MBM1 (Table II) triblock copolymer solution (10wt%) at 25°C in (a) CHCl₃; (b) *o*-xylene; (c) *o*-xylene heated at 80°C (Reproduced with permission from Ref.25, Copyright 1996 Elsevier Science)

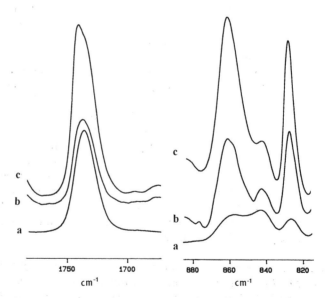

Figure 2. FTIR spectrum of a 10wt% solution in *o*-xylene at 25°C for (a) homo sPMMA M1; (b) MB diblock copolymer; (c) MBM1 triblock copolymer (Table II) (Reproduced with permission from Ref.25, Copyright 1996 Elsevier Science)

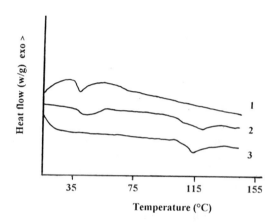

Figure 3. DSC traces of 10wt% solutions in *o*-xylene of (1)MBM1; (2) MBM1/iPMMA (10/1, wt/wt) mixture; (3)MBM1/iPMMA(2/1, wt/wt) mixture (Reproduced from Ref.27, Copyright 1996 American Chemical Society)

ratio. Curve 1 shows a melting endotherm for the MBM1 gel characteristic of the self-aggregation of the sPMMA outer blocks. The melting region extends over ca. 20°C with a maximum at 39°C. When the MBM1 copolymer is added with a minor amount of iPMMA with respect to the sPMMA blocks (curve 2), an additional endotherm is observed at a much higher temperature (113°C), which is the signature of the sPMMA/iPMMA stereocomplexation. When the s/i mixing ratio is decreased from 10/1 to 2/1, which is actually the optimum mixing ratio for the PMMA stereocomplexation (40), the melting endotherm for the self-aggregation of the sPMMA blocks completely disappears, in contrast to the melting of the stereocomplex which is much better defined (curve 3). Therefore, when MBM triblock copolymers are blended with iPMMA in o-xylene, two types of intermolecular association could occur, i.e. self-association of the sPMMA outer blocks and stereocomplexation of the sPMMA blocks with iPMMA.

Time dependent gelation

Figure 4 illustrates how the shear storage (G') and loss (G") moduli depend on time at a constant frequency of 1Hz at 10, 20 and 25°C, respectively, for a 8wt% MBM1 (Table II) solution in o-xylene. Originally, the loss modulus G" is higher than G', which is characteristic of liquids. Both the moduli increase with time as result of an increasing aggregation of the sPMMA blocks. G' however rises faster than G", so that the two curves intersect at G' = G" (crossing point) for a time known as the crossing time. When the temperature is changed, the general behavior remains unchanged excepted for the magnitude of the moduli which decreases with temperature and the crossing time increases. This observation is in agreement with the thermal dissociation of the sPMMA aggregates that stabilize the chain network.

It has been reported that the crossing point is independent of frequency at least for some chemical gels (9). This behavior has been checked for the physical gels under consideration. Figure 5a shows the time-dependence of G' and G" at three frequencies: 0.05, 0.2, and 1Hz for a 7wt% MBM1 gel at 24°C (sample G1, Table III). At each constant frequency, a crossing point is observed at different crossing times. Clearly, the crossing time is frequency-dependent for the MBM1 solution in o-xylene at 24°C. Actually, it increases with the frequency and thus with the apparent moduli. Figure 5b illustrates the time-dependence of G' and G" at three frequencies, 0.08, 0.4 and 1Hz for the MBM1 copolymer added with a small amount of iPMMA (s/i=20/1) (sample B2, Table IV). Although the general behavior of the original MBM1 copolymer (sample G1) is kept unchanged, the crossing times are shorter, all the other conditions being the same. For instance, the crossing time at 1Hz (1200 s) is more than six times smaller compared to the neat copolymer (7800 s), which indicates that a small amount of iPMMA makes the gelation much faster as result of stereocomplexation in addition to the self-association of the sPMMA blocks. The modulus at the crossing point at 1Hz (5.1 Pa) is higher compared to the G1 sample (3.2 Pa). The gelation rate is increased as the relative amount of iPMMA is larger. A decrease in the mixing ratio by a factor of two (s/i=10/1, B4, Table IV) results in a quasi instantaneous gelation, since as soon as the first measurement is carried out, G'

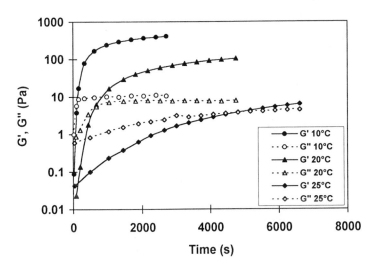

Figure 4. Shear storage (G') and loss (G'') moduli vs. time at 1Hz for a 8wt% solution at various temperatures (Reproduced with permission from Ref.25, Copyright 1996 Elsevier Science)

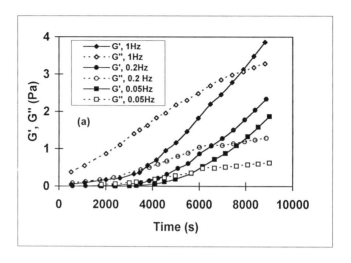

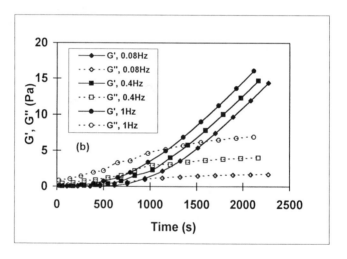

Figure 5. Shear storage (G') and loss (G") moduli vs. time at 24°C and various frequencies for 7wt% solutions in *o*-xylene: (a) MBM1 triblock; (b) MBM1/iPMMA (20/1, wt/wt) mixture. (Reproduced from Ref.27, Copyright 1996 American Chemical Society)

is higher than G" and the crossing point goes unobserved. G" is time-independent, whereas G' increases with time in a way which does not strongly depend on frequency. At mixing ratios smaller than 10/1, gelation is so fast that the time-dependence of G' and G" is no longer observed. Actually, the gel is formed before filling the measurement cell. In order to collect information for the 2/1 mixing ratio, which is the optimum ratio for stereocomplexation (*40*), solution of a lower concentration (1wt%) has been investigated at 10°C (B5, Table IV). Figure 6 shows that G" is close to zero, whereas a rapid initial increase in G' is observed, which corresponds to the rapid stereocomplexation of the sPMMA blocks of MBM1 and iPMMA. Such a fast gelation is quite surprising for a polymer concentration as low as 1wt%. It is worth noting that G' is independent of frequency in the 0.1 to 1 Hz range.

Table III. Gelation time (tg) and scaling exponents for solutions of sPMMA and MXM block copolymers in *o*-xylene

sample	polymer	wt%	T(°C)	tg(s)[a]	Δ[b]	t	s
G1	MBM1	7	24	4200	0.72	1.95	0.69
G2	MBM1	7	22	2300	0.70	1.84	0.67
G3	MBM1	7	20	1100	0.69		
G4	MBM1	7	17	350	0.67		
G5	MBM1	5	15	1360	0.69		
G6	MBM1	2	8	10800	0.70		
G7	MBM2	7	15	2930	0.70		
G8	MBM3	7	15	400	0.69		
G9	MBM4	2	8	820	0.70		
G10	MBM5	2	8	700	0.72		
G11	MBM6	12	10	300	0.70		
G12	MBM7	10	10	3660	0.70	1.96	0.67
G13	MEBM1	7	17	800	0.70		
G14	MSBSM	7	15	2000	0.72	1.84	0.67
G15	MSEBSM	7	15	28000	0.75	2.05	0.69
G16	M2	17	8	1180	0.67		

[a] time required for tan δ to be independent of frequency, [b] scaling exponent defined by eqs. 2 and 3

Table IV. Gelation time (tg) and scaling exponent (Δ) for solutions of copolymers in o-xylene added with homo PMMA

Sample	Block copolymer	Homopolymer	conc. (wt%)	b/h[a]	T (°C)	tg[b] (s)	Δ[c]
B1	MBM1	iPMMA	7	30/1	24	1330	0.72
B2	MBM1	iPMMA	7	20/1	24	650	0.74
B3	MBM1	iPMMA	7	15/1	24	180	0.75
B4	MBM1	iPMMA	7	10/1	24	/	/
B5	MBM1	iPMMA	1	2/1	10	/	/
B6	MBM1	sPMMA(M1)	7	2/1	24	6440	0.73

[a] weight ratio of the sPMMA outer blocks with respect to homopolymer, [b] time characteristic of the point where tan d is independent of frequency, [c] scaling exponent defined by eqs. 2 and 3

Scaling properties

Although the crossing point of G' and G''(G'=G'') is designated as the gel point by some authors (41), a more rigorous definition of the gel point has been given by Winter et al. (8) as the point where tan δ (=G''/G') is independent of frequency (eq. 3). This point should be referred to as the gel point (GP) and the related curing time as the gel time (tg). It is known that the statistic structure and the relaxation modes of the polymer at the gel point are self-similar and that the dynamic mechanical behavior at the gel point fits a power law for the frequency dependence of the moduli (eqs. 2 and 3) (8, 9). Although these dynamic mechanical properties are characteristic of polymers at the gel point of chemically cross-linked systems, there is now evidence for their validity in some physically cross-linked systems (10, 22). The percolation theory predicts slightly different Δ values, i.e. 0.67 when a percolation-like cluster structure and screened hydrodynamic interaction are assumed to prevail, and 0.72 on the basis of an analogy between gelation dynamics and electrical quantities in random resistor networks (for review, see ref.16). These predictions have been experimentally confirmed in case of chemical gelation, e.g. Δ=0.70 for gelation of epoxy resins (19) and Δ=0.69 for gelation of polyurethanes (20, 44). When physical gelation is concerned, Winter et al. have observed the same behavior at the gel point with Δ=0.125 (10) and 0.11 (22) for the crystallization-induced gelation of poly(propylene) and poly(β-hydroxyoctanoate), respectively; while Δ values of 0.62 (23) and 0.71 (24) have been observed for gelatin and pectin, respectively.

The scaling properties have been studied for mixtures of triblock copolymers and o-xylene. In order to determine the gel point, tan δ has been plotted against the curing time at different frequencies, as shown in Figure 7 for sample G1 (Table III). The log(tan δ) vs. time dependencies at different frequencies (from 0.05Hz to 1Hz) intersect at the same time which is the gel time (tg) as stated in the introduction. For times shorter than tg, tan δ decreases as the frequency is increased, which is typical of

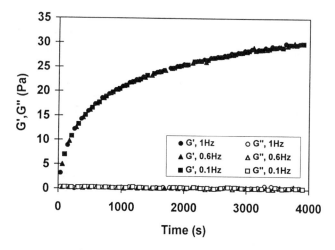

Figure 6. Shear storage (G') and loss (G") moduli vs. time at 10°C and various frequencies for 1wt% solution of the MBM1/iPMMA(2/1, wt/wt) mixture in *o*-xylene (Reproduced from Ref.27, Copyright 1996 American Chemical Society)

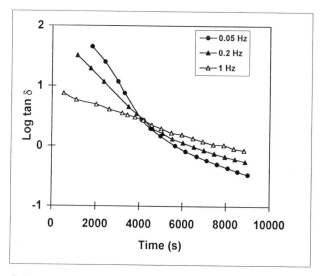

Figure 7. Loss tangent vs. time at 24°C and various frequencies for 7wt% MBM1 solution in *o*-xylene (Reproduced from Ref.27, Copyright 1996 American Chemical Society)

a viscoelastic liquid. At the gel point, the sample changes from a viscoelastic liquid to a viscoelastic solid, since beyond that point, tan δ increases with frequency. This observation indicates that the scaling law holds for the MBM1 triblock copolymer in o-xylene. The relaxation exponent Δ at the gel point can be extracted from eq.3, where δ_c is δ at the intersection of all the tan δ vs. time curves. The value of Δ (0.72, Table III) is found in the same range as for chemical gels (Table I).

Although Winter et al. expected a universal value for the scaling exponent (*8, 9, 42*), they found that Δ was changing with the structure and chemical composition of PDMS chemical gels (*43*). In case of polycaprolactone critical gels (*14*), Δ was found to decrease with increasing concentration and increasing molecular weight. Physical gelation of bacterial elastomers resulted in a strong dependence of viscoelasticity on temperature (*22*). In order to know whether Δ has a unique value or not for MBM gels in o-xylene, a series of experiments have been carried out at different temperatures and polymer concentrations for copolymers of various structures. The experimental gelation time, t_g, and the calculated scaling exponent, Δ, are listed in Table III.

Effect of temperature and copolymer concentration. The effect of temperature has been studied in case of the 7wt% MBM1 solution (samples G1 to G4, Table III). t_g decreases with decreasing temperatures from 4200s at 24°C to 350 s at 17°C, whereas the scaling exponent Δ is kept unchanged. The same conclusion holds for lower concentrations at lower temperature, as shown in Table III for samples G5 (5wt%) and G6 (2wt%) (Table III) at 15° and 8°C, respectively. It seems thus that temperature and polymer concentration have no significant effect on Δ at least in the studied ranges.

Effect of the copolymer composition and molecular weight. Gelation of MBM copolymers (MBM2 to MBM5, Table II) of various sPMMA contents (38% to 72%) and block molecular weights: 36 000<M_n(PBD)<100 000, 25 000<M_n(sPMMA)<51 000, has been studied at different concentrations (2 wt% to 7wt%). Critical gelation is observed in all cases (samples G7 to G10, Table III), and the t_g and Δ values are listed in Table III. In spite of large differences in M_n(PBD) and M_n(PMMA) and in copolymer concentration, the scaling exponent does not change significantly, 0.69<Δ<0.72 (Table III), although t_g decreases with increasing M_n(sPMMA). Mn(PBD) has no significant effect on tg (G9 and G10, Table III).

Effect of the midblock. In order to check whether the Δ value found for MXM triblock copolymer gels is dependent on the chemical nature of the midblock, additional experiments have been carried out for a series of block copolymers of the MXM type, where X is polybutadiene (PBD) of different microstructures (MBM6, MBM7), hydrogenated polybutadiene (PEB), poly(styrene-b-butadiene-b-styrene) (SBS) and hydrogenated poly(styrene-b-butadiene-b-styrene) (SEBS), respectively (Table II). Solutions in o-xylene of all these copolymers have been observed to gelify at suitable concentration and temperature, and the gelation process fits the scaling law at the gel point (samples G11 to G15, Table III). The experimental gelation time, tg, and the calculated scaling exponent, Δ, are reported in Table III. In spite of differences in the nature of the midblock, the copolymer molecular weight and the experimental conditions, the value of Δ is quite similar for all the gel samples in Table III. In contrast, the gelation time is quite different, which proves it to be dependent on the copolymer concentration, molecular weight and curing temperature.

Comparison of the G4/G13 and G14/G15 copolymer pairs shows that the gelation time of a hydrogenated block copolymer is much longer compared to the original copolymer under the same experimental conditions. Indeed the gelation time for the hydrogenated sample G13 is more than two times higher than the parent triblock copolymer G4, whereas a ten fold increase is observed as result of hydrogenation of the pentablock copolymer G14. Nevertheless, the scaling exponent Δ for all these samples remains in the 0.65-0.75 range, as it was previously observed for the MBM triblock copolymers in *o*-xylene. So, the scaling exponent Δ for MXM triblock copolymer gels appears to be essentially independent of the chemical nature of the midblock.

It is worth noting that the scaling law holds also for homo sPMMA polymer gelation in *o*-xylene (sample G16, Table III), the scaling exponent Δ is quite the same as for block copolymer, although a higher concentration is needed due to the weaker association of homo sPMMA.

Effect of addition of homo PMMA. In order to ascertain that the scaling behavior is maintained in the presence of iPMMA which decreases the crossing time and increases the thermal stability of the aged gels, gelation of MBM1/iPMMA mixtures of different s/i ratios (B1, B2 and B3 in Table IV) has been investigated. A typical example is shown in Figure 8, which indicates that the scaling law still holds for the MBM1/iPMMA blends in *o*-xylene. The experimental gelation time, tg, and the calculated relaxation exponent, Δ, are listed in Table IV.

As shown in Table IV, the gel point of the MBM1 gel (G1, Table III) is reached after 4200 s and as earlier as increasing amounts of iPMMA are added to the triblock. tg is 1330 s for sample B1 and 650 s for sample B2. Although stereocomplexation accelerates the gelation process, the scaling exponent Δ, which is characteristic for the critical conditions of gelation, is not significantly affected by the addition of iPMMA, since Δ is in the range of 0.70 to 0.75 while the s/i mixing ratio is changed from 30/1 to 15/1. This observation may be surprising, since two different mechanisms, self-aggregation of sPMMA and stereocomplexation of iPMMA/sPMMA, contribute to the gelation with quite different kinetics.

It is worth noting that addition of homo sPMMA does not accelerate the gelation (sample B6, Table IV); tg is actually and Δ remains unchanged.

Characterization of the sol-gel transition

Although the main percolation theories (Rouse approximation and electrical analogy) predict a similar Δ value for the power law (eq.2) at the gel point, they predict completely different exponents for the modulus or viscosity increase with reaction

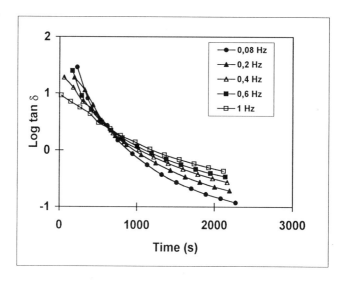

Figure 8. Loss tangent vs. time at 24°C and various frequencies for 7wt%
solution MBM1/iPMMA (20/1, wt/wt) mixture in *o*-xylene (Reproduced from
Ref.27, Copyright 1996 American Chemical Society)

extent. Static exponents have been accurately calculated by numerical methods for the two models: t=1.94, s=0.75 and Δ=0.72 for the electrical analogy and t=2.67 , s=1.32 and Δ=0.72 for the Rouse model.

The experimental determination of the exponents s and t from the divergence of the static viscosity, η_0, and the static elastic modulus, G_0, near the gel point is not very accurate since the measurements are carried out at low (but not zero) shear rates. Furthermore, in case of physical gelation, the non-covalent nature of the cross-links prevents the gelation reaction from being stopped in order to characterize the static properties at various gelation times. Static exponents can also be determined from eqs.4 and 5. In this case, τ_z and J_e^0 can be determined experimentally (23) or numerically by building up a master curve as result of the vertical and horizontal shifts of the individual frequency-dependent data (18). This second approach has been used in this work. Figure 9 shows the collapsed curves for the copolymer G1 (Table III) at tg= 4200 s and for the copolymer G15 (Table III) at tg=28,000 s, where the empirical horizontal and vertical shift factors are abbreviated as a_h and a_v, respectively. Figure 9 shows that the renormalized G' and G" moduli scale with the renormalized frequencies when Δ is ca. 0.72, in good agreement with the value obtained from eq.3. This value for the exponent Δ is consistent with the theoretical predictions by both the electrical and the Rouse models.

The renormalized horizontal and vertical factors used to build up the master curve follows the longest relaxation time, τ_z, and the steady-state creep compliance J_e^0, respectively (6, 19, 45). Therefore, the scaling relationships expressed by eqs.4 and 5 can be used to calculate the static exponents s and t. Figure 10 shows the log-log plot of a_h and a_v versus log ε, in case of sample G1 (Table III), for which t=1.95 and s=0.69. The values of s, t and Δ are listed in Table III not for all the samples, but only when enough experimental data were available in the sol-gel transition so as to construct accurately the master curves.

Comparison of the critical exponents reported in Table III shows a good agreement with the predictions by the electrical analogy model. The observations of static and dynamic exponents quite comparable for all the samples suggest that an unique gelation mechanism is operative, i.e., that the midblok of the MXM triblock does not interfere in the percolation process. According to the electrical analogy near the percolation threshold, gels formed by the triblock copolymers would consist of sPMMA outer blocks, randomly connected and weakly entangled, with an elasticity of a scalar nature. It is worth noting that similar conclusions have been drawn for physical gels based on pectin biopolymers (s=0.82, t=1.93, Δ=0.70) (24), for semi-dilute solutions of polydiacetylene in the red phase (s=1.0, t=1.9) (46), and for semi-dilute solutions of polyacrylamide cross-linked by complexation with Cr(III) (s=0.9, t=1.9) (47).

Conclusion

Thermoreversible gelation has been studied in o-xylene for block copolymers of the MXM type, where M is sPMMA and X is either polybutadiene (PBD), hydrogenated

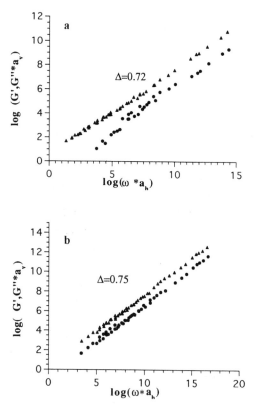

Figure 9. Log G' (●) and log G" (▲) versus log reduced frequency constructed from data available near the gel threshold for the G1 (a) and G14 (b) samples (Table III) (Reproduced from Ref.28, Copyright 1996 American Chemical Society)

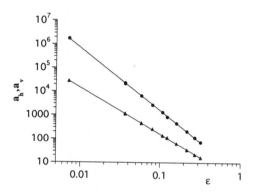

Figure 10. Log-log scaling plot of the horizontal shift a_h (●) and vertical shift a_v (▲) in fig.9 versus the reduced extent of reaction, ε, in the region near the gel point for the G1 sample (Reproduced from Ref.28, Copyright 1996 American Chemical Society)

PBD (PEB), poly(styrene-b-butadiene-b-styrene) (SBS) triblock or the hydrogenated version of this triblock (SEBS). Aggregation of the sPMMA block of the copolymers dissolved in o-xylene is responsible for the formation of a three-dimensional structure, in such a way that a scaling law behavior is observed at the gel point. The scaling exponent Δ in the 0.65-0.75 range is in good agreement with the predictions of the scalar percolation theory and with experimental measurements for chemical gelation, although functionality and size of the cross-linking entities are quite different in chemical and physical gels. It is remarkable that the exponent Δ has been found to be independent of molecular weight, chemical composition and midblock X of the MXM copolymers, and experimental conditions such as concentration and temperature. Addition of homo syndio or iso PMMA does not change the Δ value, although the gelation time is greatly affected.

Modulus-frequency master curves have been constructed by applying appropriate time dependent renormalisation factors to the frequency and modulus individual data. From the scaling of these factors with reaction time, the static scaling exponents t and s have been calculated and observed to be independent of the chemical nature of the midblock, suggesting a unique gelation mechanism. For all the samples, $1.84 < t < 2.01$ and $0.65 < s < 0.70$ have been observed in a good agreement with the scalar elasticity percolation model.

Acknowledgment

The authors are very much indebted to the IWT (Flemish Institute for the Promotion of Science-Technological Research in Industry) for the financial support of a joint research program with Raychem N.V. (Kessel-Lo, Belgium) and the Katholieke Universiteit Leuven (Prof. H. Berghmans and H. Reynaers). They warmly thank Dr. N. Overbergh (Raychem, Kessel-Lo) and Dr. Ph.Hammond (Raychem, Swindon) for stimulating discussions. They are grateful to the "Services Fédéraux des Affaires Scientifiques, Techniques et Culturelles" for general support in the frame of the "Poles d'Attraction Interuniversitaires".

References

1. de Gennes, P.G. Scaling Concepts in Polymer Physics; Cornell University Press: Ithaca, NY, 1979
2. Flory, P.J. J.Am. Chem.Soc. 1941, 63, 3083
3. Stauffer, D; Coniglio, A; Adam, M. Adv. Polym.Sci. 1982, 44, 103
4. Stauffer, D. Introduction to Percolation Theory, Taylor & Francis, London and Philadelphia,1985
5. Hess, W.; Vilgis, T.A; Winter, H.H. Macromolecules 1988, 21, 2356
6. Martin, J.E; Adolf,D; Wilcox, J.P. Phys.Rev.A 1989, 39, 1325
7. Rubinstein, M.; Colby, R.H.; Gillmor, J.R. Polym. Prepr. (Am. Chem. Soc., Div. Polym. Chem.) 1989, 30(1), 81
8. Chambon, F; Winter, H.H. J.Rheol. 1987, 31, 683
9. Winter, H; Chambon, F. J.Rheol. 1986, 30, 367
10. Lin, Y.G; Mallin, D.T; Chien, J.C.W; Winter, H.H. Macromolecules, 1991, 24, 850

11. Cuvelier, G; Launay, B. Makromol.Chem.,Macromol.Symp. 1990, 40, 23
12. Te Nijenhuis, F.; Winter, H.H. Macromolecules, 1989, 22, 411
13. Djabourov, M.; Leblond, J.; Papon, P. J.Phys. France, 1988, 49, 333
14. Izuka, A.; Winter, H.H. and Hashimoto, T. Macromolecules, 1992, 25, 2422
15. Ferry, D., Viscoelastic Properties of Polymers, Wiley, New York,1980
16. Adam, M. and lairez in The physical Properties of Polymeric Gels, Ed. by Cohen Addad J.P., John Wiley & Sons, 1996, p.87
17. Herrmann, H.J.; Derruda B.; Vannimenus, J. Phys.Rev. B 1984, 30, 4080
18. Hodgson, P; Amis, E.J. Macromolecules, 1990, 23, 2512
19. Adolf, D; Martin, J.E; Wilcoxon, J.P. Macromolecules, 1990, 23, 527
20. Durand, D.; Delsanti, M; Adam, M; Luck, J.M. Europhys. Lett.,1987, 3, 297
21. Matricardi, P.; Dentini, M and Crescenzi, V. Macromolecules, 1993, 26, 4386
22. Richtering, H.W.; Gagnon, K.D; Lenz, R.W; Fuller, R.C. and Winter, H.H. Macromolecules, 1992, 25, 2429
23. Peyrelasse, J.; Lamarque, M.; Habas, J.P. and Bounia, N.E. Phys. Rev. E, 1996, 53, 6126
24. Axelos, M.A.; Kolb, M. Phys.Rev.Lett.1990,64,12,1457
25. Yu, J. M; Jérôme, R.; Teyssié, Ph. Polymer, 1997, 38, 347
26. Yu, J. M.; Dubois, Ph.; Teyssié, Ph.; Jérôme, R.; Blacher, S.; Brouers, F. and L'Homme, G. Macromolecules, 1996, 29, 5384
27. Yu, J. M; Jérôme, R. Macromolecules, 1996, 29, 8371
28. Yu, J. M.; Blacher, S. .; Brouers, F. and L'Homme, G.; Jérôme, R Macromolecules, 1997, 30, 4619
29. Yu, J.M; Dubois, Ph.; Teyssié, Ph.; Jérôme, R. Macromolecules, 1996, 29, 6069
30. Yu, J.M; Yu, Y.; Dubois, Ph.; Teyssié, Ph.; Jérôme, R. Polymer, 1997, 38, 3091
31. Varshney, S.K.; Hautekeer, J.P.; Fayt, R.; Jérôme, R. and Teyssié, Ph. Macromolecules, 1990, 23, 2681
32. Hatada, K; Ute, K; Tanaka, K; kitayama, T and Okamoto, Y. Polym. J. 1985, 17, 977
33. Spevacek, J. and Schneider, B. Adv. Colloid. Interface Sci., 1987, 27, 81
34. Berghmans, H.; Donkas, A.; Frenay, L.; Stoks, W.; De Schryver, F. E.; Moldenaers, P. and Mewis, J. Polymer, 1987, 28, 97
35. Kusuyama, H.; Takase, M.; Higashihata, Y.; Tseng, H.T.; Chatani, Y. and Tadokoro, H. Polymer, 1982, 23, 1256
36. Kusuyama, H.; Miyamoto, N.; Chatani, Y. and Tadokoro, H. Polym. comun. 1983, 24, 119
37. Spevacek, J.; Schneider, B.; Baldrian, J.; Dybal, J. and Stokv, J. Polym. Bull. 1983, 9, 495
38. Spevacek,J.; Schneider, B.; Dybal, J.; Stokv, J. and Baldrian, J. Polym. Sci. Phys. Ed. 1984, 22, 617
39. Berghmans, W; Thijs, S.; Cornette, M.; Berghmans, H.; De Schryver, F.C.; Moldenaers, P. and Mewis, J. Macromolecules, 1994, 27, 7669
40. Yu, J.M.; Yu, Y.; Dubois, Ph.; Teyssié, Ph. and Jérôme, R.; Polymer, 1997, 38, 2143

41. Tung, C.Y.M.; Dynes, P.J. J.Appl.Polym.Sci. 1982, 27, 569
42. Chambon, F; Winter, H.H. Polym.Bull. 1985, 13, 499
43. Scanlan, J.C; Winter, H.H. Macromolecules, 1991, 24, 47
44. Prochazka, F.; Nicolai, T. and Durand, D. Macromolecules, 1996, 29, 2260
45. Martin, J.E; Adolf, D. and Wilcoxon, J.P. Phys.Rev Lett, 1988, Vol22, 2620
46. Adam, M.; Aimé, J. P. J. Phys. II France, 1991, 1, 1277
47. Allain, C.; Salomé, L. a) Polym. Com. 1987, 28, 109; b) Macromolecules, 1987, 20, 2958, c) Macromolecules, 1990, 23, 981

Chapter 20

Synthetic Design of "Responsive" Surfaces

David E. Bergbreiter

Department of Chemistry, Texas A&M University, College Station, TX 77843

Responsive polymer-modified surfaces whose properties and character changes in response to external stimuli such as temperature, pH or solvent can be prepared by a number of procedures. Such surfaces can be monolayers of functional groups or multilayers of groups. Using synthetic organic polymer chemistry, it is possible to design such interfaces such that predictable changes in reactivity, wettability or permeability can be effected. Examples of such chemistry using surface oxidation or surface graft polymerization are described.

Chemistry to modify surfaces to change the properties of a material or to engender new properties for a material is of interest in many disciplines.[1-3] In this paper, I describe and review some of the work our group has done to modify surfaces that can then respond to their environment. This work has implications in polymer chemistry, in the design of biocompatible surfaces, in catalysis chemistry, corrosion passivation and sensor chemistry. This review highlights the general ideas underlying this work. Readers interested in specific applications and in specific materials should consult the original papers cited where we and our collaborators detail our original ideas.

To begin with, it is important to define what exactly is meant by the term 'responsive surface'. There are two parts to this term, both of which require some clarification. In general, 'responsive' surfaces are surfaces of organic polymers or inorganic materials that have been modified in some way so that their properties change in response to an external stimulus like pH or temperature. In this context, responsive surfaces are much like other so-called intelligent materials.[4,5] The second part of this term refers to a substance's 'surface'. This term too needs to be clarified. In the context of our work, we have construed the term 'surface' to include more than the top few Å of material in contact with a gas. We generally use the term surface to loosely define the interfacial portion of a surface functionalized polymer or solid in which functional groups readily react with soluble reagents that would not normally penetrate the underlying polymer or solid. Thus, functional groups in a self-assembled monolayer, functional groups in a spun-cast polymer or functional groups at polymer-solution interfaces all would be construed as functional groups at a 'surface'.

Modification of existing polymers surfaces to make so-called 'smart' surfaces has precedent both in our work and in that of others.[6-9] One example of this effect would be changes in wettability in surface functionalized polymers on thermal treatment in vacuum or water.[6,10] For example, in our studies on sulfonated polyethylene surfaces,[6] we were able to show that reversible changes in hydrophobicity could be seen both for PE-[CO$_2$H] and PE-[SO$_3$H] (surfaces prepared by CrO$_3$/H$_2$SO$_4$ etching and SO$_3$/H$_2$SO$_4$ etching, respectively). As shown in Figure 1, the PE-[CO$_2$H] surface becomes more hydrophobic on heating in vacuum and more hydrophilic on heating in water. In contrast, the PE-[SO$_3$H] surface becomes more hydrophilic on heating in vacuum and more hydrophobic on heating in water. However, these changes in hydrophobicity are modest and the changes are relatively slow (vacuum heating requires ca. 1 h). Moreover, prolonged treatment leads to an irreversible change to a more hydrophobic surface in all cases - the normal reconstruction seen with many functionalized polymer surfaces. Similar work has also been described by Ferguson.[9] In this latter work, the changes were more reversible, more rapid and more substantial.

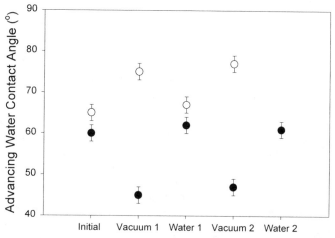

Figure 1. Changes in wettability of PE-[CO$_2$H] (O) or PE-[SO$_3$H] (●)on heating in vacuum or water in comparison to initial oxidized surface wettability (Initial).

Reversibly responsive polymer surfaces can also be prepared by other strategies. One strategy our group has favored is a process we call entrapment functionalization. This process has the advantage of allowing us to prepare sophisticated, labeled polyethylene oligomers that we have characterized by solution state analyses (principally NMR spectroscopy).[11] The resulting oligomers can then be mixed with virgin high density polyethylene - polyethylene that is free of additives. Such mixtures typically have 1% or less of the oligomer by weight. Codissolution of the virgin polymer and the functionalized oligomer then produces a homogeneous solution that can be used to solvent cast a functionalized polyethylene film.[12] Our experience shows that the oligomer's functional groups are located at the "surface" of the film derived from this casting process. Two different examples illustrate how we can use this process to prepare a responsive surface.

In the first example, we used the chemistry shown in equation 1 to prepare a pyrenylphenylmethyl-terminated polyethylene oligomer.[13] This oligomer in turn was

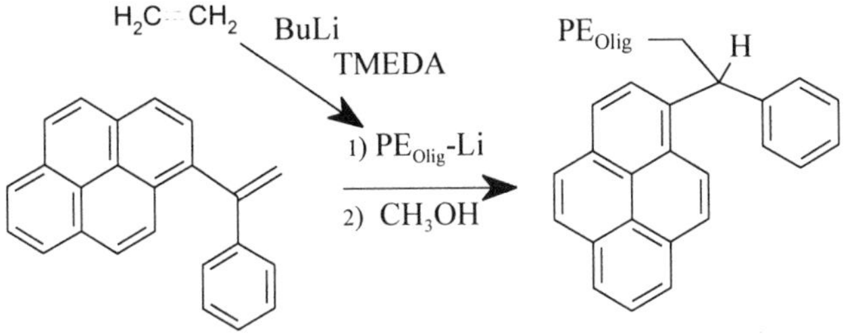

(1)

mixed with virgin polyethylene (1:100, w:w) and dissolved in *o*-dichlorobenzene at 120 °C. The resulting solution was then poured into a flat dish and the solvent was evaporated in an explosion-proof oven to form the film **1** shown in equation 2.

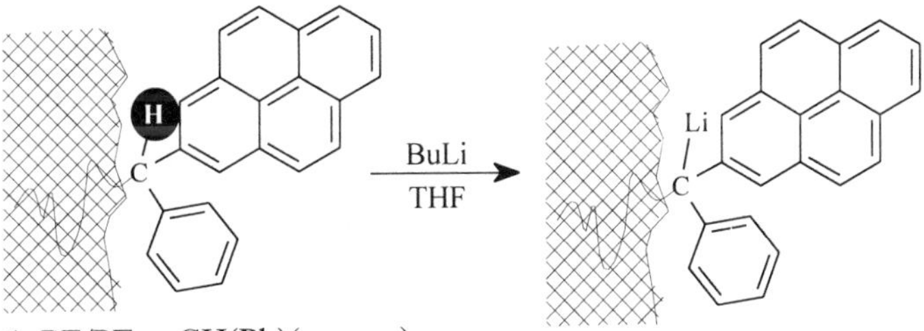

1, PE/PE$_{Olig}$-CH(Ph)(pyrene)

(2)

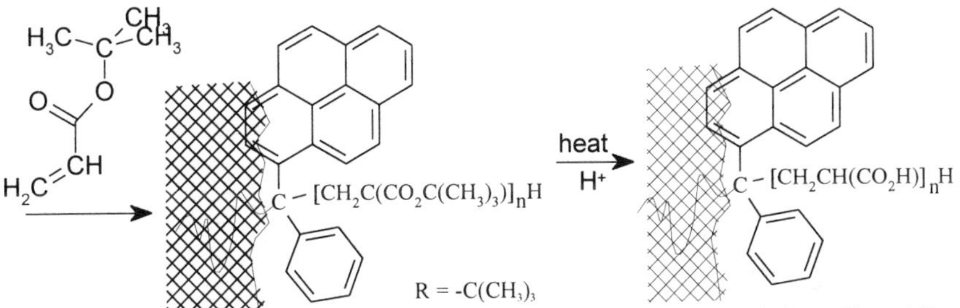

2, PE-*g*-poly(*tert*-butylacrylate)

R = -C(CH$_3$)$_3$

3, PE-*g*-poly(acrylic acid)

The pyrene-labeled polyethylene film **1** contains a weakly acidic C-H bond (highlighted **H**) that can be converted into a lithiated initiator on treatment with a

304

strong base. Thus, this film was then deprotonated with *n*-BuLi and allowed to react with *tert*-butyl acrylate to form a poly(*tert*-butyl acrylate graft) that was in turn hydrolyzed with acid to form a poly(acrylic acid) graft.[8] The surface-functionalized polyethylene so-formed contains a pyrene fluorescent label at the base of an amphoteric poly(acrylic acid) graft. Exposure of this grafted surface to a series of buffers shows that the buffer has no effect on the fluorescence intensity of the pyrene in this film. However, if 1 M NaI were present in the buffer solution, the amount of pyrenes quenched by the iodide significantly increased in the pH region of 6.2-6.5. This pH-induced change in the amount of surface-bound pyrene quenched by the soluble iodide quencher is reversible. We have interpreted this change to suggest that the surface structure changes with the polyelectrolyte graft becoming completely deprotonated and solvent swollen in the indicated pH range, thus exposing the underlying pyrene groups. When similar studies were carried out with an amine quencher (Et$_3$N, Figure 2), the amount of pyrene quenched was again sensitive to pH. However, in this instance the significant quenching was only seen above the pKa of Et$_3$NH$^+$ (i.e., above a pH of 9 where the Et$_3$N concentration becomes significant).

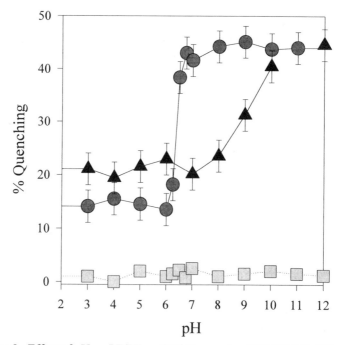

Figure 2. Effect of pH on PE/PE$_{Olig}$-C(Ph)(pyrene)-*g*-(CH$_2$CH(CO$_2$H))$_n$ quenching based on changes in the intensity of the I$_3$ peak in fluores-cence spectra of; ●, 1 M NaI in buffered aqueous solutions; □, buffer solutions without added quencher; Δ, 1 M Et$_3$N in a buffer solution.

Responsive surfaces like the modified polyethylene film **3** suggest that it may be possible to change the character of a surface significantly by changing pH. Two recent papers further illustrate this idea. The first of these papers is work by Ikada's group in which poly(acrylic acid) grafted surfaces are used as responsive membranes. In this work, the porosity of a poly(acrylic acid)-grafted membrane is significantly changed by virtue of changes in the size of the poly(acrylic acid) graft as a function of pH.[14] In this example, the presumption of changes in size of the pores as a function of ionization of the graft was supported by AFM studies.

The second example of a responsive surface with a poly(acrylic acid) graft is work from the Crooks/Bergbreiter groups. In this instance, we prepared hyperbranched graft surfaces using a forgiving synthetic approach to design a highly functionalized reactive surface.[15,16] This approach to synthesis of a functionalized responsive does not rely on a grafting from the surface strategy. Instead, we use soluble polymers (NH_2-terminated poly(*tert*-butyl acrylate)) to graft a poly(acrylic acid) precursor onto the surface. In order to compensate for synthetic vagaries associated with even simple reactions at surfaces, our two groups designed a new strategy for grafting that used repetitions of this simple step to create dendritic, hyperbranched poly(acrylic acid grafts). Unlike our earlier polyethylene work, this chemistry is considerably simpler and much more versatile. The grafts prepared in this chemistry can be as thick as desired with thicknesses of up to 1000 Å being accessible in six or seven synthetic stages.

The resulting grafted surfaces illustrate responsiveness much like that seen by Ikada. In this case, we find that a so-called 3 PAA surface (a surface grafted through three stages of activation ($ClCO_2R$), coupling ($H_2N-PTBA-NH_2$) and hydrolysis) has a thickness of 300-400 Å depending on the molecular weight of the diamino poly(*tert*-butylacrylate) ($H_2N-PTBA-NH_2$) polymer used in the functionalization step. This thickness, which was measured after immersion of

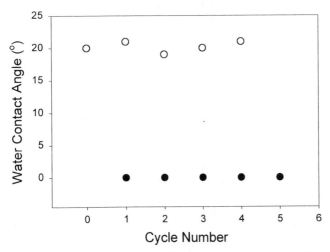

Figure 3. Changes in hydrophilicity of a hyperbranched 3-PAA graft with acid (0.1 N HCl, ○) or base (0.1 N NaOH, ●) treatment.

the surface in dilute acid and drying with an ethanol rinse and N_2 stream, reversibly expands to 40-50% after immersion in aqueous base. The reversibility of these thickness changes is correlated with a similar reversible change in surface hydrophilicity as is illustrated by the data shown in Figure 3.[16] While polymer surfaces sometimes experience some hysteresis with surface changes attenuating as the number of cycles increases or with temporary changes becoming permanent after long term treatment, these surfaces seem more robust. It is also worth noting that both the dry thickness of the acid form and the dry thickness of the basic form of these 3-PAA surfaces are likely different from the thickness of these same surfaces *in situ*. Indeed, recent work by our joint groups has indicated that these surfaces thicknesses expand to over 1000 Å in their basic form when then are completely solvated by an aqueous basic buffer.[17]

Recent work by the Crooks group has shown the potential for surface modification with dendrimers alone.[18] Our group in conjunction with the Crooks group has recently extended this chemistry by coupling the chemistry of dendrimers to the chemistry of surface chemisorbed reactive polymer brushes.[19,20] The result has been that we are able to prepare surfaces with thicknesses between 100 - 500 Å whose permeability changes significantly and reversibly with pH. In this instance, we significantly extend the idea of using a single polyelectrolyte bonded to a surface by covalently assembling a matrix of a polybase in a polyacid. The polyacid in this case is a poly(maleic anhydride)-*c*-poly(methyl vinyl ether) copolymer commercially available as Gantrez. After reaction with amines or after reaction with an alcohol, this polymer forms half acid-half amide (amic acid) groups or half acid-half ester groups and is thus a polyacid. The polybase in our procedure is a fourth generation PAMAM dendrimer assembled by sequential reactions of ethylene diamine with an acrylate moiety. This dendrimer contains tertiary amine groups in its interior and 64 - NH_2 groups on its periphery. When an appropriately functionalized surface (an amine-functionalized SAM on a gold-coated Si wafer) is modified by covalent deposition of Gantrez, the resultant surface contains a reactive anhydride brush that can react with a methanolic solution of this PAMAM dendrimer. The result is a covalent multilayer assembly of partially functionalized dendrimers in a random coil carboxylic acid-containing polymer matrix. This mixture of groups is precluded from phase separating by the amic acid crosslinks. Moreover, the initial Gantrez layer (Gz1) which is then modified by a dendrimer (D1) can be further modified because the dendrimer treatment leaves the surface as an amine-rich surface - a surface that resembles functionally the starting surface. Thus, repetition of the Gantrez and dendrimer treatments leads to increasingly thicker films.

The supported nanocomposite membrane formed by two such Gantrez/dendrimer treatment cycles is termed a D2 surface.[19,20] When this surface is examined electrochemically in the presence of anionic or cationic electroactive species, interesting pH-dependent permselective behavior is seen because of how the surface structure changes with pH. These changes in surface structure are illustrated by the schematic structures shown in Figure 4 below. At pH 3, the amine groups of the dendrimer are protonated and the carboxylic acid groups are neutral. The result is a cationic surface-supported membrane as illustrated in

Figure 4. Such a membrane is permeable to $Fe(CN)_6^{3-}$. However, a cationic redox species, $Ru(NH_3)_6^{3+}$, cannot permeate through this cationic supported membrane and this species is electrochemically silent. In contrast at pH 11, the carboxylic acid groups are carboxylate groups and anionic and the amine groups are neutral. The result is reversed activity toward the charged redox species with oxidation/reduction being facile for $Ru(NH_3)_6^{3+}$ and $Fe(CN)_6^{3-}$ being electrochemically silent. At intermediate pH values, oxidation/reduction occurs for both species. These electrochemical experiments carried out in 3M NaCl show that responsive surfaces can affect the reactivity of their underlying support toward soluble reagents.

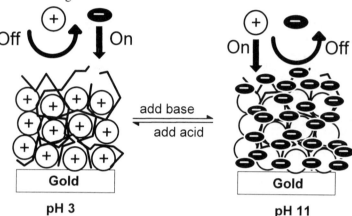

Figure 4. Reversible changes in a D2 Gantrez/PAMAM nanocomposite membrane with changes in pH.

One example of how we have prepared materials that usefully change their reactivity in response to temperature changes is our use of poly(N-isopropylacrylamide) as a support for reactants in catalysis and as a support for catalysts.[21-23] The changes in reactivity of a nitroarene bound to poly(N-isopropylacrylamide) via an amide bond toward a heterogeneous hydrogenation catalyst (Pt/C) exemplify the concept of using a soluble polymer to imbue into a reaction an anti-Arrhenius response to temperature.[24] While reactions normally run increasingly faster with temperature, this substrate on this polymer in water has reactivity that changes little with temperature in the 0 – 33 °C range. More notably, above 39 °C, this nitroarene is unreactive because above this temperature the polymer (and thus the substrate) is insoluble.

Temperature-responsive surface chemistry like that seen for 'smart' substrates can also be seen for appropriate surface-functionalized polyethylene samples.[7] In other work, we had noted that polyethylene surfaces can be functionalized with PEG oligomers or with PEG oligomers that are tagged with a pyrene fluorophore.[25-26] When a pyrene fluorophore was used, we successfully showed that the resulting polymer/water interface's environment changed in response to changes in temperature. These studies specifically showed that normal and inverse solvation of these interfaces occur with increasing temperature.

Our studies also showed that functional groups at thermally responsive interfaces could have reactivity that has an unusual temperature dependence like that of the 'smart' substrates described above.[7] Since the pyrene ester groups and PEG chains collapse at higher temperature in water suspension in a manner qualitatively illustrated in eq. 3, the reactivity of the end groups should be affected. Kinetic studies of ester hydrolysis at these surfaces show that hydrolysis rate of the surface bound pyrene esters decreases slightly at higher temperatures.[7,26] However, the reaction does not stop as it did in the case of soluble polymer-supported substrates. We speculate that the polyethylene film/solvent interface in this instance changes enough to suppress the hydrolysis reaction rate but that the less hydrophilic interface is still

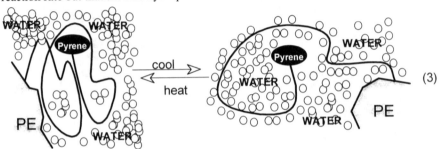

$$(3)$$

reactive at higher temperature. It is possible that other thicker functional surfaces could exhibit a more pronounced change in solvation and reactivity. Regardless, the changes in reactivity are notable in that they imply that there could be "surface" reactions that would best be run at low temperature because they could be kinetically slow at high temperature.

Conclusion

Surfaces that respond to temperature can be designed to serve several roles. Recent work both from our group and others shows that surface hydrophilicity and surface hydration can change with temperature in a manner that resembles the better known changes of soluble polymer solvation with temperature. Moreover, such changes in surface solvation can have an effect on surface reactivity – an effect that can be used to molecularly engineer useful responsive reactivity into a substrate. Likewise pH can significantly affect surface solvation and reactivity. In the case of a simple polyelectrolyte, reversible changes in hydrophilicity and reactivity of the surface induced by pH can be seen. In the case of membranes, changes in permeability as a result of either surface size changes or as a result of surface charge changes can be seen.

Since the range of what can be done synthetically at surfaces is now developing rapidly, it seems likely even more sorts of responsive surfaces will be developed. Such materials should find useful roles in areas like adhesion, permeability, biocompatibility, sensor chemistry and corrosion passivation.

Acknowledgment

Generous support of our work in surface chemistry and catalysis from the National Science Foundation and the Robert A. Welch Foundation is gratefully acknowledged.

References

1. Feast, W. J.; Munro, H. S. (Eds.), *Polymer Surfaces and Interfaces II*, Wiley, Chichester, 1993.
2. Ferguson, G. S.; Whitesides, G. M. in *Modern Approaches to Wettability: Theory and Applications*, Schrader, M. E.; Loeb, G. I., Eds., Plenum, New York, 1992.
3. David E. Bergbreiter, *Prog. Polym. Sci.*, **1994**, *19*, 529-60.
4. Gray, H. N.; Bergbreiter, D. E. *Environmental Health Perspectives*, **1997**, *105 (Supplement 1)*, 55-63.
5. Yanagida, H. *Angew. Chem. Int. Ed. Engl.*, **1988**, *27*, 1389-92.
6. Bergbreiter, D. E.; Kabza, K. *Industrial Eng. Chem. Res.*, **1995**, 2733-2739. Bergbreiter, D. E.; Kabza, K. *J. Am. Chem. Soc.*, **1991**, *113*, 1447-8.
7. Bergbreiter, D. E.; Ponder, B. C.; Aguilar, G.; Srinivas, B. *Chem. Mater.*, **1997**, *9*, 472-477.
8. Bergbreiter, D. E.; Bandella, A. *J. Am. Chem. Soc.*, **1995**, 10589-90.
9. Carey, D. H.; Ferguson, G. S. *J. Am. Chem. Soc.*, **1996**, *118*, 9780-1.
10. Holmes-Farley, S. R.; Reamey, R. H.; Nuzzo, R.; McCarthy, T. J.; Whitesides, G. M. *Langmuir*, **1987**, *3*, 799-815.
11. Bergbreiter, D. E.; Blanton, J. R.; Chandran, R.; Hein, M. D.; Huang, K. J.; Treadwell, D. R.; Walker, S. A. *J. Polym. Sci., Part A: Polym. Chem.*, **1989**, *27*, 4205-26.
12. Bergbreiter, D. E.; Hu, H. P.; Hein, M. D. *Macromolecules*, **1989**, *22*, 654-62.
13. Bergbreiter, D. E.; Srinivas, B.; Gray, H. N. *Macromolecules*, **1993**, *26*, 3245.
14. Ito, Y.; Park, Y. S.; Imanishi, Y. *J. Am. Chem. Soc.*, **1997**, *119*, 2739-40. Ito, Y.; Ochiai, Y.; Park, Y. S.; Imanishi, Y. *J. Am. Chem. Soc.*, **1997**, *119*, 1619-23.
15. Zhou, Y.; Bruening, M. L.; Bergbreiter, D. E.; Crooks, R. M.; Wells, M. *J. Am. Chem. Soc.*, **1996**, *118*, 3773-4.
16. Bruening, M. L.; Zhou, Y.; Aguilar, G.; Agee, R.; Bergbreiter, D. E.; Crooks, R. M. *Langmuir*, **1997**, *13*, 770-8.
17. Peez, R. F.; Dermody, D. L.; Franchina, J. G.; Jones, S. J.; Bruening, M. L.; Bergbreiter, D. E.; Crooks, R. M. *J. Am. Chem. Soc.*, submitted for publication
18. Wells, M.; Crooks, R. M. *J. Am. Chem. Soc.*, **1996**, *118*, 3988-9.
19. Liu, Y.; Zhao, M.; Bergbreiter, D. E.; Crooks, R. M. *J. Am. Chem. Soc.,* **1997** *119*, 8720-8721.
20. Liu, Y.; Bruening, M. L.; Bergbreiter, D. E.; Crooks, R. M. *Angew. Chem. Int. Ed. Engl.*, **1997**, *36*, 2114-2116.

310

21. Bergbreiter, D. E.; Zhang, L.; Mariagnanam, V. M. *J. Am. Chem. Soc.,* **1993,** *115*, 9295-6. Bergbreiter, D. E.; Mariagnanam, V. M.; Zhang, L. *Advanced Materials,* **1995,** *7*, 69-71
22. Bergbreiter, D. E.; Liu, Y.-S. *Tetrahedron Lett.*, **1997,** *38*, 7843-6.
23. Bergbreiter, D. E.; Case, B. *Tetrahedron Lett.*, submitted for publication.
24. Bergbreiter, D. E.; Caraway, J. W. *J. Am. Chem. Soc.*, **1996,** *118*, 6092-3.
25. Bergbreiter, D. E.; Srinivas, B. *Macromolecules,* **1992,** *25*, 6360-5.
26. Bergbreiter, D. E.; Aguilar, G. *J. Polym. Sci., Polym. Chem. Ed.*, **1995,** *33*, 1209-1217.

Chapter 21

Thermosensitive Properties of Flat Poly(tetrafluoroethylene) Plates Surface-Grafted with *N*-Isopropylacrylamide and 2-(Dimethylamino)Ethyl Methacrylate

Kiyomi Matsuda, Naoki Shibata, Kazunori Yamada, and Mitsuo Hirata

Department of Industrial Chemistry, College of Industrial Technology, Nihon University, Izumi-cho, Narashino-shi, Chiba 275, Japan

Thermosensitive properties of the surfaces of flat poly(tetrafluoroethylene) (fPTFE) plates photografted with *N*-isopropylacrylamide (NIPAAm) and 2-(dimethylamino)ethyl methacrylate (DMA) were examined by the measurement of contact angles. The surfaces of the fPTFE plates became hydrophilic with an increase in the N1s/C1s value, representing the degree of the coverage of fPTFE plate surface with grafted polymer chains. The wettabilities of NIPAAm-grafted fPTFE plate surfaces toward water and DMA-grafted fPTFE plate surfaces toward a buffer solution of pH 10 were clearly decreased between 30 and 35°C, and between 25 and 30°C, respectively. These results show that the surfaces of grafted fPTFE plates change from hydrophilic to hydrophobic around the LCSTs (lower critical solution temperatures) of these polymer solutions.

It has been examined to introduce hydrophilic groups to surfaces of some hydrophobic polymers such as polyethylene (PE) and polytetrafluoroethylene (PTFE) while leaving their bulk properties unchanged (*1,2*). For example, PE surfaces treated with oxidative agents or grafted with hydrophilic monomers showed an increase in wettabilities with treatment time or in adhesivities with the grafted amount (*3-10*). The hydrophilic PTFE surfaces prepared by the combined use of the oxygen-plasma treatment and the photografting of hydrophilic monomers also showed markedly enhanced wettabilities and adhesivities with the grafted amount (*2*).

It would be of interest to add some functionalities such as properties in response to changes in pH, temperature, electrical potential and so on as well as hydrophilic properties to PTFE surfaces by the combined technique just described. Poly(*N*-isopropylacrylamide) (PNIPAAm) is a well-known thermo-responsive polymer whose hydrogel undergoes a volume change around 31 °C in an aqueous solution. And further, poly 2-(dimethylamino)ethyl methacrylate (PDMA) in a pH 10 buffer solution exhibits a drastic decrease in the transmittance around 28 °C, which is considered to come from a phase transition of PDMA chains (*11*). Therefore, we have tried to prepare *N*-isopropylacrylamide (NIPAAm) or 2-(dimethylamino)ethyl methacrylate (DMA) grafted flat PTFE (NIPAAm- or DMA-g-fPTFE) surfaces by the above combined technique and then to follow up their thermosensitive properties.

The surface compositions and wettabilities of monomer-g-fPTFE were analyzed by means of ESCA and estimated on the basis of the contact angles toward water or an aqueous solution, respectively.

Experimental

Materials. The polymer substrates used for grafting were fPTFE plates (thickness = 1.0mm) obtained from Furon Kogyo Co. Ltd. (Japan). The fPTFE plates were cut into 2 x 6 cm^2 and washed in water, methanol and acetone for 3 hours in an ultrasonic washing machine, respectively, and then dried under reduced pressure. NIPAAm was purified by recrystallization and DMA was purified by distillation under reduced pressure. N,N'-methylenebisacrylamide (Bis; Kanto Kagaku Co. Ltd.), N,N,N',N'-tetramethyl ethylenediamine (TEMED; MERCK Co. Ltd.), N,N'-azobisisobutylonitrile(AIBN; Wako Chemical Co. Ltd.) and benzophenone(Wako Chemical Co.Ltd.) were used without purification.

Preparation of Grafted fPTFE Plates. The fPTFE plates were first treated with oxygen plasmas for 120 s at an output of 200 W and frequency of 15 kHz using a Shimadzu LCVD 20-type plasma polymerization apparatus. The plasma-treated fPTFE plates were coated with benzophenone as a sensitizer. Then, NIPAAm or DMA monomers were photografted onto the plasma-treated fPTFE plates at 60 °C by applying UV rays emitted from a 400 W high pressure mercury lamp in the individual aqueous monomer solutions at a monomer concentration of 1.0 mol/L (2).

NIPAAm-g-fPTFE plates or DMA-g-fPTFE plates were washed with cool water or with water and 50%(v/v) methanol to remove monomers and homopolymers. DMA-g-fPTFE plates were soaked in a NaOH solution of pH12 for deprotonating grafted dimethylamino groups of PDMA chains.

Surface Analysis of Grafted fPTFE Plates by means of ESCA. The photoelectron spectra on C1s, O1s, N1s and F1s for grafted fPTFE plates were recorded on a Shimadzu ESCA 750-type spectroscope with MgKα (1253.6 eV) operating at 8 kV and 30 mA. A vacuum of at least 5 x 10^{-5} Pa in the chamber was maintained for all measurements. Samples were mounted onto the probe tips with double-sided Scotch tape. Binding energies were adjusted to Au4f7/2 = 84.0 eV (12). The intensity ratios, O1s/C1s and N1s/C1s, were calculated using the equation described in previous papers (2, 11) to evaluate the coverage of g-fPTFE surface with grafted polymers.

Contact Angle Measurements. Contact angles were measured with a TYP-QI-type goniometer (Kyowa Interface Science Co. Ltd.) with an environmental chamber under an atmosphere of saturated water vapor in the range between 15 and 45 °C. The volume of the aqueous solution drops used was always 0.5μL. The aqueous solutions used were water for NIPAAm-g-fPTFE and buffer solutions of pH4 and 10 for DMA-g-fPTFE. All reported values are the average of at least 20 measurements taken at different locations on the surfaces and have a maximum error of ±1°.

Results and Discussion

Surface Analysis of NIPAAm Grafted fPTFE Plates. Figure 1 shows the results of surface analysis of untreated fPTFE, plasma-treated fPTFE, NIPAAm-g-fPTFE and PNIPAAm plates by means of ESCA. For untreated fPTFE plate, the C1s peak at 292 eV assigned to $-\underline{C}F_2-CF_2-$ and the F1s peak at 692 eV to $-C\underline{F}_2$ of PTFE plates were investigated (13,14). After the oxygen-plasma treatment for fPTFE, the broad O1s peak around 533 eV assigned to oxygen-containing groups such as peroxides was observed and the C1s peak at 292 eV assigned to $-\underline{C}F_2-CF_2-$ and the F1s peak at 692 eV to $-C\underline{F}_2$ of PTFE plates decreased together. These results show that fluorine atoms at the surface of fPTFE were partly removed and oxygen-containing groups such as peroxides were newly formed on the fPTFE surface. For NIPAAm-g-fPTFE plates, the C1s peak at 292 eV assigned to $-\underline{C}F_2-CF_2-$ and the F1s peak at 692 eV to $-C\underline{F}_2$ of PTFE plates decreased significantly and the new peaks of C1s at 285 eV assigned to $-\underline{C}-C-$ and $-\underline{C}H$ (13), and 288 eV assigned to $-\underline{C}(=O)N-$ (15) were observed. The O1s peak around 533 eV assigned to oxygen-containing groups such as peroxides became sharp one according to the peak assigned to $-C(=\underline{O})N-$ (15). Furthermore, the N1s peak around 400 eV corresponding to the amine signal $(-C(=O)\underline{N}H)$ (16) was observed on the surface of NIPAAm-g-fPTFE. The N1s peak around 400 eV corresponding to the amine signal and the sharp O1s peak at 533 eV assigned to $-C(=\underline{O})N-$ were also observed on the surface of PNIPAAm plates. Therefore, the results from ESCA show that PNIPAAm chains are recognized on the grafted PTFE surfaces.

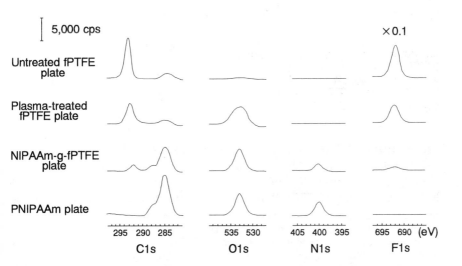

Figure 1. ESCA spectra on C1s, O1s, N1s and F1s of fPTFE plate, plasma-treated fPTFE for 120s, NIPAAm-g-fPTFE plate (N1s/C1s=0.111) and PNIPAAm plate

Figure 2 shows the intensity ratios F1s/C1s and N1s/C1s of plasma-treated fPTFE and various NIPAAm-g-fPTFE plates as a function of UV irradiation time. The intensity ratio of F1s/C1s decreased rapidly with irradiation time but never disappeared within the UV irradiation time. On the other hand, the intensity ratios of N1s/C1s increased gradually with UV irradiation time. The decrease of F1s/C1s and the gradual increase of N1s/C1s are regarded as the formation of PNIPAAm chains on the surface of fPTFE. The intensity ratio of N1s/C1s is here estimated as the degree of grafting of PNIPAAm since no gravimetric change in the grafted amount was observed.

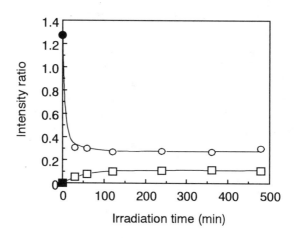

Figure 2.　Changes in intensity ratio at take off angle of 90° by ESCA for grafted fPTFE plates with irradiation time
fPTFE plate plasma-treated for 120s-
●: F1s/C1s, ▲: O1s/C1s, ■: N1s/C1s
NIPAAm grafted fPTFE plates-
○: F1s/C1s, △: O1s/C1s, □: N1s/C1s

Wettabilities of NIPAAm Grafted fPTFE Plates.　Figure 3 represents the wettabilities (cos θ) of fPTFE, plasma-treated fPTFE, various degrees of grafting of NIPAAm-g-fPTFE and PNIPAAm plates as a function of temperature. The hydrophobic surface of PTFE plate (cos θ = −0.42) was turned to slightly hydrophilic properties after the plasma treatment　(cos θ = −0.30).　The increase in the wettabilities of plasma treated fPTFE plate can be considered due to the introduced oxygen-containing groups into the surface of PTFE.　NIPAAm-g-fPTFE surfaces showed enhanced wettabilities with the degree of grafting (cos θ = 0.15~0.35). Especially, it is to be noted that abrupt drops in wettabilities of NIPAAm-g-fPTFE surfaces were observed between 30 and 35 °C (cos θ = 0.10~0.20).　The PNIPAAm plate exhibited the most wettable properties (cos θ = 0.80 below 30 °C) and the most clear drop in wettabilities between 30 and 35 °C of all.　Clearly the thermosensitive behavior of PNIPAAm plates comes from the well-known hydrophobic interaction among isopropyl groups of PNIPAAm chains. Therefore, PNIPAAm chains grafted on PTFE surfaces can show more thermosensitive properties with an increase in the degree of grafting.

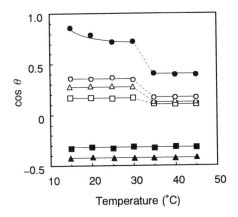

Figure 3. Contact angles (cos θ) of untreated (▲), plasma-treated for 120s (■) , NIPAAm grafted (N1s/C1s = 0.029, 0.059, 0.111) (□, △, ○) fPTFE plates, PNIPAAm plate (●) toward water plotted against temperature

Figure 4 represents cos θ toward water as a function of the degree of grafting of NIPAAm at 25 and 40°C. The wettabilities of NIPAAm-g-fPTFE surfaces at 25 °C increased with a rise in the degree of grafting of NIPAAm and were larger than those at 40 °C. The reason that the wettabilities of NIPAAm-g-fPTFE surfaces at 40 °C did not show a considerable increase with a rise in the degree of grafting of NIPAAm is attributed to the formation of dense layers resulting from the hydrophobic interaction among isopropyl groups of grafted PNIPAAm chains in the most upper region. The dense layers resulting from the hydrophobic interaction among isopropyl groups were examined with NIPAAm hydrogels containing much water. The changes in equilibrium water contents and cos θ of NIPAAm hydrogel crosslinked with Bis (Bis-NIPAAm gel) plates are shown in Figure 5 as a function of temperature. Both equilibrium water contents and cos θ of NIPAAm gels decreased remarkably between 30 and 35°C. Therefore, the changes in the wettabilities of NIPAAm gel plates correspond to those in the equilibrium water contents. Nevertheless, the equilibrium water contents of NIPAAm gels were still kept 30%(w/w) even above 35 °C (*17*), while cos θ of NIPAAm gel above 35 °C was nearly equal to PNIPAAm plates. Therefore, this indicates that the surface of NIPAAm gel plates is covered with the dense layers resulting from the hydrophobic interaction among PNIPAAm chains and the wettabilities of the surfaces of NIPAAm gel plates are not affected by polar water in the inner layer. From the results of contact angle measurements and F1s/C1s values of NIPAAm-g-fPTFE plates in Figure 2, it is apparent that the chains of PNIPAAm and PTFE are mixed at NIPAAm-g-fPTFE plate surfaces.

316

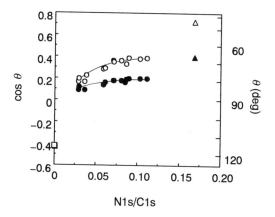

Figure 4. Contact angles (cos θ) of untreated ($\square$,$\blacksquare$), NIPAAm-grafted fPTFE ($\bigcirc$,$\bullet$) and PNIPAAm ($\triangle$,$\blacktriangle$) plates plotted against N1s/C1s opened: 25°C, shaded: 40°C

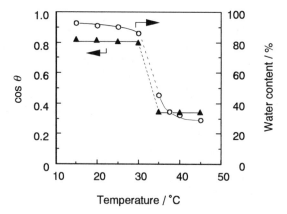

Figure 5. Changes in contact angles ($\blacktriangle$) and equilibrium water contents ($\bigcirc$) of Bis-NIPAAm gel as a function of temperature (Equilibrium water contents reproduced from our previous paper[8]).

Surface Analysis of DMA Grafted fPTFE Plates. Figure 6 shows ESCA spectra of the surfaces of untreated fPTFE, plasma treated fPTFE, DMA-g-fPTFE and PDMA plates. The N1s peaks around 400 eV corresponding to the amine signal were observed on the surfaces of DMA-g-fPTFE and PDMA plates as well as the surfaces of NIPAAm-g-fPTFE and PNIPAAm plates. Figure 7 shows the intensity ratios of F1s/C1s and N1s/C1s of DMA series as a function of irradiation time. The decrease in F1s/C1s and the increase in N1s/C1s are analogous to those of NIPAAm series.

Wettabilities of DMA Grafted fPTFE Plates. Figures 8 (a) and (b) represent the $\cos \theta$ of DMA-g-fPTFE surfaces toward pH 4 and 10 buffer solutions as a function of temperature, respectively. The $\cos \theta$ of DMA-g-fPTFE surfaces toward the pH 4 buffer solution increased in a rise of N1s/C1s but showed no change with temperature. However, the $\cos \theta$ of DMA-g-fPTFE surfaces toward the pH 10 buffer solution dropped between 25 and 30 °C in similar to that of NIPAAm-g-fPTFE plates between 30 and 35 °C. The hydrophobic interaction among dimethylamino groups deprotonized at pH 10 is considered to be responsible for the decrease in wettabilities of DMA-g-fPTFE surfaces.

The wettabilities of DMA-g-fPTFE surfaces toward pH 4 and 10 buffer solutions are shown in Figures 9 (a) and (b). The thermosensitive properties of DMA-g-fPTFE surfaces were depicted in Figures 8 (b) and 9 (b) by curves similar to those of NIPAAm-g-fPTFE plates in Figure 4. Therefore, we can make the same point as above, that is, the DMA-g-fPTFE surfaces consist of a mixture of chains of grafted PDMA and bulk PTFE and those toward pH 10 aqueous solutions exhibit the thermosensitive properties.

Evaluation of Wettabilities of NIPAAm-g-fPTFE Plates Based on Surface Free Energies. In order to make the hydrophobic surface properties at high temperature shown in Figure 4 clearer, we followed up the changes in the component of hydrogen bonding of surface free energies. The values of the dispersion, polar and hydrogen bonding components of free energies of water, 1-bromonaphthalene and diiodomethane calculated from the surface tensions based on Fowkes equation (*18*) at 25 and 40°C are listed in Table I for calculation of these surface free energies. The values of three components (γ_L^d, γ_L^p and γ_L^h) of free energies were calculated from the contact angles measured on PE and PTFE plates with these liquids. It is here assumed that the surface free energy of PE plate consists of only γ_S^d and that of PTFE plate consists of γ_S^d and γ_S^p. These values were calculated with Kitazaki-Hata's method (*19*).

Table II shows the list of the surface free energies γ_S of untreated fPTFE, plasma-treated fPTFE, NIPAAm-g-fPTFE, PNIPAAm and NIPAAm gel plates at 25 and 40°C calculated on the basis of the modified Fowkes equation (Kitazaki-Hata's method) (*19*).

$$\gamma_{SL} = \gamma_S + \gamma_L - \left(2\sqrt{\gamma_S^d \cdot \gamma_L^d} + 2\sqrt{\gamma_S^p \cdot \gamma_L^p} + 2\sqrt{\gamma_S^h \cdot \gamma_L^h} \right)$$

The surface free energies γ_S and the components of the dispersion γ_S^d, polar γ_S^p and hydrogen bonding γ_S^h of untreated fPTFE and plasma-treated fPTFE plates showed little change at 25 and 40°C. However, γ_S of NIPAAm-g-fPTFE (N1s/C1s=0.111), PNIPAAm and NIPAAm gel plates dropped from 40.6, 61.6 and 60.5 mJ/m^2 (at 25°C) to 37.7, 48.4 and 40.8 mJ/m^2 (at 40°C), respectively. It is considered that the

318

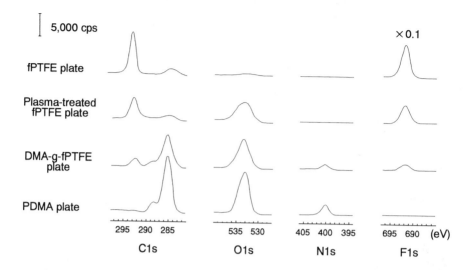

Figure 6. ESCA spectra on C1s, O1s, N1s and F1s of fPTFE plate, plasma-treated fPTFE plate for 120s, DMA-grafted fPTFE plate and PDMA plate

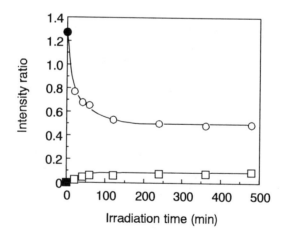

Figure 7. Changes in intensity ratio at take off angles of 90° by ESCA for grafted fPTFE plates with irradiation time
fPTFE plate plasma-treated for 120s-
●: F1s/C1s,■: N1s/C1s
DMA grafted fPTFE plates
○: F1s/C1s,□: N1s/C1s

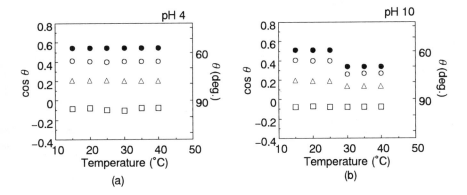

Figure 8. Contact angles (cos θ) of DMA-grafted fPTFE($\bigcirc,\triangle,\square$) and PDMA ($\bullet$) plates for (a) pH 4 ($CH_3COOH/CH_3COONa$, I=0.01) and (b) pH 10 ($NaOH/NaHCO_3$, I=0.01) buffer solutions plotted against temperature
N1s/C1s-$\bigcirc$: 0.037,$\triangle$: 0.010,$\square$: 0.006

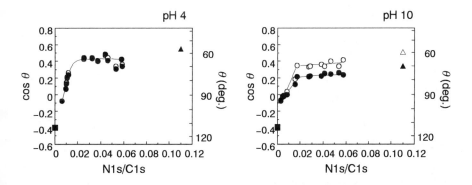

Figure 9. Contact angles (cos θ) of untreated ($\square,\blacksquare$), DMA-grafted fPTFE ($\bigcirc,\bullet$) and PDMA ($\triangle,\blacktriangle$) plates for (a) pH 4 (CH_3COOH/CH_3COONa, I=0.01) and (b) pH 10 ($NaOH/NaHCO_3$, I=0.01) buffer solutions plotted against N1s/C1s
opened: 20°C, shaded: 35°C

decrease in γ_s at 40 °C, which comes from a decrease in γ_s^h (see Table II), contributes to these decreases in wettabilities (12). Therefore, the decrease in wettabilities for the surface of NIPAAm-g-fPTFE plate may be ascribed to the loss of hydrogen bonding between grafted PNIPAAm chains and water molecules of the liquid drop.

Table I. The dispersion, polar and hydrogen bonding components of free energies for pure liquids[a]

Used liquids		γ_L^d	γ_L^p	γ_L^h	γ_L
Water	(25 °C)	28.1	3.0	41.4	72.5
	(40 °C)	26.8	2.3	41.2	70.3
1-Bromonaphthalene	(25 °C)	43.9	0.0	0.0	43.9
	(40 °C)	42.7	0.0	0.0	42.7
Diiodomethane	(25 °C)	45.6	4.2	0.0	49.9
	(40 °C)	44.0	4.2	0.0	48.2

a) All surface free energies are recorded in mJ/m^2.

Table II. The surface free energies of the used plates calculated from Kitazaki-Hata's method [a]

Plate	25 °C				40 °C			
	γ_s^d	γ_s^p	γ_s^h	γ_s	γ_s^d	γ_s^p	γ_s^h	γ_s
Untreated PTFE	19.6	1.1	0.0	20.7	19.6	1.4	0.0	21.0
Plasma-treated PTFE[b]	12.8	9.2	0.0	22.0	12.6	9.4	0.0	22.0
NIPAAm-g-PTFE[c]	32.9	0.0	9.4	42.3	34.4	0.0	2.5	37.4
PNIPAAm	39.0	0.0	22.4	60.9	40.5	0.0	6.4	46.9
NIPAAm gel	27.5	0.0	36.6	63.6	32.2	0.9	6.4	39.5

a) All surface free energies are recorded in mJ/m^2.
b) Plasma-treated for 120 s.
c) N1s/C1s=0.111.

Conclusions

1. Inert PTFE surfaces have been made more wettable by the combine use of oxygen-plasma treatment and photografting of NIPAAm or DMA.
2. Both NIPAAm- and DMA-g-fPTFE surfaces exhibit unique thermosensitive properties due to the hydrophobic interaction among grafted polymer chains.
3. The decrease in wettabilities for NIPAAm-g-fPTFE surfaces is ascribed to the loss of hydrogen bonding between grafted PNIPAAm chains and water molecules from the results obtained on the basis of the surface free energies.

Literature Cited

1. Yamada, K.; Tsutaya, H.; Tatekawa, S.; Hirata, M. *J. Appl. Polym. Sci.*, **1992**, *46*, 1065.
2. Yamada, K.; Ebihara, T.; Gondo, T.; Sakasegawa, K.; Hirata, M. *J. Appl. Polym. Sci.*, **1996**, *61*, 1899.
3. Hirata, M.; Tokumoto, Y.; Ichikawa, M.; Izumi, T. *Nippon Kagaku Kaishi* [*J. Chem. Soc. Jpn., Chem. Ind. Chem.*], **1985**, *1985*, 661 [in Japanese].
4. Yamada, K.; Kimura, T.; Tsutaya, H.; Hirata, M. *J. Appl. Polym. Sci.*, **1992**, *44*, 993.
5. Bergbreiter, D.E.; Kabza, K., *Industrial Eng. Chem. Res.*, **1995**, *34*, 2733
6. Bergbreiter, D.E.; Kabza, K., *J. Am. Chem. Soc.*, **1991**, *113*, 1447
7. Holmes-Farley, S. R.; Reamey, R. H.,; Nuzzo, R.; McCarthy, T. J.; Whitesides, G. M., *Langmuir*, **1987**, *3*, 799
8. Carey, D. H.; Ferguson, G. S., *J. Am. Chem. Soc.*, **1996**, *118*, 9780
9. Bergbreiter, D.E.; Ponder, B. C.; Anguilar, G.; Srinivas, B., *Chem. Mater.*, **1997**, *9*, 472
10. Bergbreiter, D.E.; Bandella, A., *J. Am. Chem. Soc.*, **1995**, *117*, 10589
11. Yamada, K.; Sato, T.; Tatekawa, S.; Hirata, M. *Polym. Gels and Networks*, **1994**, *2*, 323.
12. Briggs, D.; Brewis, D. M.; Konieczo, M. B. *J. Mater. Sci.*, **1976**, *11*, 1270.
13. Clark, D. T.; Feast, W. J.; Kilcast, D.; Musgrave, W. K. R. *J. Polym. Sci., Polym. Chem. Ed.*, **1973**, *11*, 389.
14. Clark, D. T.; Feast, W. J.; Ritchie, I.; Musgrave, W. K. R. *J. Polym. Sci., Polym. Chem. Ed.*, **1974**, *12*, 1049.
15. Clark, D. T.; Cromarty, B. J.; Dilks, A. *J. Polym. Sci., Polym. Chem. Ed.*, **1978**, *16*, 3173.
16. Clark, D. T.; Harisson, A. *J. Polym. Sci., Polym. Chem. Ed.*, **1981**, *19*, 1945.
17. Matsuda, K.; Matsudo, H.; Shimizu, K.; Yamada, K.; Hirata, M. *Nippon Kagaku Kaishi* [*J. Chem. Soc. Jpn., Chem. Ind. Chem.*] **1993**, *1993*, 380 [in Japanese].
18. Fowkes, F. M. *J. Phys. Chem.*, **1962**, *66*, 382.
19. Kitazaki, Y.; Hata, T. *J. Adhesion Soc. Jpn*, **1972**, *8*, 131; **1974**, *10*, 174 [in Japanese].

Chapter 22

Temperature- and pH-Responsive Polymers for Controlled Polypeptide Drug Delivery

C. Ramkissoon-Ganorkar, F. Liu, M. Baudyš, and S. W. Kim

Department of Pharmaceutics and Pharmaceutical Chemistry (CCCD),
University of Utah, Salt Lake City, UT 84112

pH- and temperature-sensitive terpolymers of N-isopropylacrylamide (NIPAAm), butylmethacrylate (BMA), and acrylic acid (AA) were used to prepare polypeptide releasing beads via loading in aqueous solution. Insulin was protected and not released from the beads at low pH, such as that in the stomach. Linear polymers with NIPAAm/BMA/AA mol feed ratio of 85/5/10 were prepared by free radical polymerization. Molecular weight (M.Wt.) of the polymers was varied using different volume percentages of benzene/tetrahydrofuran (THF) as solvent. The weight average M.Wt. of the polymer synthesized in THF was 33,000 and 990,000 for the polymer synthesized in benzene. The lower critical solution temperature (LCST) was determined by cloud point measurement and was 22 °C at pH 2.0 and 85 °C at pH 7.4 for the polymers. Beads were loaded with the polypeptide by preparing a 10% polymer, 0.2% polypeptide solution in 20 mM glycine/HCl buffer, pH 2.0, 0.15 M NaCl at 4 °C and adding the solution dropwise into an oil bath at 35 °C. Loading efficiency was between 20 to 100% depending on the M. Wt. of the polymer and M. Wt. of the polypeptide. Release studies were performed at pH 2.0 and pH 7.4, and 37 °C, simulating in vivo conditions. Low M.Wt. polymer beads displayed a dump-like release profile and high M. Wt. polymer beads released drug slowly over a period of eight hours. The loaded and released polypeptide drugs were bioactive as shown by in vivo studies. These polymers allow aqueous phase loading of proteins in beads, protect polypeptides from the harsh acidic environment in the stomach, and can provide oral controlled release of polypeptide drugs. The low M.Wt. polymer beads can be used for immediate release of polypeptide in the duodenum while the high M. Wt. polymer beads can be used for slow release of polypeptide in the colon.

Stimuli-sensitive polymers which change their structure and physical properties in response to external stimuli have an incredible potential for applications in pharmaceutical technology, biotechnology industry and other industrial applications (1-14). In spite of the recent tremendous progress in recombinant DNA technology and conventional chemical synthesis that have contributed many polypeptides and proteins for the treatment of diseases, administration and delivery of protein drugs remain a major problem. Presently, most protein formulations are reconstituted at the time of administration and are given parenterally by injections or infusions. Thus, chronic treatment is usually painful and unpleasant. Naturally, alternative administration routes have been sought in the last decade, among them the oral route would be preferable for treatment of chronic diseases. Problems associated with oral polypeptide drug delivery include degradation in the acidic environment of the stomach, enzymatic degradation in the intestine and poor absorption across the gastrointestinal (GI) mucosa. A drug delivery system, using temperature-sensitive and pH-sensitive polymers may be applied to enable aqueous protein drug loading, provide protection and enhance GI absorption to make oral polypeptide delivery feasible.

We propose the use of "smart" polymers that are pH/temperature-sensitive, specifically statistical terpolymers of N-isopropylacrylamide (NIPAAm), butylmethacrylate (BMA) and acrylic acid (AA) in the form of pH/temperature-sensitive polymer beads as a drug delivery system for the oral delivery of polypeptide drugs. NIPAAm-based polymers are not yet approved for human use, but they show promising results based on in vitro studies done on various types of cells, which remained viable in the presence of NIPAAm polymers (6,7,14). NIPAAm confers the polymer temperature sensitivity, which is characterized by a lower critical solution temperature (LCST). The polymer is soluble below the LCST and precipitates above the LCST. Consequently, hydrogels made of temperature-sensitive polymers show reversible high/low swelling based on temperature change (15). pH-sensitivity is obtained by the use of AA in the polymer. At low pH, almost no carboxylic group will be ionized, permitting greater interaction between the polymer chains through hydrogen bonding, leading to shifting of the LCST to a lower temperature, and hence the polymer will be in the unswollen (insoluble) state (16). Whereas, at high pH, due to ionization of the carboxylic groups, the polymer will take up water, thereby swelling (dissolving parallely). The incorporation of acrylic acid increases the LCST at neutral pH due to the hydrophilicity of the charged AA group (17,18). BMA was also incorporated in the polymer, it improves the mechanical strength of the beads and also affects the polymer LCST due to its hydrophobicity. Thus, these polymer beads based on their pH- and temperature-sensitive properties are insoluble in acidic pH and 37 °C and protect the drug from gastric degradation. The polymer becomes soluble at neutral pH and can release the drug in the intestinal tract where most of the absorption takes place.

In this study, pH/thermo-sensitive polymer beads were prepared using statistical terpolymers of NIPAAm, BMA and AA and investigated for oral delivery of polypeptide drugs. Loading efficiency into the beads was studied as a function of M.

Wt. of polypeptide drug, using insulin, calcitonin, trypsin inhibitor and angiotensin II as model polypeptides. Release rates from these systems were controlled by varying the molecular weight (M. Wt.) of the polymers (19), and insulin was used as a model polypeptide drug.

Materials And Methods

Materials. NIPAAm obtained from Fisher Scientific Inc. (Fair Lawn, NJ), was recrystallized from hexane. BMA and AA were purchased from Aldrich Chemical Company (Milwaukee, WI). BMA and AA were purified by vacuum distillation at 57 °C / 17 mm Hg and 39 °C / 10 mm Hg respectively. 2,2'-azobis-isobutyronitrile (AIBN), purchased from Eastman Kodak Company (Rochester, NY), was recrystallized from methanol. Bovine insulin, trypsin inhibitor and angiotensin II were purchased from Sigma Chemical Company (St. Louis, MO). Human calcitonin was a gift from Suntory Ltd. (Tokyo, Japan). Heavy white mineral oil and decane were purchased from Aldrich Chemical Company (Milwaukee, WI). Acetonitrile, HPLC grade, was purchased from Fisher Scientific Inc. (Fair Lawn, NJ). All other chemicals were reagent grade.

Polymer synthesis. The synthesis of linear terpolymers of NIPAAm, BMA and AA, with feed ratio of NIPAAm/BMA/AA = 85/5/10 or 80/10/10, was carried out in varying volume percent of benzene and tetrahydrofuran (THF) as solvent. AIBN was used as free radical initiator (7.41 mmol AIBN per mol monomer). Dried N_2 gas was bubbled through the solution for 20 min to remove dissolved oxygen. The solution was polymerized for 24 hours at 60 °C under N_2 atmosphere. The synthesized terpolymers were recovered by precipitation in diethylether. The polymers were filtered and dried under vacuum overnight.

Polymer characterization. The polymers were characterized by M. Wt. and LCST. M. Wt. was estimated by Size Exclusion Chromatography. N-(2-hydroxypropyl)methacrylamide (HPMA) polymers or Poly(ethyleneoxide) (PEO) (Waters Corporation, MA) were used as standards. An FPLC system (Pharmacia) connected to a refractometer was used. A Superose 6 HR 10/30 column (Pharmacia) and 0.05 M Tris buffer, pH 8.0, 0.5 M NaCl as eluent or an ultrahydrogel 2000 column (Waters Corporation, MA) and 10 mM phosphate buffer, pH 7.4, 0.15 M NaCl as eluent, were used. A flow rate of 0.4 ml/min was used. The M. Wts. and M. Wt. distributions were obtained relative to standard M. Wt. HPMA or PEO. The LCST of the polymers was determined by cloud point measurement. The polymer solutions (1% w/v) were prepared in 20 mM glycine/HCl buffer, pH 2.0, 0.15 M NaCl and 10 mM phosphate buffer, pH 7.4, 0.15 M NaCl. The temperature of the solutions was raised from 10 °C to 80 °C in 2 °C increments every 5 min and the absorbance at 450 nm was measured using a Perkin-Elmer Lambda 7 UV/Vis spectrophotometer. The LCST was defined as the temperature at the inflection point in the absorbance versus temperature curve.

Bead preparation. Drug-loaded beads were prepared from an aqueous solution (20 mM glycine buffer, pH 2.0, 0.15 M NaCl) containing pH/temperature-sensitive polymer (7 or 10% w/v) and polypeptide drug (0.2% w/v) as described previously (20,21). The solution was kept at 4 °C overnight to allow solubilization of the polymer. The drug/polymer solution at 4 °C was added dropwise using a syringe and 25G needle into 50 ml of mineral oil kept at a temperature of 35 °C which is above the LCST of the polymers. The mineral oil was covered with 5 ml decane to reduce surface tension and to aid in the penetration of the solution drop at the air/oil interface. The formed beads were washed with hexane, dried at 35 °C for one hour in the open air and then dried in a rotary evaporator with aspiration for half hour.

Scanning Electron Microscopy. The structural morphology of beads was evaluated using scanning electron microscopy (SEM). Cross-sections of the beads were coated with gold and were examined with a scanning electron microscope (Cambridge, S240 SEM), using a LaB_6 filament and a secondary electron (SE) detector at 15-25 kV acceleration voltage. Cross-sections were obtained by breaking the beads in liquid nitrogen.

Dissolution/Swelling Studies. Beads were placed in 20 mM glycine/HCl buffer, pH 2.0, 0.15 M NaCl at 37 °C or in 10 mM phosphate buffer, pH 7.4, 0.15 M NaCl at 37 °C so as to study the dissolution and swelling behavior of the beads under different pH conditions. At different time intervals, pictures of the beads were taken using a Nikon Eclipse E800 microscope connected to a Pentium Probe 200 MHz computer using Image ProPlus software for image processing.

Loading efficiency. The content of polypeptide in the beads was determined after complete dissolution of the beads at 4 °C. The dried beads containing protein drug were dissolved in 2 ml of 20 mM glycine/HCl buffer, pH 2, 0.15 M NaCl at 4 °C overnight. The resulting solution was heated up to 50 °C for 10 min. After the polymer solution precipitated, the supernatant was collected and assayed for polypeptide by reversed phase high performance liquid chromatography (HPLC). This procedure was repeated three times until a negligible amount of polypeptide could be detected. The HPLC apparatus consisted of a Waters gradient system (automated gradient controller, model 680 and HPLC pump, model 501, Waters Corporation, MA). The samples were injected via an intelligent sample processor (WISP, model 712, Waters) connected to a UV detector (model 484, Waters) and an integrator (model 745, Waters). The C_4 column (5 μm, 4.6 x 250 mm, Vydac, Hesperia, CA) was equilibrated with eluent A (water, 0.1% trifluoroacetic acid) and eluent B (acetonitrile, 0.1% trifluoroacetic acid), the percentages of which depended on the polypeptide assayed. A flow rate of 1 ml/min was used and a gradient of 2% B/min was used in all cases. The absorbance of the eluent was recorded at 276 nm. The column was calibrated with polypeptide solutions of known concentration.

Release studies. Drug-loaded beads were placed in 10 ml glycine/HCl buffer, pH 2.0, 0.15 M NaCl at 37 °C for two hours and then in 10 ml isotonic PBS, pH 7.4 at 37 °C so as to mimic in vivo conditions in the GI tract. At different time points, one ml of the release medium was collected and replaced by the same volume of buffer. Drug released was assayed by reversed phase HPLC as described above.

Bioactivity of insulin. The bioactivity of insulin released from the polymeric beads in vitro was studied by measuring blood glucose level depression in male Sprague-Dawley rats (200-350 g, Sasco, Omaha, NEB). As a reference, a fresh insulin solution was prepared in 10 mM PBS, pH 7.4. Insulin solutions (0.5 IU/ml/kg) were injected intravenously through the tail vein. Blood samples were taken from the jugular vein before injection and 15, 30, 60, 90, 120, 180 and 240 min after injection. Glucose levels were measured with an Accu-Check III blood glucose monitor (Boehringer Mannheim, Indianapolis, IN). In order to remove soluble polymer, the release medium was adjusted to pH 2.0 at room temperature so that the polymer precipitated out. The supernatant containing insulin was appropriately diluted and the pH was adjusted to neutrality before injection.

Results And Discussion

The M. Wt., polydispersity and LCST of the polymers are shown in Table I. The lowest M. Wt. obtained was 33,000 and the highest M. Wt. was 990,000. As the volume percent of THF in the solvent mixture increases, the M. Wt. decreases. THF acts as a chain transfer agent, thus decreasing the size of the polymer chains (19,22). The polydispersity increases as the M. Wt. of the polymer increases.

Table I. Molecular Weight, Polydispersity and LCST of Different
NIPAAm/BMA/AA Polymers

Polymer Composition NIPAAm/BMA/AA[a]	Weight Averaged M. Wt.	Polydispersity	LCST (°C) pH 2.0	LCST (°C) pH 7.4
80/10/10[b]	37,000	2.2	14	77
85/5/10[b]	33,000	1.8	23	84
85/5/10[c]	450,000	5.6	24	85
85/5/10[d]	990,000	> 7.0	24	>88

[a] - Feed ratio (mol%);
[b,c,d] - Prepared by free radical polymerization in a mixture of benzene and THF
([b] 0:100; [c] 90:10; [d] 100:0; benzene:THF volume ratio).

At acidic pH, the LCST is low as compared to neutral pH. Lower BMA content for the last three polymers causes LCST to increase slightly. At neutral pH, LCST rises sharply. This observation can be explained by the effect of ionizable groups on thermosensitive polymers (16,17,18). At low pH, the acrylic acid is uncharged and hence, the polymers have a low LCST. At high pH, the acrylic acid is charged, and thus, the polymers exhibit a high LCST. The effect of M. Wt. on LCST was only marginal.

The formation of beads is explained by the formation of a thin skin layer due to instant polymer precipitation at the oil-water interface (20). The beads have diameter ranging from 0.9 to 1.8 mm, with the low M. Wt. beads having the least diameter and the high M. Wt. beads having the maximum diameter. The diameter of the beads was determined by optical microscopy. The polypeptide drug gets entrapped inside the polymer which collapses once the temperature is above the LCST.

The SEM pictures of the interior of the low and high M. Wt. polymer beads (85/5/10 polymer) are shown in Figure 1. The interior of the low M. Wt. polymer beads is more porous than the high M. Wt. polymer beads. It was observed that as M. Wt. increases, porosity decreases, whereas, compactness and density of the beads increase. It is proposed that the extent of polymer chains entanglement is the physical basis for these differences. Entanglement depends on the concentration and M. Wt. of polymer (23). Thus, the low M. Wt. polymer has short chains that provide a low degree of entanglement, so that the beads are porous and less compact. Whereas, the long polymer chains of the high M. Wt. polymer have a high degree of entanglement, thus, producing beads with higher mechanical strength and lower porosity.

At pH 2.0 and 37 °C, beads do not swell or dissolve. They stay intact indefinitely. However, at pH 7.4 and 37 °C, beads behave differently depending on M. Wt. of the polymer. Low M. Wt. beads show no swelling, but disintegrate within one hour (Figure 2). High M. Wt beads swell gradually and even after a period of eight hours are still undergoing rate-limiting swelling and have not disintegrated yet (Figure 3). This difference in swelling/dissolution behavior is believed to be intimately linked with M. Wt. and the degree of entanglement as discussed above. The high M. Wt. polymer chains form physical crosslinks that produces a network showing swelling properties similar to chemically crosslinked hydrogels.

Loading efficiency decreases as the M. Wt. of the polypeptide drug decreases (Table II). This is due to the porosity of the beads which allows drug loss in the aqueous phase as the polymer collapses. Insulin, with a M. Wt. of 5800 Da, has a very high loading efficiency. At the loading concentration used, insulin exists mostly in hexameric form (24) with an apparent M. Wt. of 36,000 Da which most probably explains such a high loading efficiency.

The dependence of release kinetics on polymer M. Wt. was investigated with the 85/5/10 polymer beads loaded with insulin (Figure 4). At pH 2 and 37 °C, only up to 10% of loaded insulin was lost after two hours. At pH 7.4 and 37 °C, the low M. Wt. polymer beads displayed a dump-like profile. In contrast, the high M. Wt. polymer beads released insulin slowly over a period of more than 8 hours. The release of insulin under these different conditions of pH and at 37 °C correlates with the dissolution/swelling behavior of the beads described above and depicted in Figures 2 and 3. At pH 2 and 37 °C, the beads are above LCST, and hence, no swelling or disintegration is seen, and consequently, there is negligible drug release. The 10%

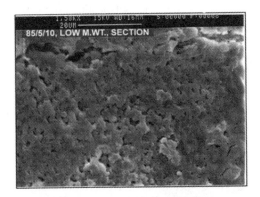

Figure 1. Scanning electron microscope cross-sectional views at 1,580 and 1,530 magnification of low M. Wt. polymer bead (left) and high M. Wt. polymer bead (right). Beads were prepared from polymers with NIPAAm/BMA/AA mol ratio of 85/5/10.

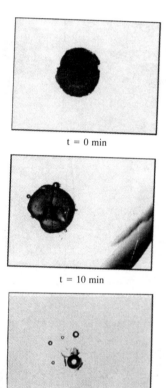

t = 0 min

t = 10 min

t = 50 min

Figure 2. Time dependence of low M. Wt. polymer bead dissolution (NiPAAm/BMA/AA = 85/5/10) at pH 7.4 and 37 °C studied by optical microscopy.

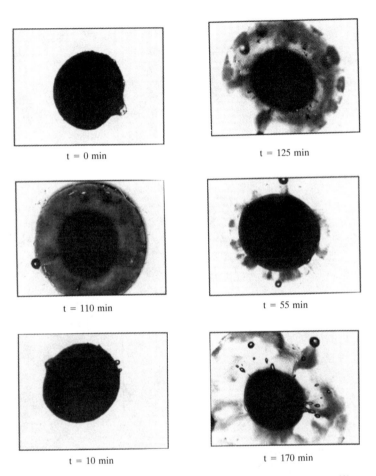

t = 0 min

t = 125 min

t = 110 min

t = 55 min

t = 10 min

t = 170 min

Figure 3. Time dependence of high M. Wt. polymer bead swelling and disintegration (NIPAAm/BMA/AA = 85/5/10) at pH 7.4 and 37 °C studied by optical microscopy.

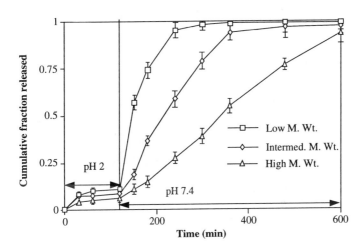

Figure 4. Modulating insulin release profile from pH-/temperature-sensitive beads made of terpolymers of constant NIPAAm/BMA/AA composition 85/5/10 and increasing molecular weight.

Table II. Loading efficiencies of polymer beads

Polypeptide	Polypeptide M. Wt.	Loading efficiency[a] (%)			
		80/10/10[b,c]	85/5/10[b,d]	85/5/10[b,e]	85/5/10[b,f]
Angiotensin II	1,046	ND	30 ± 9	ND	40 ± 7
Calcitonin	3,400	ND	20 ± 3	33 ± 4	48 ± 5
Insulin	5,800	92 ± 6	90 ± 2	92 ± 3	95 ± 2
Trypsin Inhibitor	9,000	33 ± 3	49 ± 3	43 ± 4	55 ± 4

ND - Not determined;
[a] - Determined after beads dissolution at 4 °C in the loading buffer;
[b] - NIPAAm/BMA/AA feed ratio (mol%);
[c] - M. Wt. - 37,000;
[d, e, f] - Statistical polymers of the same composition but increasing M. Wt. - 33,000; 450,000; 990,000.

loss must be due to surface-located drug. In contrast, at pH 7.4 and 37 °C, the beads are below LCST, and thereby, disintegrate or swell based on the polymer M. Wt. It is again observed that the release profile follows the dissolution or swelling behavior illustrated in Figures 2 and 3. The low M. Wt. beads show a fast release rate, the high M. Wt. polymer shows a slow release profile and the intermediate M. Wt. polymer shows a release profile in between these two. Scheme 1 illustrates the relationship between the release profile and dissolution/swelling characteristics observed for the different M. Wts. polymer beads. The effect of M. Wt. of polypeptide drug on the release profile is being investigated.

The bioactivity of the released insulin is shown in Figure 5. These results demonstrate that the beads protect insulin from the harsh acidic conditions and that the activity of insulin is retained after the loading procedure and on contact of the insulin with the polymer.

Conclusion

pH/Temperature-sensitive polymers poly(NIPAAm-co-BMA-co-AA) were used to prepare polypeptide (insulin) releasing beads via loading in aqueous solution. After prolonged contact with the hydrophilic polymers, insulin was bioactive. At the low pH of the stomach, insulin was protected and not released from the beads. At pH 7.4, the low M. Wt. hydrophilic polymer beads displayed a dump-like profile and dissolved within 1 hour, (bead dissolution controlled release mechanism), while the high M. Wt. hydrophilic polymer swelled only and released insulin slowly over a period of 8 hours, (swelling and diffusion controlled release mechanism). Thus, the unique properties of the pH/temperature-sensitive polymer beads make it a viable

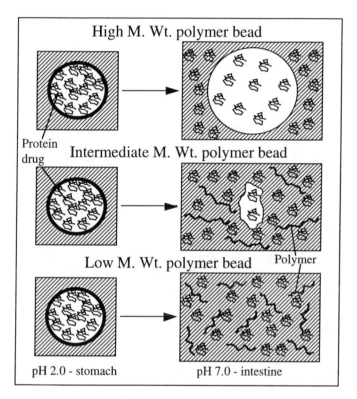

Scheme 1. Mechanism of protein drug release from pH/thermo-sensitive beads made of polymers with increasing molecular weight. Top - swelling and drug diffusion mechanism results in slow release suited for colon targeting. Bottom - bead dissolution results in fast release suited for duodenal delivery. Center - combination of both mechanisms results in intermediate release suited for lower small intestine targeting.

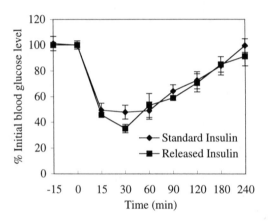

Figure 5. Blood glucose depression activity of released insulin from polymer beads determined by IV injection in rats (n=4).

system for oral drug delivery of peptide and protein drugs at a specific site in the gastrointestinal tract. The low M. Wt. hydrophilic polymer beads can serve as an oral delivery system for duodenal targeting, the intermediate M. Wt. hydrophilic polymer beads can serve as an oral delivery system for the lower small intestine targeting, while the high M. Wt. hydrophilic polymer beads can be used to target polypeptide drugs predominantly to the colon.

Acknowledgment

This work was supported by MacroMed, Inc., Salt Lake City, USA, and Suntory Ltd., Tokyo, Japan. We thank Drs. Y. H. Bae, A. Gutowska, A. Serres, and V. Menard for valuable discussions and Don Mix for his assistance.

References

1. Bae, Y. H., Okano, T., Kim, S. W. *J. Control. Rel.* **1989**, *9*, 271.
2. Hoffman, A. S. *Artificial Organs* **1995**, *19*, 458.
3. Chen, G., Hoffman, A. S. *Nature* **1995**, *373*, 49.
4. Gutowska, A., Bae, Y. H., Jacobs, H., Feijen, J., Kim, S. W. *Macromolecules* **1994**, *27*, 4167.
5. Liu, F., Zhuo, R. X. *Biotechnol. Appl. Biochem.* **1993**, *18*, 57.
6. Vernon, B., Gutowska, A., Kim, S. W., Bae, Y. H. *Macromol. Symp.* **1996**, *109*, 155.
7. Iwata, H., Amemiya, H., Akutsu, T. *Artif. Organs* **1990**, *14, Suppl 3*, 7.
8. Edelman, E. R., Kost, J., Bobeck, H., Langer, R. *J. Biomed. Mater. Res.* **1985**, *19*, 67.
9. Okahata, Y., Noguchi, H., Seki, T. *Macromolecules* **1987**, *20*, 15.
10. Suzuki, A., Tanaka, T. *Nature* **1990**, *346*, 345.
11. Kwon, I. C., Bae, Y. H., Kim, S. W. *Nature* **1991**, *354*, 291.
12. Dagani, R. *Chem. Eng. News* **1997**, *Jun. 9*, 26.
13. D'Emanuele, A. *Clin. Pharmacokinet.* **1996**, *31*, 241.
14. Okano, T., Yamada, N., Okuhara, M., Sakai, H., Sakurai, Y. *Biomaterials* **1995**, *16*, 297.
15. Schild, H. G., Tirrell, D. A. *J. Phys. Chem.* **1990**, *94*, 4352.
16. Dong, L.-C., Hoffman, A. S. *J. Control. Rel.* **1991**, *15*, 141.
17. Feil, H., Bae, Y. H., Feijen, J., Kim, S. W. *Macromolecules* **1993**, *26*, 2496.
18. Feil, H., Bae, Y. H., Kim, S. W. *Macromolecules* **1992**, *25*, 5528.
19. Schild, H. G. *Prog. Polym. Sci.* **1992**, *17*, 163.
20. Kim, Y.-H., Bae, Y. H., Kim. S. W. *J. Control. Rel.* **1994**, *28*, 143.
21. Serres, A., Baudyš, M., Kim, S. W. *Pharm. Res.* **1996**, *13*, 196.
22. Rodriguez, F. *Principles of Polymer Systems;* Publisher: McGraw-Hill Book Company, 1970.
23. Grossman, P. D., Soane, D. S. *J. Chromatogr.* **1991**, *559*, 257.
24. Blundel, T. L., Dodson, G. G., Hodgkin, D. C., Mercola, D. *Adv. Protein Chem.* **1972**, *26*, 279.

INDEXES

Author Index

Subject Index